Methods in Polyphenol Analysis

Methods in Polyphenol Analysis

Edited by

Celestino Santos-Buelga
University of Salamanca, Spain

Gary Williamson
Nestlé Research Center, Lausanne, Switzerland

RS•C
advancing the chemical sciences

ISBN 0-85404-580-5

A catalogue record for this book is available from the British Library

Published by The Royal Society of Chemistry,
Thomas Graham House, Science Park, Milton Road,
Cambridge CB4 0WF, UK

Registered Charity Number 207890

For further information see our web site at www.rsc.org

Typeset by Keytec Typesetting Ltd., Bridport, Dorset, UK
Printed in the United Kingdom at the University Press, Cambridge

Preface

Polyphenols in the diet are becoming increasingly recognised as important in long term health and reduction in the risk of chronic disease. Also, we have not forgotten that they have been known for a long time for their roles in colour, tanning, taste and astringency, making essential contributions to the unique sensation of the "British cup of tea" the popularity of "Chinese green tea" or the characteristic colour of fruits and derived products, such as red wine. Polyphenols are found widely in plant foods, notably onions, cocoa, apples, broccoli, tea, coffee, red wine, berries, and even in Chinese herbal medicines.

This book is a product of the European Union Framework V project on polyphenols bioavailability and colon cancer, affectionately called "POLYBIND". We decided to try to fill the gap in this area with a book that reflected the increased interest in polyphenols and their health benefits. It has been many years since the great "bibles" on polyphenols by the late Prof. Jeffrey Harborne were printed. We cannot compete with these great works, but this book is intended as an update on the latest techniques to analyse for polyphenols and is based on work by experts (our colleagues and friends) in this area.

Owing to the chemical complexity of polyphenols, analysing this class of compounds remains as challenging as ever. However, with the potential for health claims attached to these compounds in the future, methods for accurate and sensitive analysis become more and more important. The challenge is magnified by many fold in the case of polyphenols which are oligomerised or polymerised into chemical species which are difficult to define at a molecular level. We have tried to cover a range of techniques, and many of the chapters clearly indicate the resurgence of MS for analysis in biology over the last decade. We also present methods for analysis in biological fluids and foods of purified and synthetic compounds, metabolites, conjugates and microbial metabolites.

Finally, thank you to all of our colleagues who have directly or indirectly contributed to this book, especially to Dr. Maite Escribano-Bailon for her assistance with the manuscripts and the preparation of the subject index, and if the reader has any improvements to the methods or tricks to overcome problems that we have not covered, then please let us know – there is always room for improvements in analysis of polyphenols!

Gary Williamson
Celestino Santos-Buelga
January 2003

Contents

Contributors

I. C. W. Arts, *RIKILT, PO Box 230, 6700 AE Wageningen, The Netherlands*
D. Barron, *Institut de Pharmacognosie et Phytochimie, Université de Lausanne, BEP, CH-1015 Lausanne, Switzerland*
T. J. Bond, *James Finlay Limited, 163 West George Street, Glasgow G2 2JJ, UK*
V. Cheynier, *Unité Mixte de Recherche Sciences pour l'Oenologie, INRA, 2 place Viala, 34060 Montpellier cedex, France*
M. N. Clifford, *Centre for Nutrition and Food Safety, School of Biomedical and Life Sciences, University of Surrey, Guildford, Surrey GU2 7XH, UK*
C. Cren-Olivé, *Chimie Organique et Macromoléculaire, Université des Sciences et Technologies de Lille, Bâtiment C4, 59655 Villeneuve d'Ascq, France*
A. P. Davies, *Unilever Research, Colworth House, Sharnbrook, Bedford MK44 2LQ, UK*
A. Davis, *Unilever Research, Colworth House, Sharnbrook, Bedford MK44 2LQ, UK* and *School of Chemistry, University of Nottingham, University Park, Nottingham NG7 2RD, UK*
S. de Pascual-Teresa, *Hugh Sinclair Unit of Human Nutrition, School of Food Biosciences, University of Reading, PO Box 226, Reading RG6 6AP, UK, and Area de Nutrición y Bromatología, Facultad de Farmacia, Universidad de Salamanca, Campus Miguel de Unamuno, 37007 Salamanca, Spain*
A. J. Day, *Procter Department of Food Science, University of Leeds, Leeds LS2 9JT, UK*
M. T. Escribano-Bailón, *Area de Tecnología de Alimentos, Escuela Politécnica Superior de Zamora, Universidad de Salamanca, Campus Viriato, 49022 Zamora, Spain*
F. Ferreres, *Departamento de Ciencia y Tecnología de Alimentos, CEBAS (CSIC), PO Box 4195, Murcia 30080, Spain*
H. Fulcrand, *Unité Mixte de Recherche Sciences pour l'Oenologie, INRA, 2 place Viala, 34060 Montpellier cedex, France*
C. García-Viguera, *Departamento de Ciencia y Tecnología de Alimentos, CEBAS (CSIC), PO Box 4195, Murcia 30080, Spain*
M. I. Gil, *Departamento de Ciencia y Tecnología de Alimentos, CEBAS (CSIC), PO Box 4195, Murcia 30080, Spain*
A. Gil-Izquierdo, *Departamento de Ciencia y Tecnología de Alimentos, CEBAS (CSIC), PO Box 4195, Murcia 30080, Spain*

J. F. Hammerstone, *Mars, Incorporated, 800 High Street, Hackettstown, NJ 07840, USA*

P. C. H. Hollman, *RIKILT, PO Box 230, 6700 AE Wageningen, The Netherlands*

K. Hostettmann, *Institut de Pharmacognosie et Phytochimie, Université de Lausanne, BEP, CH-1015 Lausanne, Switzerland*

M. A. Kelm, *Mars, Incorporated, 800 High Street, Hackettstown, NJ 07840, USA*

S. A. Lazarus, *Masterfoods Australia/New Zealand, 4 Corella Close, Wyong, NSW 2259, Australia*

J. R. Lewis, *Unilever Research, Colworth House, Sharnbrook, Bedford MK44 2LQ, UK*

C. Manach, *U3M, INRA Clermont-Theix, 63122 St Genès Champanelle, France*

M. R. A. Morgan, *Procter Department of Food Science, University of Leeds, Leeds LS2 9JT, UK*

P. W. Needs, *Institute of Food Research, Norwich Research Park, Colney, Norwich NR4 7UA, UK*

K. Ndjoko, *Institut de Pharmacognosie et Phytochimie, Université de Lausanne, BEP, CH-1015 Lausanne, Switzerland*

K. O'Leary, *Faculty of Medicine, Hammersmith Campus, Ducane Road, London W12 0NN, UK*

G. W. Plumb, *Institute of Food Research, Norwich Research Park, Colney, Norwich NR4 7UA, UK*

J. C. Rivas-Gonzalo, *Area de Nutrición y Bromatología, Facultad de Farmacia, Universidad de Salamanca, Campus Miguel de Unamuno, 37007 Salamanca, Spain*

C. Rolando, *Chimie Organique et Macromoléculaire, Université des Sciences et Technologies de Lille, Bâtiment C4, 59655 Villeneuve d'Ascq, France*

C. Santos-Buelga, *Area de Nutrición y Bromatología, Facultad de Farmacia, Universidad de Salamanca, Campus Miguel de Unamuno, 37007 Salamanca, Spain*

H. H. Schmitz, *Mars, Incorporated, 800 High Street, Hackettstown, NJ 07840, USA*

F. Tomás-Barberán, *Departamento de Ciencia y Tecnología de Alimentos, CEBAS (CSIC), PO Box 4195, Murcia 30080, Spain*

D. P. Venema, *RIKILT, PO Box 230, 6700 AE Wageningen, The Netherlands*

G. A. Wächter, *Mars, Incorporated, 800 High Street, Hackettstown, NJ 07840, USA*

G. Williamson, *Nestlé Research Centre, PO Box 44, Vers Chez Les Blanc, Lausanne 26, CH-1000, Switzerland*

J.-L. Wolfender, *Institut de Pharmacognosie et Phytochimie, Université de Lausanne, BEP, CH-1015 Lausanne, Switzerland*

CHAPTER 1

Polyphenol Extraction from Foods

MARÍA TERESA ESCRIBANO-BAILÓN AND CELESTINO SANTOS-BUELGA

1 Introduction

The objective in extracting phenolic compounds from their plant sources is to liberate these compounds from the vacuolar structures where they are found, either through rupturing plant tissue or through a process of diffusion. The first case is accomplished by carrying out particle reduction generally by using a homogenizer in which the plant substance to be treated is interposed with the extraction solvent that will be used later. In the second case, nothing more than steeping is required.

In this chapter we will present a review of the most frequently used methods for extracting phenolic compounds, for analytical purposes from their plant sources, though it must be borne in mind that there is no single extraction protocol which can be considered optimally for all types of samples. There are three principal techniques that may be used: (1) extraction using solvents, (2) solid-phase extraction and (3) supercritical extraction. The latter method may also be considered as a type of solvent extraction in which the solvent is a fluid in a supercritical state, though it should be considered separately because of its peculiarities in technique and equipment.

2 Solvent Extraction

This is a process designed to separate soluble phenolic compounds by diffusion from a solid matrix (plant tissue) using a liquid matrix (solvent). The process can be divided into two stages:

1. *Initial stage*. Swelling of the particles or of the solid fragments is observed due to sorption of the solvent in the solid phase. This sorption is caused by osmotic forces, by capillarity and by solvation of the ions in the cells. In this stage, a certain

percentage of the polyphenols in the cells damaged in previous cutting, grinding or freezing of the product are extracted directly by washing. At the same time, the soluble components are dissolved. In some extractions there may also occur a solubilization through hydrolysis of a fraction naturally insoluble.

2. *Diffusion stage*. Diffusion takes place in two steps; an internal step within the solid phase and another external step through the outer layers that surround the particles or the solid fragments. In the extraction of coloured phenolic compounds, such as anthocyanin pigments, this stage is immediately perceived by the colour of the solution.

Factors That Influence the Efficiency of Solvent Extraction

The factors that contribute to the efficiency of solvent extraction are: type of solvent, pH, temperature, number of steps and volume of solvent, and particle size in the sample.

- Nature of the solvent. The most widely used solvent for extracting phenolic substances is methanol and methanol/water mixtures. Other solvents such as acetone, ethyl acetate and solvent mixtures have also been utilized, but they usually provide lower yields. Supercritical fluids have special properties that will be discussed later.
- pH of the extraction medium. This determines the degree of solubility for soluble compounds and also influences the possible solubilization of the hydrolysable fraction.
- Temperature. High temperatures improve the efficiency of the extraction since heat renders the cell walls permeable, increases solubility and diffusion coefficients of the compounds to be extracted and decreases the viscosity of the solvent, thus facilitating its passage through the solid substrate mass and subsequent separating processes (filtering or sedimentation). However, excessive temperature may degrade polyphenolic compounds so that the use of temperatures higher than 25 °C is uncommon. For example, Careri and coworkers,[1] in order to extract flavanones from orange juice, adds methanol and heats the mixture to 55 °C for 15 min to increase hesperidin solubility.
- Number of extraction steps and volume of solvent. The efficiency of the extraction increases along with the number of extraction steps. In this sense, it is more efficient, for example, to carry out four extractions with 50 ml of solvent than one with 200 ml. Quantitative yields are obtained only when 3–5 sequential extractions of the original plant material are carried out.
- Particle size and shape. Homogenization favours the extraction process and can be carried out in contact with the extraction solvent.

Sample Preparation

It is advisable to complete the extraction using dry, frozen or lyophilized samples since some phenolics are unstable or can be degraded by enzyme action in undried plant material. Oven drying is always unadvisable for it may decrease the extractability of some polyphenols (*e.g.* catechins), which would remain linked to fibre or proteins.[2] Furthermore, thermal degradation may also occur. However, the elimination of water through lyophilization generally does not affect the phenolic compounds excessively, and allows samples to be kept for longer periods.[3] Freezing the sample prior to extraction is also advisable since ice crystals produce lesions in the cellular structure and consequently facilitate the exit of cellular components and thus the process of extraction. If phenolic compounds quantification is the objective of the subsequent analysis, snap freezing in liquid nitrogen immediately after harvesting is advisable.

As stated above, reducing particle size in the sample to be extracted eases the process and allows for greater yields. A powder is obtained by crushing the dry, frozen or freeze-dried material in the presence of liquid nitrogen since oxidation is not a problem when working at such a low temperature. If the material cannot be crushed, it may be macerated with the solvent to be used for extraction. In this case, the alcoholic extraction solvent denatures plant enzymes, thus avoiding problems due to enzyme activity.

Matrices containing high levels of lipidic compounds usually require defatting in a Soxhlet apparatus prior to phenolic extraction.[4]

The Extraction Process

The extraction procedure is determined by the types of phenolic compounds to be extracted and whether the objective is quantitative or qualitative.

Ultrasound–assisted extraction is often used for the extraction of plant material using liquid solvents. This extraction process is faster and more complete in comparison with traditional methods such as maceration/stirring, because the surface area in contact between the solid and liquid phases is much greater due to particle disruption taking place.[5] Extraction time varies and depends on the previous preparation of the sample. Some very complete extractions may be accomplished in as little as 30 min.[5,6] The extraction can be also performed in a Soxhlet apparatus, thus combining percolation and immersion techniques.[7]

It is usual to incorporate an antioxidant (*tert*-butylhydroquinone, BHT, ascorbic acid or sulfites) to prevent phenolic oxidation,[1,8–18] but for anthocyanidins the addition of ascorbic acid to the extraction medium is not advisable due to the degradative properties of this antioxidant on the anthocyanins. It has also been indicated that using ascorbic acid during flavanol extraction may produce a certain degradation of proanthocyanidins.[19]

Soluble phenolic compounds are generally extracted using water, methanol, ethanol or acetone. The presence of attached sugars tends to render the phenolic compounds more water soluble, and combinations of the above solvents with water are thus better solvents for glycosides. In contrast, less polar aglycones such as isoflavones, flavanones and highly methoxylated flavones and flavonols tend to be more soluble in non-aqueous solvents.[19–21]

Methanol is the solvent most commonly employed. It has been used to extract flavanones,[1] flavones and flavone-glycosides,[22] methoxyflavones[23] and flavone dimers.[24] One problem that has been observed with methanol extraction in some juices after centrifugation is the appearance of turbidity in the last extractions. This can be prevented by adding of a small percentage of water to the extraction solvent.[25]

Methanol 70–80% has produced good yields in extracting hydroxycinnamic derivatives, flavones, flavonols and catechins from fruits,[6,26,27] legumes,[28] grape seeds[29] and wine pomace.[30]

Most flavonoids occur naturally as glycosides. As previously indicated, flavonoid glycosides show enhanced solubility in water compared with the corresponding aglycones. The use of methanol/water (50:50) is usually enough to produce a good extraction of glycosides from most plant material. However, due to the variety of heterosidic combinations, certain groups of flavonoids, such as flavones and flavonols, are not generally characterized as such intact compounds, but in the form of their aglycones.[1,18,31,32] For that reason, a hydrolysis procedure to break the glycoside bonds prior to or during extraction is required. Treatment with 1.5M HCl in a methanol–water solution 50:50 containing BHT as an antioxidant at reflux at 90 °C for 1 h has been used for the extraction of flavonol aglycones from plant products with good yields.[31] Basile *et al.*[33] reported complete extractions of flavones, flavonols and flavonones from *Castannea sativa* leaves using an aqueous solution of sulfuric acid at 70 °C and then extracting with ethyl acetate.

Selective Extraction Using Different pH Values

Fractionated extraction at different pH values has been used for selective extraction of free, sterified and insoluble-bound phenolic acids.[4] For that purpose, a sample previously defatted was extracted with 60% acetone, acidified to pH 3 with trichloroacetic acid and then centrifuged. After centrifugation the supernatant was evaporated under vacuum to remove acetone and the extract was treated with 4M NaOH under N_2 for 4 h to hydrolyse the sterified phenolic acids. The solution was then acidified to pH 2 with HCl. For extraction of the insoluble-bound phenolic acids, the residual after acetone extraction was treated with 4M NaOH under N_2 for 4 h and further acidified with HCl to pH 2 and centrifuged. The supernatant was combined with the acidified extract obtained earlier. The combined extracts were extracted with ethyl ether/ethyl acetate (1:1), the organic layers co-evaporated to dryness and all of the phenolic acids then re-dissolved in methanol.

Separation of neutral from acidic phenolics can be achieved through successive extraction with ethyl acetate at pH 7 and pH 2. This method accomplishes catechin and dihydrochalcone extraction with good accuracy and reproducibility, but recoveries are not as good as for procyanidins and flavonols.[34–37]

Selective extraction of wine phenolics can also be carried out at different pH values. The first step is to remove alcohol in a vacuum-rotating evaporator at a temperature lower than 30 °C to preserve polyphenols. Adjusting the sample to pH 7 followed by extraction with $CHCl_3$ permits obtaining hydrophobic compounds in the organic layer. Hydrophilic compounds can be recovered from the water phase by further adjustment to pH 6 followed by extraction with $CHCl_3$.[38] Adjustment to pH 2 with 0.1M HCl followed by extraction with diethyl ether has been used[39] to separate a variety of wine polyphenols: gallic acid, protocatechuic acid, protocatechuicaldehyde, (+)-catechin, 2,5-dihydroxybenzaldehyde, vanillic acid, caffeic acid, syringic acid, (−)-epicatechin, syringaldehyde, *p*-coumaric acid, ferulic acid, *trans*-resveratrol, myricetin quercetin and kaempferol.

Solvent Extraction of Anthocyanins

Extraction of anthocyanins is commonly carried out under cold conditions with methanol containing a small amount of acid. The acid most commonly employed is hydrochloric acid; acetic acid and formic acid are less advisable since anthocyanin acylation and artefact formation may occur. The use of acid is very important in determining extractant effectiveness and it is necessary for obtaining the flavylium cation form, which is red and stable in a highly acid medium. However, acid may cause partial hydrolysis of the acyl moieties in acylated anthocyanins, especially in anthocyanins acylated with dicarboxylic acids such as malonic acid. It has been reported that solvents containing up to $0.12\ mol\,l^{-1}$ hydrochloric acid (*i.e.* 1% HCl in methanol) cause partial hydrolysis of acylated anthocyanins from red grapes.[40] Thus replacing HCl with weaker acids such as tartaric[41] or citric acid[42–44] may be advisable for the extraction of anthocyanins with the dicarboxylic substituents intact. Also, small amounts of more volatile stronger acids, such as trifluoroacetic acid (0.5–3%), have been used for extractions of complex polyacylated anthocyanins.[45] In our experience, the use of methanol containing 0.1% HCl (*i.e.* 0.012N HCl in methanol) or methanol:1N HCl (95:5) (*i.e.* 0.05N HCl in methanol) do not cause significant degradation of the most usual monoacylated anthocyanins. For complex labile pigments, these same solvents can be used, but it is advisable to carry out the extraction at low temperature (*e.g.* keeping the samples in maceration in a freezer) and under argon or nitrogen.

Acetone has also been used to extract anthocyanins from several plant sources. According to Garcia-Viguera *et al.*,[46] in comparison with acidified methanol this solvent is more efficient and allows for more reproducible yields for anthocyanins extracted from red fruit, since it avoids problems with pectins and permits a much lower temperature for sample concentration.

Less common 3-deoxyanthocyanidins may occur in nature as aglycones, *e.g.* carajurin present in leaves of *Arrabidaea chica*; these low polarity anthocyanidins are best extracted with chloroform.[47]

Solvent Extraction of Flavan-3-ols

Extraction of these compounds (catechins, proanthocyanidins, condensed tannins) is more problematic than extraction of anthocyanins. The composition of the extractable fraction of flavanols from a sample may vary in accordance with the extraction procedure employed. Different solvents have thus been proposed, the principal ones being methanol, acetone, ethanol, ethyl acetate and water. Purification of the extracts is usually carried out by solvent partition (*e.g.* in ethyl acetate)[48–52] and/or selective precipitation with NaCl or caffein.[53]

Methanol and aqueous methanol are the most common solvents for extracting low-degree polymerization catechins from food, although it is assumed that the extraction may not be complete, depending on the characteristics of the matrix. According to Arts and Hollman,[54] both acetone and methanol (although methanol is more agreeable to work with) give similar maximum catechin yields, but the extraction is influenced by type and concentration of the solvent, which affect the yield of catechins. These authors found that 90% methanol was adequate for apple and grape (although extraction was sufficient for as low as 60% methanol, yields decreased notably below this percentage) and 80% for beans (as low as 40% was adequate). The differences in extraction yields may be explained by the action of polyphenol oxidase, the activity of which is reduced by methanol. A low methanol percentage extraction does not completely inactivate polyphenol oxidase in fresh fruits, thus reducing catechin yields. This does not apply to heat-treated products.

In some samples, treatment with water previous to methanol extraction increases the amount of extractable tannins. For instance, the extractability of flavanols from powdered lentils improves after fermentation in water. Increased flavanol extraction is highest in the methanol extracts following water treatment. When methanol is used directly, only about one third of the flavanols are found, as compared with when the water series is used. Enzymes may be involved in supporting the solubility of flavanols, since methanol is known to be an effective enzyme inhibitor. The question remains as to whether the solubility of flavanols is enhanced by a physical process or *via* enzyme activity during the extraction procedure.

Acetone is another widely used solvent.[10,55–57] Comparative studies have been carried out to establish its extractive efficiency compared with methanol. Acetone and methanol seem to have distinct specificities in the extraction of polyphenolic substances. It has been indicated that methanol is the best solvent for catechin extraction, whereas a better yield for procyanidins is obtained with 70% acetone. It has also been indicated that aqueous methanol, due to its polarity, extracts polyphenols linked to polar fibrous matrices more effectively, while acetone/water mixtures are more useful for extracting polyphenols from proteic matrices, since they appear to degrade the polyphenol-protein complexes.[13,58–60]

The combination of various extraction solvents, in order to take advantage of the specificity of each, is an alternative employed for achieving more complete extractions of proanthocyanidins. The extraction sequence: 80% methanol, 50% methanol, water and 75% acetone has been used for grapes,[61–63] whereas successive extractions with chloroform (monomers and dimers), acetone (trimers) and methanol (tetramers) have been employed for the latex of *Croton lechleri.*[64]

According to Haslam,[65] whichever the method chosen, the quantities of tannins extracted are believed to be small in comparison with those that remain unextracted.

Auxiliary Operations

Once the extraction to separate extract from plant material is achieved, centrifugation for 15–20 min at 20000–25000*g* is recommended. Filtration is less advisable since the retention of certain phenolic compounds (such as anthocyanins) in the filtration membrane may occur. A short gentle heating in warm water (*e.g.* 55 °C, 15 min) prior to centrifugation has produced good results for the extraction of flavones and flavanones[25] (in fact, hesperidin is difficult to extract without heating), but it is unadvisable for other temperature-sensitive flavonoids such as proanthocyanidins or anthocyanins.

The water–alcohol extract obtained often contains numerous non-phenolic substances such as sugars, fats, terpenes, pigments, *etc.* which can interfere with later analysis. Consequently, a purification step is necessary. The aqueous extract obtained after removal of the organic solvent is often washed with hexane to eliminate liposoluble components. It has been indicated that this extraction may be improved by the addition of 1–2% metaphosphoric acid and 10–20% ethanol,[66] although care should be taken that this does not cause a loss of certain phenolic molecules. Another purification method consists of extracting the defatted/depigmented aqueous extract with ethyl acetate to which ammonium sulfate (20%), metaphosphoric acid (2%) and ethanol (20%) have been added. These conditions are necessary to enable quantitative extraction of the main phenolic compounds by ethyl acetate. It should be noted that anthocyanins, which are practically insoluble in ethyl acetate, remain in the aqueous phase. This method for purification has been used to study phenolic compounds in various fruits.[26,67–69]

Whichever extraction method is used, the final extracts are usually concentrated. To achieve this, vacuum evaporation at low temperature (<30 °C) is recommended, although it is not advisable to dry the extract completely, since further dissolution of compounds from the residue may be more difficult and some degradation may occur. The addition of water prior to the complete evaporation of the solvent and further freeze-drying of the aqueous extract obtained is recommended.

Microwave-assisted Extraction

Microwave-assisted extraction (MAE) is a new extraction technique that combines microwave and traditional solvent extraction. Several studies[70–73] show that MAE has many advantages over conventional extraction methods that include shorter time, less solvent used, or higher extraction rate. Traditional solid–liquid extraction (SLE) methods typically take several hours, while MAE only takes a few minutes. MAE is a simple, cheap procedure that can be applied to more materials than SLE and with less polarity limitation for the extractant.

MAE has been shown to be an efficient method for extracting phenolic compounds from tea leaves[70] and grape seeds.[72] The methodology includes mixing the sample with an appropriate solvent, a ratio of 20:1 ($ml\,g^{-1}$) being sufficient. The extraction rate improves proportionately with the degree of grinding. After that, the sample is irradiated with microwaves for 4–12 min. In all cases, the irradiation is not constant so as to avoid temperature elevation. So, after an irradiation period of 45–60 s, the sample is cooled at room temperature or by cooling water. A pre-leaching time of 90 min at room temperature before MAE seems to increase polyphenol extraction.[72]

Because solvents with high dielectric constants can absorb more microwave energy, the polarity of the solvent is very important in microwave extraction. Although there is no uniform opinion on this, polar solvents are usually believed better than non-polar ones. However, there exists an opposite opinion, the 'broken cell-wall theory',[73] according to which microwave-transparent solvents are better than microwave-absorbing ones. Using a microwave-transparent solvent, all of the microwave energy is absorbed by the plant material; the water inside the cellular structures absorbs microwave energy very quickly, which suddenly increases the temperature inside cells and finally results in breaking the cell walls and releasing compounds into the surrounding solvent. That would explain why higher extraction levels of polyphenols have been found with acetone than with methanol, water or ethanol.[70] However, it has also been reported that when the solvent polarity was modified by the addition of water, increased yields are obtained.[70,72]

3 Solid-phase Extraction (SPE)

This is a rapid, easy, and economical alternative to solvent extraction since it significantly reduces the volume of organic solvent required. It is used to extract compounds from a liquid matrix or as a complement to solvent extraction. SPE may be also used as a fractionation/purification method or for pre-concentration of compounds. Nevertheless this extraction method does not always allow quantitative extraction so that the phenolic compound levels may be underestimated. For instance, with juices, SPE is ineffective for extracting flavonoids located in the solids found in suspension and which may represent an important fraction of the polyphenols present. For this type of matrix, solvent extraction appears to be the best alternative.

Fractionation of Phenolic Compounds

Phenolic compounds can be fractionated into neutral and acidic groups by means of a SPE method. Prior to the extraction process, juice is often centrifuged. For alcoholic samples, such as wine or cider, the alcohol much be removed by evaporation under vacuum. To avoid oxidation of polyphenols, an antioxidant can also be added.

Solid-phase extraction with C_{18} cartridges has been extensively employed for the selective extraction of phenolic acids and flavonoids from red wines,[74] orange juice,[75] cranberry juice,[76] grapes,[11] apple,[77,78] musts and ciders.[79] There are important differences between C_{18} solid-phase supports in relation to physical characteristics such as carbon loading and pore size.[79] A higher average pore size increases the retention capacity due to stronger interactions between the non-polar surface and the analyte, and also, a close linear relationship between carbon loading and the capacity factor is found, so the higher carbon loading accounts for a greater retention of the ionized form of phenolic acids. Suàrez *et al.*[79] achieved high reproducibility and recovery of phenolic analytes by using Extra-Sep C_{18} columns (1000 mg; Lida Kenosha, WI, USA).

A method for extracting neutral and acidic phenolic compounds with C_{18} cartridges is described below, although some optimization may be required as a function of the sample and support used. The cartridges must be preconditioned, one for neutral phenolics by sequentially passing through it 8 ml of methanol and 4 ml of water adjusted to pH 7.0 and the other one for acidics by passing 0.01M HCl instead of water through it. A suitable volume of sample is to load 8 mg of total polyphenols per gram of solid phase. The sample is adjusted to pH 7 with NaOH and loaded onto the C_{18} neutral cartridge and washed with 10 ml of pH 7 water. The washing volume can be modified for different matrices in order to eliminate phenolic acids remaining as interferents in the neutral extract. The washing step is critical and should not provoke losses of neutral phenols – catechin, especially, is often affected. The effluent portion is adjusted to pH 2.0 with 2M HCl, passed through the preconditioned acidic column and washed with 5 ml of 0.01M HCl to eliminate any residual sugars. This last step should be avoided in ciders or wines since these samples hardly contain sugars, and losses of some phenolic acids have been observed. Finally, the adsorbed fractions are eluted from the cartridges with 12 ml of methanol.[79]

Extraction of free phenolic acids can also be achieved using anion-exchange cartridges.[7] The process requires adjusting the solution containing the phenolic acids to pH 7.0–7.2 with an aqueous solution of sodium bicarbonate and bringing the cartridge to this same pH. Phenolic acids, as weakly acidic compounds (pK_a ~4–5) are extracted most effectively by anion-exchange sorbents (*i.e.* quaternary amine columns) about 2 pH units above their pK_a so they are retained on the sorbent bed and can be further eluted with methanol acidified with 0.2M H_3PO_4 (1:1).

Solid-phase Extraction of Anthocyanins

Several stationary phases have been used for preparing anthocyanins. Amberlite ion exchange resins can be used to eliminate polar non-phenolic compounds from crude anthocyanin extracts.[80,81] Polyvynilpyrrolidone (Polyclar AT) has been employed to separate individual anthocyanins, although irreversible adsorption to the phase may occur.[82] Separation can be improved with mixed stationary phases composed of PVP (20%) and Silicagel G (80%); washing with water to eliminate sugars and acids, and further elution using EtOH:H_2O gradients (containing 0.1% of 1N HCl) permits obtaining simpler anthocyanin fractions or even pure compounds.[83] Toyopearl gel HW-40(s) (*i.e.* Fractogel HW-40s) is another column support used, and has been useful for separating anthocyanins and anthocyanin-derived pigments.[84,85]

A procedure for separating anthocyanins from other sample components using

Polyclar AT (PVP) has been used with good results in our laboratory. Polyclar AT is poured into a beaker containing water and left to settle in order to discard the finest particles remaining in the supernatant. It is then packed into the column and equilibrated by passage of water. The anthocyanin extract is carefully deposited onto the column, washed with water acidified with 1% of 0.1N HCl and the anthocyanins by further elution with a mixture of methanol/water/HCl (70:30:1, v/v/v) until no more colour is eluted from the column. The stationary phase can be regenerated by passing 1N NaOH through it and then washing it with water until neutral again.

In general, separation with these phases is not complete and further purification is usually required using the same support or by HPLC, to obtain pure anthocyanins. Reversed phases are more selective for anthocyanin isolation, especially if they are applied to pre-purified anthocyanin fractions rather than to crude extracts. Anthocyanin elution follows similar patterns to those in analytical HPLC (see Chapter 5). For the preparation of anthocyanins and anthocyanin fractions, acetonitrile and formic acid (the solvents most usually employed for analytical HPLC) are not advisable, due to the theoretical possibility of acylation and the difficult evaporation of the acetonitrile. In our laboratory, satisfactory separations of anthocyanins and anthocyanin-derived pigments using reversed-phase C_{18} have been obtained both at medium and high pressure using methanol with either acetic or trifluoroacetic acid as acid modifiers.[86]

Solid-phase Extraction of Proanthocyanidins

The separation of proanthocyanidins poses more difficulties than that of anthocyanins. Sephadex LH-20 has been extensively used with good results for fractionating proanthocyanidin mixtures as well as separating procyanidins from other phenolics in different plant material and in beverages.[9,13,14,17,58,87–90] This is an exclusion gel that also acts by adsorption and partition in direct phase. The most usual solvents are ethanol, methanol and acetone and their water mixtures. The separation is based on the establishment of hydrogen bonds between phenolic hydrogens or carboxylic groups and acceptors in the gel. The strength of the adsorption depends on the number of phenolic hydrogens per molecule; polymeric polyphenols, like condensed tannins, are adsorbed more readily than monomers, such as catechins. Ethanol, a frequently employed solvent, is not very efficient for displacing polymeric polyphenols. Acetone is a better desorbent, since carbonyl oxygen acts as a strong acceptor for hydrogen bonding and is capable of displacing the polymers of the gel. It is important that the elution be carried out at low flow rate in order to give time for the exchanges to take place and to achieve good separations.[17] Other stationary phases used for the separation of flavanols are Sephadex LH-25,[91] Sep-Pak C_{18} cartridges,[11,63,74,92] Fractogel TSK,[93–97] Biogel P-2,[98] Polyamide[61,99] and Amberlite.[100,101]

Generally, the fractions obtained have to be submitted to a further separation, either by column chromatography or HPLC, when compound isolation is required.

Further information about HPLC separation of proanthocyanidins can be found in Chapter 5.

4 Supercritical Fluid Extraction (SFE)

Supercritical fluid extraction was developed in the 1960s and, in recent years, has acquired some relevance for the extraction of polyphenols from plant sources. The main advantage of SFE is that it combines the characteristics of gases and liquids for extraction. The low viscosity of the supercritical fluids confers a high capacity for diffusion and improves access to phenolic compounds bound to the cell wall. Moreover, its relatively high density confers a high solvation power, which greatly facilitates the extraction process. Furthermore, it minimizes any possible degradation processes, such as oxidations or isomerizations (*e.g.* isomerization of the naturally occurring *trans*-resveratrol to *cis*-resveratrol) that may occur with other more conventional extraction techniques, because it reduces extraction time and because the process can be carried out in the absence of light and air.

Supercritical carbon dioxide is the most widely used extraction solvent. It has certain advantages that are attracting increasing interest among researchers: it is chemically inert, has a low toxicity, presents no problems in terms of pollution, and has a short concentration time. However, supercritical carbon dioxide is an apolar solvent and, to allow for the extraction of polar compounds such as polyphenols, other more polar solvents have to be added as modifiers, or else the compounds to be extracted must be first derivatized to decrease their polarity.[102–113]

SFE occurs in two steps. In the first step, the phenolic compounds are extracted from the matrix after being solubilized by the supercritical CO_2 In the second step, the vent valve on the extractor is opened immediately and carefully, and the soluble compounds are trapped either in a liquid solvent (*i.e.* methanol) or on an inert solid surface. One advantage of solid trapping over liquid trapping is that much less solvent is used to elute analytes from the solid trap than is required for liquid trapping.[104] Many optimization studies have been performed on both the extraction and the trapping steps. The variables to be set up include: CO_2 density, modifier type, modifier percentage, extraction temperature, dynamic extraction time, CO_2 flow rate, trap temperature, trap solid phase, trap rinse solvent and trap rinse solvent flow rate.

To extract the phenolic compounds, solid or semi-solid samples must be pulverized and packed into a sample cartridge and the modifier added. Depending on the polarity of the compound to be extracted, ethanol, methanol, aqueous methanol or ethyl acetate may be used as modifiers. The high polarity of polyphenols usually requires high density and a high percentage of modifier. Extraction temperature also has to be set up. Low temperatures are recommended because most phenolic compounds are unstable and easily oxidized at high temperatures. Finally, liquid carbon dioxide at high pressure is allowed to flow into the sample cartridge. Approximately 20–25 ml of solvent are used for 1 g of material, and the extraction process is run at least three times.

The solubility of the compounds to be extracted in the supercritical fluid is probably the most important property that must be determined in order to efficiently design an extraction process based on supercritical solvents. Unfortunately, solubility data on polyphenols are still scarce in the literature.[107–109,113–115]

For the extraction step, Lin *et al.*[6] developed optimal conditions for supercritical fluid extraction of flavones from *Scutellaria baicalensis* radix. As expected, the yield obtained with pure supercritical carbon dioxide was not satisfactory and changes in pressure and temperature led to only negligible improvements. Good results were obtained only by increasing the polarity of the extraction solvent; thus, optimal extraction was obtained using supercritical carbon dioxide–MeOH–water (20:2.1:0.9), at a temperature of 50 °C and a pressure of 200 bar.

For the trapping step, one of the most influential variables affecting recovery is the type of trapping material. After using octadecylsilica (C_{18}) and ethylvinylbenzene–divinylbenzene (EVB–DVB) as solid phases to trap several phenolic compounds (cinnamic acids, benzoic acids, benzoic aldehydes, resveratrol and catechin), Palma and Taylor[104] concluded that C_{18} was the best trapping material.

5 Pressurized Liquid Extraction (PLE)

This technique has been recently introduced for phenolic compound extraction.[116] In PLE, high temperature and high pressure are used to accelerate the extraction. Pressure increases contact between the extracting fluid and the sample and the high temperature breaks the phenolic-matrix bonds. The hot solvent denatures the cells due to the coagulation of lipoproteins, making the permeability of the cell wall less and less selective. Through reheating, the volume of the internal liquid phase increases, leading to increased pressure, which in turn causes centrifugal circulation of the solutions through the pores.

PLE also offers the possibility of performing extractions in an inert atmosphere protected from light, which represents an advantage since phenolic compounds are very sensitive to these two factors. The stability of phenolic compounds during PLE using methanol under pressure at temperatures ranging from 40 to 150 °C as the extracting fluid, was studied by Palma *et al.*[116] It was determined that at temperatures of about 100 °C the maximum degree of degradation suffered by phenolic compounds was 10%. These same authors showed that temperatures of between 50 and 100 °C produce the best results for the extraction of phenolics from grape skins, but higher temperatures (150 °C) were necessary for a suitable extraction from grape seeds. The preliminary results obtained allow those authors to consider PLE as a promising alternative for polyphenol extraction.

References

1. M. Careri, L. Elviri, A. Mangia and M. Musci, *J. Chromatogr. A*, 2000, **881**, 449.

2. F. S. Jackson, T. N. Barry, C. Lascano and B. Palmer, *J. Sci. Food Agric.*, 1996, **71**, 103.
3. I. C. W. Arts, P. C. H. Hollman and D. Kromhout, *The Lancet*, 1999, **354**, 488.
4. L. Xu and L. L. Diosady, *Food Res. Int.*, 1997, **30**, 571.
5. M. Palma and C. G. Barroso, *Anal. Chim. Acta*, 2002, **458**.
6. M. C. Lin, M. J. Tsai and K. C. Wen, *J. Chromatogr. A*, 1999, **830**, 387.
7. K. Glowniak, G. Zgorka and M. Kozyra, *J. Chromatogr. A*, 1996, **730**, 25.
8. Z. Czochanska, L. Y. Foo, R. H. Newman, L. J. Poter and W. A. Thomas. *J. Chem. Soc. Chem. Commun.*, 1979, 375.
9. A. E. Hagerman and L. G. Butler, *J. Agric. Food Chem.*, 1980, **28**, 944.
10. J. Jerumanis, *J. Inst. Brew.*, 1985, **91**, 250
11. A. W. Jaworski and C. Y. Lee, *J. Agric. Food Chem.*, 1987, **35**, 257.
12. J. Field, G. Cettinga and L. Habets, *J. Ferment. Bioeng.*, 1990, **69**, 148.
13. L. Hussein, M. A. Fattah and E. Salem, *J. Agric. Food Chem.*, 1990, **38**, 95.
14. M. T. Escribano-Bailon, Y. Gutierrez-Fernandez, J. C. Rivas-Gonzalo and C. Santos-Buelga, *J. Agric. Food Chem.*, 1992, **16**, 79.
15. M. Azar, E. Verette and S. Brun, *J. Food Sci.*, 1987, **52**, 1255.
16. J. Oszmianski and J. Sapis, *J. Agric. Food Chem.*, 1989, **37**, 1293.
17. K. Kantz and V. L. Singleton, *Am. J. Enol. Vitic.* 1990, **34**, 3208.
18. S. H. Hakkinen and A. R. Torronen, *Food Res. Int.*, 2000, **33**, 517.
19. M. P. Bradshaw, P. D. Prenzler and G. R. Scollary, *J. Agric. Food Chem.*, 2001, **49**, 934.
20. K. R. Markham, '*Techniques of Flavonoid Identification*', Academic Press, London, 1982.
21. K. R. Markham and S. J. Bloor, in '*Flavonoids in Health and Disease*', eds C. A. Rice-Evans and L. Packer, Marcel Dekker Inc., New York, 1998, p. 3.
22. S. L. Yang, M. F. Roberts, M. J. O'Neill, F. Bucar and D. Phillipson, *Phytochemistry*, 1995, **38**, 255.
23. C. W. Huck, C. G. Huber, K. H. Ongania and G. K. Bonn, *J. Chromatogr. A*, 2000, **870**, 453.
24. I. C. Moreira, M. G. Carvalho, A. B. Bastos and R. Braz-Filho, *Phytochemistry*, 1999, **51**, 833.
25. W. E. Bronner and G. R. Beecher, *J. Chromatogr. A*, 1995, **705**, 247.
26. J. Billot, C. Hartmann, J. J. Macheix and J. Rateau, *Physiol. Veg.*, 1978, **16**, 693.
27. C. Melin, J. Billot and J. F. Dupin, *Physiol. Veg.*, 1979, **17**, 557.
28. A. Escarpa, M. D. Morales and M. C. Gonzalez, *Anal. Chim. Acta*, 2002, **460**, 61.
29. J. Oszmianski and C. Y. Lee, *Am. J. Enol. Vitic*, 1990, **41**, 204.
30. A. Gonzalez-Paramas, S. C. de Pascual-Teresa, J. C. Rivas-Gonzalo and C. Santos-Buelga, in '*Polyphenols Communications 2000*', Proceedings of the XXth International Conference on Polyphenols, Freising-Weihenstephan, Germany, 2000, p. 143.
31. M. G. L. Hertog, P. C. H. Hollman and M. B. Katan, *J. Agric. Food Chem.*, 1992, **40**.

32. S. Häkkinen and S. Auriola, *J. Chromatogr.*, 1998, **829**, 91.
33. A. Basile, S. Sorbo, S. Giordano, L. Ricciardi, S. Ferrara, D. Montesano, R. Castaldo-Cobianchi, M. L. Vuotto and L. Ferrara, *Fitoterapia,* 2000, **71**, 110.
34. E. Delage, G. Bohuon, A. Baron and J. F. Drilleau, *J. Chromatog.*, 1991, **555** 125.
35. B. Suarez, J. Santamaria, J. J. Mangas and D. Blanco, *J. Agric. Food Chem.*, 1994, **42**, 2732.
36. M. H. Salagoity-Auguste and A. Bertrand, *J. Sci. Food Agric.*, 1984, **35**, 1241.
37. A. Schieber, W. Ullrich and R. Carle, *Inn. Food Sci. Emerg. Technol.*, 2000, **1**, 161.
38. Z. Kovacs and Z. Dinya, *Microchem. J.*, 2000, **67**, 57.
39. M. A. Rodríguez-Delgado, S. Malovana, J. P. Perez and T. Borges, *J. Chromatogr. A*, 2001, **912**, 249.
40. E. Revilla, J. M. Ryan and G. Martin Ortega, *J. Agric. Food Chem.*, 1998, **46**, 4592.
41. T. Philip, *J. Food Sci.*, 1974, **39**, 859.
42. J. H. Main, F. M. Clydesdale and F. J. Francis, *J. Food Sci.*, 1978, **43**, 1693.
43. J. P. Calvi and F. J. Francis, *J. Food Sci.*, 1978, **43**, 1448.
44. F. M. Clydesdale, J. H. Main, F. J. Francis and R. A. Damon, *J. Food Sci.*, 1978, **43**, 1687.
45. T. Goto, T. Kondo and H. Tamura, *Tetrahedron Lett.*, 1984, **25**, 6021.
46. C. Garcia Viguera, P. Zafrilla, F. Artes, F. Romero, P. Abellan and F. A. Tomas-Barberan, *J. Sci. Food Agric.*, 1998, **78**, 565.
47. B. Devia, G. Llabres, J. Wouters, L. Dupon, M. T. Escribano-Bailón, S. C. de Pascual-Teresa, L. Angenot and M. Tits, *Phytochem. Anal.*, 2002, **13**, 114.
48. L. Foo and J. J. Karchesy, *Phytochemistry*, 1989, **28**, 3185.
49. R. K. Gupta and E. Haslam, *J. Chem. Soc. Perkin Trans. 1*, 1978, 892.
50. E. Haslam, *Phytochemistry*, 1980, **19**, 2577.
51. H. Kolodziej, M. K. Sakar, J. F. W. Burger, R. Engelshowe and D. Ferreira, *Phytochemistry*, 1991, **30**, 2041.
52. S. Pascual-Teresa, Y. Gutierrez-Fernandez, J. C. Rivas-Gonzalo and C. Santos-Buelga. *Phytochem. Anal.*, 1998, **9**, 21.
53. A. E. Hagerman and L. G. Butler, in '*Herbivores: Their Interactions with Secondary Plant Metabolites*', Vol I, eds G. A. Rosenthal and M. R. Berenbam, Academic Press, London, 1991, 355.
54. I. C. W. Arts and P. C. H. Hollman, *J. Agric. Food Chem.*, 1998, **46**, 5156.
55. R. B. Broadhurst and W. T. Jones, *J. Sci. Food Agric.*, 1978, **29**, 788.
56. Y. Kashiwada, M. Morita, G. Nonaka and Y. Nishioka, *Chem. Phar. Bull.*, 1990, **38**, 856.
57. G. Nonaka, O. Kawahara and Y. Nishioka, *Chem. Pharm. Bull.*, 1983, **31**, 3906.
58. G. C. Galleti and R. Self, *Ann. Chim. (Rome)*, 1986, **76**, 195.
59. M. N. Clifford and J. R. Ramírez Martínez, *Food Chem.*, 1991, **40**, 191.

60. S. Kallithraka, C. García Viguera, P. Bridle, and J. Bakker, *Phytochem. Anal.*, 1995, **6**, 265.
61. M. Bourzeix, D. Weyland and N. Heredia, *Bull. OIV*, 1986, **29**, 788.
62. M. Bourzeix, M. Clarens and N. Heredia, *Bulg. Acad. Sci.: Comm. Dep. Chem.*, 1987, **20**, 3.
63. E. Revilla, M. Bourzeix and E. Alonso, *Chromatographia*, 1991, **31**, 465.
64. Y. Cai, F. J. Evans, M. F. Roberts, J. D. Phillipson, M. H. Zenk and Y. Y. Gleba, *Phytochemistry*, 1991, **30**, 2033.
65. E. Haslam, '*Plant Polyphenols. Vegetable Tannins Revisited*'. Cambridge University Press, Cambridge, 1989.
66. J. L. Poessel, Thèse 3eme Cycle, Montpellier, France, 1983, cited in '*Fruit Phenolics*', eds J. J. Macheix, A. Fleuriet and J. Billot, CRC Press Inc, Boca Raton, Florida, 1990.
67. M. J. Amiot, A. Fleuriet and J. J. Macheix, *J. Agric. Food Chem.*, 1986, **34**, 823.
68. A. Fleuriet and J. J. Macheix, *J. Chromatogr.*, 1972, **74**, 339.
69. B. Möller and K. Herrmann, *Phytochemistry*, 1983, **22**, 477.
70. X. Pan, G. Niu and H. Liu, *Chem. Eng. Proc.*, 2003, **42**, 121.
71. G. A. Csiktusnàdi Kiss, E. Forgàcs, T. Cserhati, T. Mota, H. Morais and A. Ramos, *J. Chromatogr. A*, 2000, **889**, 41.
72. N. Hong, V. A. Yaylayan, G. S. Raghavan, J. R. Pare and J. M. Bélanger, *Nat. Prod. Lett.*, 2001, **15**, 197.
73. H. Jinyu, H. Wei, H. Shunde, X. Boyong and D. Xiu, *Sep. Purifi. Technol.*, 2002, in press.
74. J. Oszmianski, T. Ramos and M. Bourzeix. *Am. J. Enol. Vitic.*, 1988, **39**, 259.
75. R. L. Rouseff, G. R. Dettweiler and R. M. Swaine, *J. Chromatogr. Sci.*, 1992, **30**, 383.
76. H. Chen, Y. Zuo and Y. Deng, *J. Chromatogr. A*, 2001, **913**, 387.
77. S. Burda, W. Oleszek and C. Y. Lee, *J. Agric. Food Chem.*, 1990, **38**, 945.
78. M. Y. Coseteng and C. Y. Lee, *Food Sci.*, 1987, **52**, 985.
79. B. Suàrez, A. Picinelli and J. J. Mangas, *J. Chromatogr. A*, 1996, **727**, 203.
80. A. Chandra, M. G. Nair and A. Iezzoni, *J. Agric. Food Chem.*, 1993, **41**, 1063.
81. H. Wang, M. G. Nair A. F. Iezzoni, G. M. Strasburg, A. M. Booren and J. I. Gray, *J. Agric. Food Chem.*, 1997, **45**, 2556.
82. C. G. Van Teeling, P. E. Cansfield and R. A. Gallop, *J. Chromatogr. Sci.*, 1971, **9**, 505.
83. E. Hebrero, C. Santos-Buelga and J. C. Rivas-Gonzalo, *Am. J. Enol. Vitic.*, 1988, **39**, 227.
84. N. Mateus, A. M. S. Silva, J. Vercauteren and V. A. P. de Freitas, *J. Agric. Food Chem.*, 2001, **49**, 4836.
85. T. Shoji, A. Yanagida and T. Kanda, *J. Agric. Food Chem.*, 1999, **47**, 2885.
86. F. J. Heredia, E. M. Francia Aricha, J. C. Rivas-Gonzalo, I. M. Vicario and C. Santos Buelga, *Food Chem.*, 1998, **63**, 491.

87. R. S. Thompson, D. Jacques, E. Haslam and R. I. N. Tanner, *J. Chem. Soc. Perkin Trans. 1*, 1972, 1387.
88. A. G. H. Lea, P. Bridle, C. F. Timberlake and V. L. Singleton, *Am. J. Enol. Vitic.*, 1979, **30**, 289.
89. R. W. Hemingway, L. Y. Foo and L. J. Porter, *J. Chem. Soc. Perkin Trans. 1*, 1982, 1209.
90. G. A. Spanos, R. E. Wrolstad and D. A. Heatherbell, *J. Agric. Food Chem.*, 1990, **38,** 1572.
91. T. C. Somers, *Nature*, 1966, **209**, 368.
92. E. Alonso, M. Bourzeix and E. Revilla, *Seed Sci. Technol.*, 1991, **19** 545.
93. M. Boukharta, M. Girardin and M. Metche, *J. Chromatogr.*, 1988, **455**, 406.
94. J. A. Delcour, E. J. Serneels, D. Ferreira and D. G. Roux, *J. Chem. Soc. Perkin Trans. 1*, 1985, 669.
95. L. J. Porter, Z. Ma and B. G. Chan, *Phytochemistry*, 1991, **30**, 1657.
96. J. M. Souquet, V. Cheynier, F. Brossaud and M. Moutounet, *Phytochemistry*, 1996, **43**, 509.
97. J. M. Ricardo da Silva, J. Rigaud, V. Cheynier, A. Cheminat and M. Moutounet, *Phytochemistry,* 1991, **30,** 1259.
98. K. Yokotsuka, S. Shinkai and T. Kushida, *J. Ferment Technol.*, 1980, **58**, 107.
99. J. M. Ricardo da Silva, J. P. Rosec, M. Bourzeix and N. Heredia, *J. Sci. Food Agric.*, 1990, **53**, 85.
100. R. Di Stefano and M. C. Cravero, *Riv. Vitic. Enol.*, 1991, **2**, 37.
101. A. Degenhardt, U. H. Engelhardt, A. S. Wendt and P. Winterhalter, *J. Agric. Food Chem.*, 2000, **48**, 5200.
102. Y. Lin, N. G. Smart and C. M. Wai, *Trends Anal. Chem.*, 1995, **14**, 123.
103. M. T. Tena, A. Rios and M. Valcarcel, *Fresenius' J. Anal. Chem.*, 1998, **361**, 143.
104. M. Palma and L. T. Taylor, *Anal. Chim. Acta*, 1999, **391**, 321.
105. M. Palma and L. T. Taylor, *J. Agric. Food Chem,* 1999, **47**, 5044.
106. M. Palma and L. T. Taylor, *J. Chromatogr.*, 1999, **849**, 117.
107. A. Berna, A. Chafer and J. B. Monton, *J. Supercrit. Fluids*, 2001, **19**, 133.
108. A. Berna, A. Chafer, J. B. Monton and S. Subirats, *J. Supercrit. Fluids*, 2001, **20**, 157.
109. A. Chafer, A. Berna, J. B. Monton and R. Muñoz, *J. Supercrit. Fluids*, 2002, In press.
110. T. Tsuda, K. Mizno, K. Ohshima, S. Kawakishi and T. Osawa, *J. Agric. Food Chem.,* 1995, **43**, 2803.
111. J. Castaneda Acosta, A. W. Cain, N. H. Fischer and F. C. Knopf, *J. Agric. Food Chem.*, 1995, **43**, 63.
112. D. D. Michael and N. S. James, *Anal. Chem.*, 1996, **68**, 3038.
113. R. Murga, R. Ruiz, S. Beltràn and J. L. Cabezas, *J. Agric. Food Chem.,* 2000, **48**, 3408.
114. H. Uchiyama, K. Mishima, S. Oka, M. Ezawa, M. Ide, T. Takai and P. W. Park, *J. Chem. Eng. Data*, 1997, **42,** 570.
115. E. S. Choi, M. J. Noh and K. P. Yoo, *J. Chem. Eng. Data*, 1998, **43**, 6.
116. M. Palma, Z. Piñeiro and C. G. Barroso, *J. Chromatogr. A*, 2001, **921**, 169.

CHAPTER 2

Methods of Polyphenol Extraction from Biological Fluids and Tissues

ANDREA J. DAY AND MICHAEL R. A. MORGAN

1 Introduction

Research on the absorption, metabolism and tissue distribution of dietary polyphenols in human and animal studies relies on efficient extraction of the polyphenol or putative polyphenol metabolites from the blood, urine or organ under investigation. Though analysis of polyphenols in food (necessarily involving extraction) has received much critical attention and review over many years, resulting in the reaching of general consensus as to optimum procedures, the more recent interest in animal fluids and tissues has moved quickly and, we suggest, without full regard for analytical efficiency. Additional factors to those relevant to extraction from plants and food matrices need to be considered when dealing with mammalian metabolism, and extraction procedures require full validation even if similar procedures are employed for the two purposes. In addition, different polyphenol classes may require different extraction techniques. It must also be remembered that sophistication of analytical detection and quantification – and there are now some highly sophisticated instrumental techniques available – cannot compensate for inefficient analyte extraction.

Our present aim is to review some of the methods that have been used for extraction of polyphenols in human and animal studies, to discuss the major problems that may occur during extraction, and to show that more attention needs to be paid to validation of analytical and extraction procedures.

It is our opinion that some of the results presented in the literature, and the conclusions derived therefrom, are incorrect because of the use of non-validated and non-optimum extraction procedures. We will not seek to validate published procedures or recommend a 'best' method, since further information is required before such an approach can be fully justified, desirable though it might be.

Primary Considerations

Regardless of the extraction procedure used, consideration first needs to be given to the chemical identity of the analyte on which information is required. In mammalian tissues and fluids, polyphenols will usually be present as conjugated metabolites (*e.g.* β-glucuronides or sulfates). Hydrolysis of such metabolites to form aglycones may be undertaken if

- it is not of interest to monitor the particular form(s) of the metabolite,
- the aglycone alone is the source of interest, or
- the sensitivity of the analysis of individual compounds is inadequate and requires the concentration effect given by hydrolysis.

However, it must be emphasised that, whatever the justification, hydrolysis will destroy biological information. Where detailed information is required as to the nature of the metabolite, then even more care is needed in choice of extraction protocol.

We will deal first with the use of solvent extraction of polyphenols (with or without hydrolysis) before covering the choice of hydrolysis procedure.

2 Polyphenol Extraction

Methods for extraction of polyphenols can be divided into two, those which use solvent extraction and those which combine solvent extraction with solid phase extraction.

Solvent Extraction

Solvents used include acetone, acetonitrile, methanol and ethanol in the presence of acid, and ethyl acetate. The nature of the solvent used for extraction is critical since, for example, solvents such as ethyl acetate and acetone are poor at extracting the most polar metabolites such as the sulfates.[1] It is imperative to characterise extraction efficiencies for the compounds and solvents of interest as well as for the matrix to be extracted. Table 1 reviews the characteristics of a number of solvent extraction methods described in the literature in recent years.

Comparing Solvents for Extraction of Quercetin Conjugates

As an illustration, the following approach has been used by us to assess the recovery of quercetin and quercetin-3-glucoside from human plasma using previously published procedures.[2–4]

Samples. Plasma samples from volunteers on a flavonol-free diet, spiked with quercetin (26 μM) or quercetin-3-glucoside (46 μM). Acetic acid (0.65 mM; 0.1 volume) used to adjust pH to 5; ascorbic acid (to a final concentration of 1 mM)

Table 1 *Methods of polyphenol extraction from animal tissue or fluids using solvent only*

Polyphenol	*Biological fluid*	*Method description*	*Solvent*	*Hydrolysis*	*% Recovery*	*Additional information*
Naringenin, naringenin glycosides[18]	Rat plasma and urine	Plasma acidified to pH 4.9 with acetic acid. Samples were enzyme hydrolysed, extracted with 2.8 vol. acetone, centrifuged and supernatants evaporated to original volume. Acidified urine (pH 4.9) was enzyme hydrolysed and extracted in 1 volume acetone.	Acetone	Enzyme. *H.P.* 6×10^6 β-glucur. + 1.8×10^5 sulf. U l^{-1}; *E.C.* 10^6 U l^{-1} β-glucur.	n/g	No aglycone present in plasma. 5 metabolites disappeared with *H.P.* but only 5–10% with *E.C.* Phenolic acids also recovered mainly as conjugates.
Quercetin and rutin were fed as a single meal or over 10 d[19]	Rat plasma	Plasma acidified to pH 4.9 with acetic acid then enzyme hydrolysed and extracted with 8.5 vol. Acetone, centrifuged and supernatants evaporated to 2 × original volume.	Acetone	Enzyme. *H.P.* β-glucur, 5×10^6 + sulf. 2.5×10^5 U l^{-1}	Recovery of pure compounds >85%.	Quercetin was present in plasma mainly as methyl glucuro-sulfate conjugates.
Quercetin and catechin as a single meal or over 10 d[9]	Rat plasma and liver	Plasma acidified to pH 4.9 with acetic acid and enzyme hydrolysed. Aglycones were extracted with 2.75 volume acetone, centrifuged and evaporated to 2 × original volume.	Acetone	Enzyme. (*plasma*: 8×10^5 U l bovine β-glucur.; 5×10^6 β-glucur. + 2.5×10^5 U l sulf. *Liver*: $1.1. \times 10^7$ U l^{-1}	n/g	Free catechin (80%) and methylcatechin (95%) observed in liver, whereas 25–30% of quercetin/ methyl-quercetin was in the liver (100% methylated). No sulfated

continued overleaf

Table 1 *(continued)*

Polyphenol	*Biological fluid*	*Method description*	*Solvent*	*Hydrolysis*	*% Recovery*	*Additional information*
		Livers were homogenised with 2 vol. ice-cold acetate buffer (0.05 M, pH5, containing 1 $g\,l^{-1}$ ascorbate). Homogenate was centrifuged (5 min, 15000 *g*) and the supernatant enzyme hydrolysed. Sample was extracted with 2.75 volumes acetone as above.		β-glucur. + 4.3×10^5 $U\,l^{-1}$ sulf.)		(methyl) catechin, most (methyl) quercetin was sulfated.
Onions, buckwheat tea, quercetin-4′-glucoside or rutin were given as a rich source of quercetin glycosides[14]	Human plasma and urine	Plasma or urine (500 μl containing ascorbic acid, 10 μl, 0.5%, w/v) was adjusted to pH 5 with acetic acid and enzyme hydrolysed. Sample extracted with acetone (600 μl), vortex-shaken (10 min), and centrifuged. Ascorbic acid (10 μl, 0.5% w/v) was added to the supernatant before evaporation. The residue was dissolved in DMF/H_2O (150 μl, 2:1, v/v).	Acetone	Enzyme. *H.P.* β-glucur./sulf.	n/g	Quercetin and glycosides were not detected, but 4–5 glucuronides were. Sulfates were not found. All metabolites were equivalent regardless of food source.
Onions fed as a rich source of quercetin[15]	Human plasma	Plasma (500 μl) was mixed with acetic acid (20 μl, 0.58 M) and acetone	Acetone	NO	n/g	No sulfates were found in plasma samples.

		(500 μl), vortex-mixed and centrifuged.				
Quercetin[20]	Rat plasma	Plasma (1 ml) was added to MeOH/acetic acid (100:5, v/v, 19 ml), then centrifuged. The supernatant evaporated, dissolved in 2% acetic acid (20 ml, pH 2.65) and extracted 3× with equal volume ethyl acetate. Then dried, dissolved in H_2O (5 ml) and washed 2 × with n-hexane and centrifuged. The aqueous phase evaporated, dissolved in aqueous MeOH (200 μl, 50% v/v containing 0.03% TFA). Plasma was enzyme hydrolysed in acetate buffer (0.1 M, pH 5, 50 min, 37 °C) before extraction as above.	Acidic MeOH and ethyl acetate	Enzyme. H.P. 25 U	n/g	No quercetin present. 5 metabolites, with 2 corresponding to synthesised glucuronides.
Onions fed as rich source of quercetin glucosides[21]	Human plasma	Plasma (50 μl) was enzyme hydrolysed (0.1 M acetate buffer, pH 5, 37 °C, 50 min), followed by extraction with MeOH-acetic acid (100:5 v/v, 900 μl). After centrifuging	Acidic MeOH	Enzyme. H.P. 25 U	95% recovery, quercetin spiked plasma	Quercetin identified only after hydrolysis of plasma. Glucosides were not present in unhydrolysed samples.

continued overleaf

Table 1 *(continued)*

Polyphenol	*Biological fluid*	*Method description*	*Solvent*	*Hydrolysis*	*% Recovery*	*Additional information*
		supernatant was evaporated and dissolved in MeOH.				
Onions were fed as a rich source of quercetin[22]	Human plasma	MeOH (1 ml, containing 2 g l^{-1} TBHQ) and HCl (10 M, 400 μl) were added to plasma (600 μl) and refluxed (90 °C, 2 h). MeOH (2 ml, containing 2 g l^{-1} TBHQ) was added and the sample centrifuged (1000 g, 15 min).	Acidic MeOH	Acid hydrolysis	Recovery of spiked quercetin was 88% ± 11%.	Conjugates were not analysed.
Onions, as a rich source of quercetin, quercetin or rutin supplements were fed[23]	Human urine and ileostomy effluent	MeOH (12.5 ml, containing 2 g l^{-1} TBHQ) and HCl (5 ml 10 mol l^{-1}) were added to urine (7.5 g), and refluxed (90 °C, 2 h), then cooled and made to 50 ml with MeOH. Effluent (0.5 g freeze dried) was added to aq. MeOH (40 ml, 62.5% v/v; containing 2 g/l TBHQ) and HCl (10 ml, 10 mol/l) and refluxed (2 h, 90 °C) then made to 100 ml with MeOH.	Acidic MeOH	Acid hydrolysis	Recovery of onion or supplements to effluent was 90–92%, and 99% for quercetin added to urine.	Various acid concentrations and time duration were assessed for optimum conditions.
Epicatechin[10]	Rat plasma	Plasma (50 μl) was added to acetate buffer (0.1 mol/l,	Acidic MeOH	Enzyme. (a) β-glucur. (50	n/g	Free, glucuronides, sulfates, and mixed

		50 μl, pH 5) and acetic acid in MeOH (900 μl, 0.83 mol/l). The mixture was centrifuged and the supernatant diluted in H_2O (1/1, v/v). Plasma (50 μl) was mixed with various enzymes (a) 37 °C, 2 h, (b) 37 °C, 4 h, (c) 37 °C, 45 min, then extracted as above.		U) type VII-A (b) sulf. (25 U) type VIII in presence of saccharic acid 1,4 lactone (21 mmol/l) as β-glucur. inhibitor (c) sulf. type H-5 (25 U sulf. + 500 U β-glucur.)		conjugates of epicatechin and methyl-epicatechin were measured in rat plasma.
Green tea was consumed as a rich source of catechins[24]	Human plasma	Plasma (1 ml) was added to HCl-MeOH (1:1 v:v, 1 ml, 0.5 M) and incubated (r.t., 30 min). The mixture was extracted with ethyl acetate (2 ml, 2×), vortex-mixed (3 min), centrifuged and dried. Then redissolved in MeOH-H_2O (3:2, 250 μl, pH 2.8 containing ortho-phosphoric acid).	Acidic MeOH	Acid hydrolysis	Recovery from spiked blank was 94–102%, or plasma 93–101%.	Optimum acid hydrolysis conditions were tested (0.5, 1, 2 M) with different times (0, 10, 30, 60 min) at r.t. or 37 °C. At high HCl concentration catechin gallates were degraded (20–80%).
Onions fed as a rich source of quercetin[25]	Human plasma and urine	Plasma (600 μl) and urine (750 μl) were hydrolysed (90 °C, 2/3 h, urine/plasma) with HCl (2 ml, 1.2 M) in aq. MeOH (50% v:v, containing 20 mM Na-diethyl-dithio-carbamate). Extracts (100 μl) before or	Acidic MeOH	Acid hydrolysis	n/g	Flavonols and glycosides were found in plasma and urine based on co-elution with standards.

continued overleaf

Table 1 *(continued)*

Polyphenol	*Biological fluid*	*Method description*	*Solvent*	*Hydrolysis*	*% Recovery*	*Additional information*
		after hydrolysis were diluted with H_2O (250 μl), adjusted with TFA (pH 2.5) and plasma extracts centrifuged.				
Quercetin glycosides[3]	Rat plasma	Plasma (180 μl) was acidified to pH 4.9 with acetic acid (20 μl) and enzyme hydrolysed. Extracted with MeOH–HCl (500 μl, 200 mM) and centrifuged. Unhydrolysed plasma (200 μl) was extracted with MeOH-HCl as above.	Acidic MeOH	Enzyme. β-glucur./sulf. *H.P.* 1×10^5 U ml^{-1} β-glucur., 5×10^3 U ml^{-1} sulf.	n/g	Quercetin glycosides were not present in plasma and metabolites.
Fruit; rich source of hydroxy-cinnamates[26]	Human urine	MeOH (2.4 ml) and HCl (3 M, 100 μl) were added to urine (1 ml) centrifuged and the supernatant dried to the aq. fraction.	Acidic MeOH	NO	n/g	Hydroxycinnamates were detected in urine.
Quercetin and rutin[27]	Rat plasma	Plasma extracted with 4 volumes ethanolic HCl (50 mmol l^{-1}). The supernatant was evaporated and a volume of acetate buffer (0.2 mol l^{-1}, pH 4.9) equivalent to original plasma	Acidic ethanol	Enzyme. *H.P.* β-glucur, 5×10^5 U l^{-1} and sulf. 2.5×10^4 U l^{-1}.	n/g	In neutral conditions >90% of Q-met. ppt with pellet, indicating strong albumin binding. When acidic EtOH was used metabolites were recovered.

		volume added. Samples were centrifuged or enzyme hydrolysed and centrifuged.				
Quercetin Intravenously administered[28]	Human plasma	Plasma (200 μl) was mixed with DMSO–MeOH (1/4, v/v; 400 μl) and centrifuged. The supernatant was mixed with an equal volume of H_2O.	DMSO-MeOH	NO	Recovery of quercetin was 93–97% from spiked samples.	MeOH and acetonitrile considered unsuitable as may co-ppt compound with proteins. MeOH–DMSO previously shown to extract protein-bound drugs.
Rats were injected with equol (isoflavone metabolite).	Rat uterine tissue, urine and plasma.	Rat uterine tissue (30 mg) was homogenised with cold buffer (1 ml, 10 mM Tris, 1.5 nM EDTA, pH 7.4).	EtOH	Enzyme. No details given.	n/g	Flavonoids were found in urine in free and conjugated form and in plasma and uterus tissue.
Humans were given soya supplements as a rich source of isoflavones[29]	Human urine and plasma	Plasma or homogenate (0.2 ml) was vortex-mixed (10 min) with EtOH (1.2 ml) and centrifuged. Supernatant (1 ml) was evaporated and re-dissolved with MeOH (0.1 ml) shaking (5 min) followed by diluted with H_2O (0.3 ml). Plasma or tissue (0.2 ml) was mixed with β-glucuronidase (50 μl) and acetate buffer (0.2 ml, 0.1 M pH 5 containing 0.1% ascorbic acid and 0.01% EDTA) and				

continued overleaf

Table 1 *(continued)*

Polyphenol	*Biological fluid*	*Method description*	*Solvent*	*Hydrolysis*	*% Recovery*	*Additional information*
		incubated (3 hrs, 37 °C) before addition of EtOH (1.2 ml) as above. Urine was diluted to equal volume with mobile phase and centrifuged.				
Onions were fed as a rich source of quercetin[30]	Human plasma and ileostomy effluent	Ileostomy fluid (1 g, freeze dried) was extracted with MeOH (3 × 10 ml). Extracts were dried and redissolved in MeOH (1 ml) and mobile phase (2 ml, 35% MeOH and 5% acetic acid). Plasma diluted by (a) equal volume MeOH and centrifuged, (b) adding acid and centrifuged,[25] or (c) by lyophilisation and extraction with MeOH.	MeOH	NO	Extraction recovery was 93%.	Effluent contained mainly quercetin. Quercetin glucoside and aglycone were absent from plasma.
Grapefruit juice was fed as a rich source of flavanones[31]	Human urine	Plasma or urine (750 μl) and NH_4PO_4 (350 mg) were extracted with ethyl acetate (4 ml), centrifuged and organic phase removed, before extraction was repeated. Organic layers	Ethyl acetate	Enzyme. Entrails provided 28000 U ml^{-1} β-glucur + 733 U ml^{-1} sulf. Bovine liver provided	Naringenin recovery from urine was 77–101%, 69–89% for naringin and 77% for	Naringin and naringin glucuronides were not found in urine, naringenin and glucuronides were found. Enzyme hydrolysis had reached a max by 4 h.

		were evaporated and re-dissolved in (a) mobile phase (150 μl) or (b) the sample was diluted with an equal volume of citrate (pH 3.8) or acetate (pH 5) buffer and enzyme hydrolysed.		28000 U ml^{-1} β-glucur.	hesperetin.	
Soy supplements were given as a rich source of isoflavones[32]	Human serum	Serum (75 μl) was extracted with acetonitrile (75 μl), vortex-mixed (10 min) and centrifuged. Aliquots (100 μl) were added to citrate buffer (0.1 M, pH 5, 1 ml) and enzyme hydrolysed. Aglycones were extracted with ethyl acetate (3 × 1 ml) dried and dissolved in MeOH-H_2O (50 μl). Blood from finger pricks (50 μl) were spotted dry on filter paper. Isoflavones were extracted with MeOH (3 × 0.5 ml) from cut-out spots. Extract evaporated and dissolved in MeOH (50 μl) and formic acid (100 μl, 0.1 %).	Acetonitrile and ethyl acetate.	Enzyme. sulf./glucur (*H.P.* containing 10^5 U ml^{-1} glucur. + 5 × 10^3 U ml^{-1} sulf.), sulf. (*Aerobacter aerogenes*, 0.84 U), or 3.24 U recombinant β-glucur.	Recovery was 86 ± 20% for genistein and 69 ± 25% for daidzein.	Glycetin, genistein and daidzein glucuronides and sulfates were detected in plasma.
Tea polyphenols[33]	Human plasma and urine	Plasma (200 μl) mixed with ascorbate-EDTA (20 μl; 0.4 M NaH_2PO_4 buffer	Methylene chloride and ethyl acetate	Enzyme. β-glucur (500 U) and sulf. (40 U),	82–88% plasma recovery of	EC found mainly as sulfates, some glucuronides, but none

continued overleaf

Table 1 *(continued)*

Polyphenol	*Biological fluid*	*Method description*	*Solvent*	*Hydrolysis*	*% Recovery*	*Additional information*
		containing 20% ascorbate–0.1% EDTA pH 3.6) and enzymes (20 μl) was incubated. Methylene chloride (500 μl) and H_2O (300 μl) were added, vortex-mixed with ethyl acetate (4 min) and centrifuged. Aqueous phase was re-extracted with ethyl acetate (700 μl). Combined extract (1.4 ml) mixed with 1% ascorbic acid (10 μl), evaporated dissolved in acetonitrile (200 μl) and centrifuged. Urine sample (50 μl) mixed with NaH_2PO_4 buffer (150 μl, pH 6.8), ascorbate-EDTA (20 μl), enzymes (20 μl), and incubated. Ethyl acetate (1 ml) was added, vortex-mixed (4 min) and centrifuged. Aq. phase was re-extracted with ethyl acetate (700 μl) as above.		source not stated.	stds. 87–95% urine recovery of stds.	free. EGC found in all forms but mainly glucuronidates. EGCG found mainly sulfated in plasma. Lower grade β-glucur. caused conversion of EGCG to EGC and other compounds, resulting in lower recovery.
Quercetin Intravenous or oral	Human plasma and urine	Plasma or urine (2 ml) and citrate buffer (1 M, pH 2,	H_2O-saturated methyl ethyl	Enzyme. (units and source	n/g	10 h gave complete enzyme hydrolysis,

dose given to humans[8]		1.5 ml) was extracted with H_2O-saturated methyl ethyl ketone (8 ml). Organic phase (6 ml) evaporated, dissolved in acetone (0.5 ml), reacted with tetraphenyl-diboroxide (0.6 mg). Complex was adjusted to pH 3 with citrate (2 ml), extracted in diisopropyl-ether (6 ml). Urine (5 ml) was adjusted to pH 5.3 with an equal vol. acetate buffer (0.2 M) enzyme hydrolysed Not clear from paper if plasma was also treated the same.	ketone and diisopropyl ether	not given).		equivalent to 30 min acid hydrolysis (high background not good for low estimates). Plasma quercetin found after i.v. but not oral administration. Quercetin excreted mainly as conjugates after i.v., but not detected after oral administration.
Naringin. Intravenous administration[34]	Rat plasma and brain tissue	Plasma (0.1 ml) was extracted with acetonitrile (0.2 ml), and centrifuged. Brain tissue was mixed with Ringer solution (2 ml g^{-1}) and homogenised. Acetonitrile (0.5 ml) extracted sample (50 μl) was centrifuged. Plasma or brain tissue homogenate (250 μl) was enzyme hydrolysed. Plasma (100 μl) or brain tissue (50 μl) was mixed with acetonitrile (0.2 or 0.5 ml	Acetonitrile	Enzyme. β-glucur. and sulf. *H.P.* units not specified).	Recovery from plasma was 103–108%, from brain 88–100%.	Naringenin and glucuronide were found in plasma and brain tissue. Aglycone was found more in the brain.

continued overleaf

Table 1 *(continued)*

Polyphenol	*Biological fluid*	*Method description*	*Solvent*	*Hydrolysis*	*% Recovery*	*Additional information*
		respectively) and centrifuged.				
Quercetin glucoside supplements[35]	Human plasma	Plasma samples were extracted with acetonitrile (2 vol) and phosphoric acid (1 vol, containing ascorbic acid, 4 $g\,l^{-1}$), before centrifuging. Plasma (25 μl) was enzyme hydrolysed in acetate (pH5) and extracted as above.	Acetonitrile	Enzyme. Bovine liver, 25 mU	Recovery of standards was 83-115%.	Quercetin glucosides were not observed but glucuronides were.
Onions fed as a rich source of quercetin glucosides[16]	Human plasma	Plasma (10 ml; containing 1 mM ascorbic acid) was acidified with acetic acid (pH 5) extracted with acetonitrile (2.5 vol) by vortex-mixing (10 min) and centrifuging. Supernatant was dried, made to 500 μl with H_2O, then MeOH (500 μl) and centrifuged. Combined extracted plasma was dried, dissolved in buffer (1 ml, pH 7). Sample (200 μl) enzyme hydrolysed, MeOH (200 μl) added and centrifuged.	Acetonitrile	Enzyme. (β-glucur., 5U, 25 μl E.C. and buffer; sulf., 0.4 U, 25 μl *Aerobacter aerogenes* and buffer 25 μl).	n/g	Quercetin glucuronides, mixed conjugates, sulfates were detected in plasma samples.

TBHQ = *tert*-butylhydroxyquinone; n/g not given; β-glucur. = β-glucuronidase; sulf. = aryl-sulfatase; *H.P.* = *Helix pomatia*; *E.C.* is *E. Coli*.

added to stabilise samples. Controls were potassium phosphate buffer (0.1 M, pH 7) treated as for plasma. All samples treated in duplicate.

Extraction solvents. (i) 3 volumes of acetone (ii) 4 volumes of methanol (iii) 2 volumes acetonitrile.

Extraction procedures. Samples vortexed mixed for 30 seconds every 2 minutes for 10 minutes. Centrifugation (13600 *g*, 4 °C, 10 minutes) to remove precipitated proteins. Supernatants dried to around 100 μl by rotary evaporation at 50 °C under vacuum, made up to 500 μl with water, methanol (500 μl) added. An additional two samples extracted with acetonitrile freeze-dried and the residue taken up in water (500 μl) and methanol (500 μl).

Analysis. Samples were centrifuged (13600 *g*, 4 °C, 2 minutes) and filtered into HPLC vials. Aliquots (20 μl) were injected onto the HPLC column for analysis.[5]

Results. Table 2 shows the recoveries observed.

The control samples provided over 80% recovery in all methods tested, though the recovery of quercetin and quercetin-3-glucoside was poor when either acetone or methanol was used to precipitate the proteins. There was little observed difference for recovery of quercetin and quercetin-3-glucoside between freeze-drying or rotary evaporation for removal of acetonitrile. Hence freeze-drying may be an efficient process for extraction of large numbers of plasma samples.

Similar experiments need to be carried with glucuronide and sulfate conjugates and with different matrices in order to establish optimum extraction conditions.

Table 2 *Recovery of quercetin from spiked plasma samples using different solvents for extraction*

Method	*Quercetin* (% recovery[†])	*Quercetin-3-glucoside* (% recovery[†])
Acetone[a]		
buffer	98, 110	107, 113
plasma	54, 74	81, 89
Methanol[b]		
buffer	80, 101	101, 108
plasma	40, 66	51, 78
Acetonitrile[c]*		
buffer	85, 90	102, 109
plasma	86, 90	99, 102
Acetonitrile[c] – freeze dried		
buffer	101, 107	111, 114
plasma	95, 98	99, 103

[a] Method adapted from ref 3.
[b] Method adapted from ref 4.
[c] Method adapted from ref 2.
* With removal of solvent using rotary evaporator.
[†] Recovery is based on comparison to standard.

Solid-phase Extraction

Solid-phase extraction, almost invariably in the form of pre-prepared, disposable cartridges, allow both concentration and a degree of purification as unwanted materials are removed from the polyphenol fraction. Solid phases of relevance to polyphenols include ion exchange, functionalised and hydrophobic sorbents. The polyphenols can be applied in aqueous environments or dissolved in organic solvents depending on the column and the separation required. Elution, for the same reasons, can also be in aqueous media or in solvents. Clearly each type of column needs to be fully characterised and optimised for different polyphenols and their metabolites. As an example of a typical application, sulfated polyphenols can be strongly bound by polyamide columns, unwanted material washed away and the sulfates eluted using aqueous methanol at alkaline pH. Optimal recoveries for all of the solid phases will only be observed if columns are washed and conditioned correctly, and if the columns are used in a reproducible manner. Erlund *et al.* tested various SPE for recovery of quercetin from plasma.[6] The authors found that Bond Elut C18 and IST ENV columns gave greater than 97% recovery, but that an acid buffer wash step was required to release quercetin bound to proteins. They also reported a solvent extraction method for the aglycone using an acidic mixture of tolulene and dichloromethane (80:20, v/v) after assessing a variety of solvents. Erlund *et al.* went on to successfully apply their methodology to a human study.[7] Table 3 summarises recent methodology for extraction of polyphenols from animal fluids and tissues using solid phase extraction.

3 Hydrolysis

Both enzymic and acid hydrolysis of samples have been employed to release aglycones. The aglycone can then be extracted by an apolar solvent such as ethyl acetate (Figure 1a). If the level of conjugate in comparison to total aglycone is required then the sample can be analysed before and after hydrolysis and the concentration of the conjugates calculated indirectly (Figure 1b).

Acid Hydrolysis

Acid hydrolysis usually takes the form of an incubation of sample with 2N HCl for 15 minutes at 95 °C. Such conditions are sufficient to hydrolyse the glucoside, glucuronide and sulfate conjugates to liberate aglycone. Acid hydrolysis has the significant advantage of being reproducible but does not confer a high degree of specificity to the analysis. An exception may be the use of partial hydrolysis with organic acids, such as 10% acetic acid. Under such conditions, identification of intermediate compounds may be helpful to structural identification since the sulfate may be hydrolysed faster than the glucoside group. Such an approach is difficult and requires considerable expertise and supporting validation. The use of acid hydrolysis, especially under harsh conditions, may result in losses of

Table 3 *Polyphenol extraction using solid-phase extraction*

Polyphenol (isolate/food)	*Biological fluid*	*Method description*	*Column-solvent*	*Hydrolysis (% recovered)*	*Additional information*
Cyanidin glucosides[36] Fed to rats and humans	Rat liver and plasma	Rat liver (500 mg) was washed with NaCl (0.15 ml l^{-1}) and homogenised in ice-cold TFA (2.5 ml, 0.44 mol l^{-1}). ODS SPE cartridge was pre-washed with MeOH (10 ml) and equilibrated with formic acid (10 ml, 1.5 mol l^{-1}). Plasma (1 ml) diluted with TFA (0.2 ml, 0.44 M) or liver homogenate (1 ml) was applied, water soluble compounds, polar and neutral lipids were eluted respectively with formic acid (10 ml 1.5 M), dichloromethane (10 ml) and benzene (10 ml). Anthocyanins were recovered with TFA in MeOH (5 ml, 0.44 M), dried and redissolved in TFA (200 μl, 0.44 M). Plasma (0.5 ml) was diluted with phosphate buffer (50 μl, 0.1 M containing 2.9 mM EDTA, pH 6.8) and enzyme hydrolysed before extracted as above.	ODS SPE column, eluted with TFA in MeOH	Enzyme. β-Glucur. (1250 U 50 μl^{-1}), or sulf. (100 U). Enzyme source not stated. (n/g)	Anthocyanins were present in plasma and liver samples after supplementation, but treatment with enzymes did not alter levels.

continued overleaf

Table 3 *(continued)*

Polyphenol (isolate/food)	*Biological fluid*	*Method description*	*Column-solvent*	*Hydrolysis (% recovered)*	*Additional information*
Delphinidin and cyanidin glucoside and rutinoside[37]	Rat and human plasma and urine.	Plasma acidified with 1/40 volume HCl (6 M), then 400 μl was diluted with oxalic acid (400 μl, 10 mM). Human (4 ml) and rat (0.4 ml) plasma was applied to a pre-washed Sep-Pak C_{18} cartridge (10 ml MeOH containing 5% TFA, equilibrated with 10 ml oxalic acid). After 10 ml oxalic acid, anthocyanins were eluted with MeOH containing 5% TFA, then evaporated and redissolved in 3% phosphoric acid (200 μl). Urine was acidified with 1/40 volume HCl (6 M) then 10 ml applied to a Sep-Pak + C_{18} ENV cartridge conditioned as above. Anthocyanins were eluted with MeOH (5 ml) containing 5% TFA, dried and redissolved in 3% phosphoric acid (200 μl).	Sep-Pak C_{18} cartridge, eluted with MeOH containing 5% TFA (urine also used C_{18} ENV).	NO. Recovery of spiked samples was 70–73%.	Detection of original compounds in plasma and urine based on UV and retention time with standards suggests these compounds are absorbed intact.
Quercetin and kaempferol[38]	Spiked human plasma	Plasma (1 ml) was diluted and acidified with phosphate buffer (pH 2, 0.33 M, 5 ml). Sep-Pak light cartridges were pre-equilibrated with MeOH (2 ml)	Sep-Pak C_{18} cartridges eluted with aq. acetone and phosphate buffer.	NO Recovery of kaempferol was 86%, and quercetin >90%.	Various Sep-Pak cartridges were tested using different conditions.

		and phosphate buffer (1 × 6 ml). The sample was slowly applied to cartridge, washed with buffer (2 ml), and flavonoids eluted with aq. acetone (4 ml, 50% v/v acetone containing 250 mM $Na_2H_2PO_4$ and 0.1% v/v diethylamine, adjusted to pH 3.5 with orthophosphoric acid), dried, redissolved in acetone (600 μl, 45% v/v acetone in 250 mM $Na_2H_2PO_4$) and centrifuged.		
Isoflavones and isoflavone glycosides[39] Incubated in tied-off rat stomachs	Rat plasma	Plasma was subjected to enzyme hydrolysis. Aglycones were extracted with methanol/acetic acid, centrifuged and supernatant diluted with water prior to analysis.	Sep-Pak C_{18} Vac cartridge, eluted with aq. MeOH.	Enzyme. H-5 sulf. containing sulf. + β-glucur.
		Plasma (4 ml) was enzyme hydrolysed. Released aglycones were extracted 3× with MeOH (80 ml total). Extracts were centrifuged and supernatant was extracted 3× with methylene chloride (20 ml). MeOH/H_2O layers were evaporated. Residue was sonicated with MeOH (30 ml) and centrifuged. Supernatant evaporated, sonicated with H_2O and applied to pre-conditioned Sep-Pak C_{18} Vac cartridge. Washed with		

continued overleaf

Table 3 *(continued)*

Polyphenol (isolate/food)	*Biological fluid*	*Method description*	*Column-solvent*	*Hydrolysis (% recovered)*	*Additional information*
		water (30 ml), then isoflaonves were eluted with 90% MeOH (8 ml). Evaporated and redissolved in MeOH.			
Blackcurrant and apple juice was fed as a rich source of quercetin[40]	Human plasma and urine	Plasma (300 μl) was diluted to 2 ml with acidic MeOH (50% v/v, 1.2 M HCl, containing 20 mM sodium diethyl-dithiocarbamate). Samples were heated in sealed vials (2 h, 90 °C).	Plasma was extracted with acidic MeOH.	*Plasma*: Acid *Urine*: Enzyme *H.P.* 5.5×10^3 β-glucur. + 2.6×10^3 U l^{-1} sulf.	Quercetin was excreted in urine at a consistent % independent of consumption, plasma did not reflect intake levels.
		Urine (15 ml) was adjusted to pH 5 with sodium acetate and enzyme hydrolysed. The sample was extracted on a Bond Elut C_{18} cartridge, pre-washed with MeOH (3 ml, 5% containing 1% formic acid), eluted with MeOH (3 ml, 90% v/v, containing 1% formic acid and 1% ascorbic acid) followed by MeOH (100%). Elutants were combined, evaporated and redissolved in MeOH (150 μl, 30% v/v, containing 1% formic acid).	Urine was extracted on a Bond Elut C_{18} cartridge eluted with acidic MeOH.		

Gingko biloba tablets were fed as a source of quercetin and kaempferol[41]	Human urine	Urine (2 ml) was acidified with HCl (1 ml, 1 M) and passed through 3 cm XAD2 resin in a Pasteur pipette. After washing with H_2O (1 ml) sample was eluted with MeOH (2 ml) and dried before derivitisation for GC–MS. Urine (2 ml) was enzyme hydrolysed in (a) acetate buffer (0.5 ml, 1 M, pH 5) or (b) phosphate buffer (0.5 ml, 1 M, pH 6.8), and extracted as before.	XAD2 resin eluted with MeOH	Enzyme. (a) Sulf. (type VIII) 100 U (b) β-glucur. (Type 1X-A) 500 U.	Free kaempferol was found in urine. Sulf. had little effect on total, but quercetin and kaempferol were released after β-glucuor. treatment
Ginkgo biloba was fed as source of phenylacetic acid metabolites from quercetin and kaempferol[42]	Human plasma and urine	Urine (1 ml) was loaded onto a SPE C_{18} cartridge preactivated with MeOH (3 ml) and H_2O (10 ml) and eluted with H_2O (3 ml), 30% MeOH (3 ml), 50% MeOH (3 ml) and 100% MeOH (3 ml). Each eluate was evaporated and redissolved in H_2O (1 ml). Urine (1 ml) was heated (100 °C, 45 min) in sealed tubes with HCl (0.2 ml, 1 M) and MeOH (0.8 ml). Sample was neutralised with NaOH (1 M), evaporated and redissolved in MeOH (1 ml, 30% v/v) before loading on SPE column as above.	SPE C_{18} cartridge elued with H_2O and MeOH.	Acid hydrolysis	Intact flavonols were not measured. Plasma samples contained no metabolites (maybe due to degradation over 5 h – shown by others).
Pure flavonones or grapefruit and orange	Human plasma and urine	Urine and plasma samples were split in two. Urine (10 ml) or	C_{18} SPE column eluted with MeOH	Enzyme. *H.P.* 10000 U of	Free flavanone was not found.

continued overleaf

Table 3 *(continued)*

Polyphenol (isolate/food)	*Biological fluid*	*Method description*	*Column-solvent*	*Hydrolysis (% recovered)*	*Additional information*
juice were fed as rich sources of flavanones[43]		plasma (3 ml) were added to acetate buffer (3 ml, 1 M, pH 4.5) and enzyme hydrolysed (20 h, 37 °C). Both hydrolysed and non-hydrolysed samples were extracted using a pre-conditioned (6 ml 95% EtOH, and 10 ml H_2O) C_{18} SPE column. After a MeOH wash (4 ml, 10% v/v), the flavonones were eluted with MeOH (3 ml) and the eluent dried to 0.5 ml.		β-glucur.	
Luteolin and glucoside absorption was investigated using rat everted-gut and rat and human *in vivo* studies[44]		Plasma (0.5 ml) was acidified with 1 vol oxalic acid (0.01 M) and applied to a Sep-Pak C_{18} cartridge. After washing with oxalic acid (0.01 M), MeOH/H_2O/0.5 M oxalic acid (25:73:2, v/v) and H_2O, the MeOH eluate was evaporated then redissolved in MeOH (100 μl) and centrifuged. Each sample was acidified with acetate buffer (1 M, pH 4.5) and enzyme hydrolysed. Oxalic acid (1 vol) was added, the mixture centrifuged and the supernatant prepared as above	Sep-Pak C18 cartridge	Enzyme. Treated with 5.4×10^2 U ml^{-1} β-glucur. + 0.1×10^2 U ml^{-1} sulf. Recovery was 95–104%.	Luteolin was found in a free form although mainly conjugated to glucuronides. The glucoside was not present.

Methodology tested for analysis of naringin and naringenin[45]	Spiked human urine	Naringin: urine was diluted with H_2O (2 ml) and applied to a strong anion exchanger Sep-Pak Accell QMA cartridge (2 g), pre-conditioned with MeOH (15 ml) and H_2O (15 ml). The cartridge was washed with H_2O (1 ml), purged with air, and eluted with formic acid in MeOH (10 ml, 1 mM) at 3 ml/min. The eluate was evaporated and redissolved in MeOH (2 ml).	Sep-Pak Accell QMA cartridge eluted with formic acid and MeOH.	Formic acid in methanol gave a good recovery (92–94% naringin, 80–81% naringenin).	Sep-Pak C_{18} cartridge was found not to effectively remove endogenous compounds in urine that interfered. The combined use of a Sep-Pak and a strong anion exchanger is often used, but when tried was equally effective as Sep-Pak Accell QMA alone.
		Naringenin: urine was diluted with H_2O (2 ml) and applied to a column pre-conditioned with MeOH (5 ml) and H_2O (5 ml). The column was washed with H_2O (2.5 ml), purged with air, and eluted with formic acid in MeOH (7 ml, 4 mM) at 3 ml/min. The eluate was evaporated and redissolved in EtOH (2 ml).			
Naringin was fed as a pure compounds[46]	Human urine	Urine (1 ml) was extracted with a Sep-Pak QMA cartridge as described in above paper.[45] Urine (0.5 ml) was enzyme hydrolysed (24 h, 37 °C), then EtOH (20 ml) was added, the sample vortex-mixed (1 min) and centrifuged. Supernatant was	Sep-Pak Accell QMA cartridge eluted with formic acid and MeOH.	Enzyme. *H.P.* β-glucur. (5 ml, 10000 U in 0.2 M acetate buffer pH 4.5)	Naringin and aglycone were found in urine, but mainly naringenin was excreted as two glucuronide forms.

continued overleaf

Table 3 *(continued)*

Polyphenol (isolate/food)	*Biological fluid*	*Method description*	*Column-solvent*	*Hydrolysis (% recovered)*	*Additional information*
		centrifuged. Supernatant was evaporated, dissolved in H_2O and extracted as above.			
Naringenin and the 7-glc was given by i.v. or oral administration[47]	Rat urine	Bond elute C_{18} cartridges were pre-conditioned with MeOH (1 ml) followed by acidic H_2O (1 ml, pH <3). Urine (100 μl) was diluted with H_2O (1 ml) and PBS (100 μl). After incubation (37 °C, 16 h) the urine was washed through column with acidic H_2O (1 ml), then eluated with MeOH (3 ml), evaporated and reconstituted in mobile phase (100 μl). For analysis of conjugates, PBS contained enzymes for hydrolysis.	Bond elute C_{18} cartridges	Enzyme. 50000 U ml^{-1} β-glucur. (type L-II active at pH 3.8 from limpet). Samples were incubated for 16 h to ensure complete hydrolysis.	Naringenin-7-glc was not found in urine, only aglycone and glucuronides. After i.v. administration some glycoside was in the urine, but less recovered than when the aglycone was administered.
Genistein[48]	Rat tissues and serum	Thawed tissue was homogenised with citrate buffer (25 mmol l. pH 5, 1 ml/100 mg, 30–60 s, r.t.) A portion (200 μl) transferred and MeOH (800 μl) added, sonicated (10 min) and mixed with citrate buffer (10 ml). Sample was enzyme hydrolysed (60 min, 37 °C). For fatty tissues, NaCl (200 mg) was added to samples and lipids	Oasis HLB cartridges	Enzyme. *H.P.* 100 μg containing 40–60 U of β-glucur + 1.5–4 U sulf.	Aglycone and conjugate were determined in all tissues.

		removed by shaking with hexane (3 × 1.5 ml). All samples were centrifuged and genistein extracted by SPE using Oasis HLB cartridges pre-activated with MeOH (1 ml), H_2O (1 ml) and citrate buffer (1 ml). The samples were loaded on to cartridges, washed with H_2O (1 ml) and eluted with MeOH (1 ml). The eluent was dried and redissolved in MeOH (25 μl) and H_2O (25 μl).			
Quercetin and rutin[7]	Human plasma.	Plasma (1 ml EDTA), acetate buffer (110 μl, 0.78 M, pH 4.8), ascorbic acid (100 μl, 0.1 M) and enzyme (40 μl) were incubated (17 h, 37 °C). Sample diluted in phosphate buffer (2 ml, 70 mM, pH 2.4) and extracted on a Bond Elut C_{18} SPE column preconditioned with MeOH (6 ml) and phosphate buffer (6 ml). Column was washed with phosphate buffer (9 ml) and H_2O (0.5 ml), eluted with MeOH (2 ml) and dried. Toluene-dichloromethane (80/20, v/v, 1 ml) and acetic acid (5.3 M) – oxalic acid (32 mM) (80/20, v/v, pH 2.4, 200 μl) were added, vortex-mixed (1 min), centrifuged and lower phase analysed.	Bond Elut C18 SPE column eluted with MeOH.	Enzyme. *H.P.* 4000 U β-glucur. and 200 U sulf. (type HP–2). Recovery of rutin was 98–106%.	For other recovery and details of method development see Erlund *et al.*[6]

continued overleaf

Table 3 *(continued)*

Polyphenol (isolate/food)	*Biological fluid*	*Method description*	*Column-solvent*	*Hydrolysis (% recovered)*	*Additional information*
		Plasma (0.5 ml), acetate buffer (55 μl, 0.78 M, pH 4.8), ascorbic acid (50 μl, 0.1 M) and phosphate buffer (2 ml) were added. Rutin was extracted using Bond Elut C_{18} extraction as before. Dried eluants were dissolved in MeOH (100 μl) and acetic acid (5.3 M) – oxalic acid (32 mM) (80/20, v/v, pH 2.4, 100 μl).			

EDTA = ethylenediaminetetraacetic acid; TFA = trifluroacetic acid; PBS = phosphate buffered saline; β-glucur. = β-glucuronidase; sulf. = aryl-sulfatase.

polyphenol through degradation and derivatisation. High background readings have also been reported by some researchers reducing the sensitivity of analysis.[8]

Enzymic Hydrolysis

Enzymic hydrolysis, when used under optimal conditions, can provide a much higher degree of specificity and sophistication than acid hydrolysis. Enzymes commonly employed are a combination of β-glucuronidase and sulfatase from *Helix pomatia,* purified fractions of these individual enzymes from the same source, β-glucuronidase from bovine liver, aryl-sulfatase from *Aerobacter aerogenes* and recombinant β-glucuronidase from *Eschericshia coli.* Cellulase may also be used as a crude preparation containing a mixture of unspecified hydrolytic enzymes. Enzymic hydrolysis is dependent on biological material that will vary in purity and activity from batch to batch (we have observed varying degrees of activity for aryl-sulfatase on standards), might contain important contaminants (such as other enzymes or even polyphenols, a point widely ignored), and which can be difficult to standardise. Sulfatase is often contaminated with β-glucosidase or β-glucuronidase; β-glucuronidase preparations also contain sulfatase or β-glucosidase activity. Many authors do not seem to be aware of these points, sometimes not even providing the biological source of the enzyme(s) used. Often, enzyme preparations are used indiscriminately to release aglycones. Rarely do researchers assess the purity of the enzymes used, such as by using the appropriate substrate controls looking for additional enzymic activity on other conjugates. It cannot be emphasised enough that the use of enzymes for these purposes has to be very carefully monitored.

When relevant enzymes are likely to be present in the cells or tissue of interest (which is almost always likely to be the case) then care needs to be taken that the enzymes do not interfere in the subsequent processes. It must be ensured that the enzymes are deactivated (which may be the case with the extraction solvents) or inhibited by the use of specific inhibitors. Often this is not done as tissue samples have been homogenised in buffer before solvent extraction.[9]

Enzymic hydrolysis has a major advantage in that it can – if used with care – be applied in order to provide structural information of the conjugate. For this purpose the enzyme preparation(s) used will need to be further purified before use, and fully characterised to show the absence of relevant activities. In this way, for example, the use of a purified sulfatase enzyme resulting in the 'disappearance' from an HPLC chromatogram of a suspected sulfate metabolite can be regarded as supporting evidence of identity (see problems with sulfate determination below). An additional approach is the use of specific inhibitors, such as saccharic acid 1,4-lactone as a β-glucuronidase inhibitor[10] (Figure 1c, 1d) in conjunction with a mixed, or unpurified, enzyme preparation. In this way an indirect determination of the β-glucuronide conjugate component can be calculated.

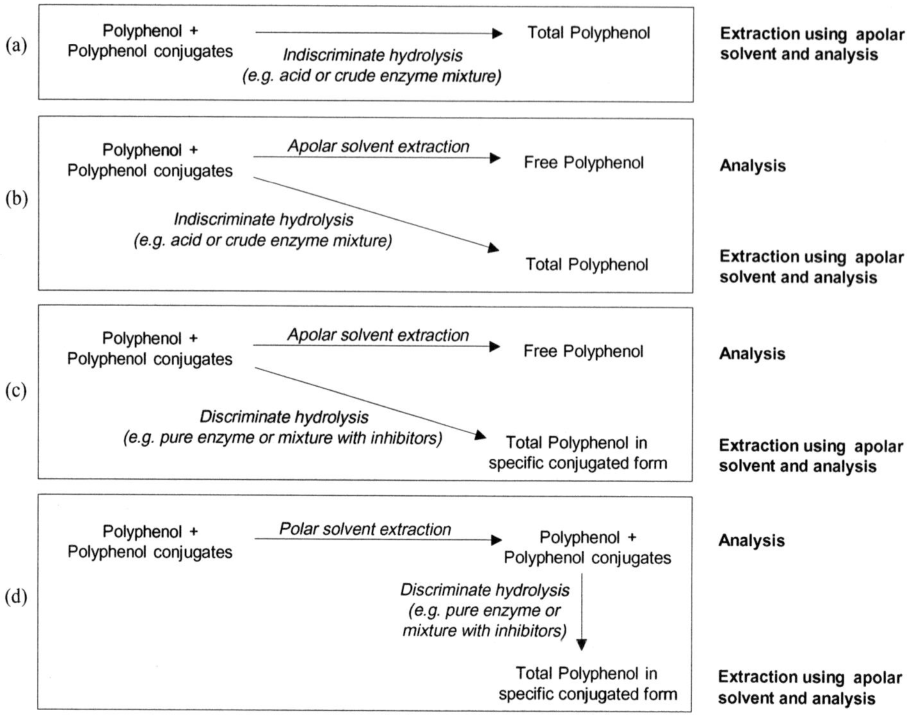

Figure 1 *Use of solvent for extraction of polyphenols*

4 Particular Problems with Sulfate Determination

Sulfates can be attached to the uronic acid group of polyphenol glucuronide conjugates in plant tissue. It is possible, though it has not been demonstrated as yet, that these conjugates form during mammalian metabolism and if so treatment with β-glucuronidase would release aglycone initially present as sulfate. It is also worth noting that sulfatase acts primarily on phenol sulfate esters, and does not cleave alcohol sulfate esters, such as the flavonol-3-sulfates, with the same efficiency. Indeed, lack of activity on the 3-sulfate has been utilised in the synthesis of 3-sulfates,[11–13] but may lead to mis-identification if aryl-sulfatase is used for characterisation of metabolites.

As has been mentioned previously, sulfates are poorly extracted by ethyl acetate and acetone, and this has given rise to problems in the literature. Several of the studies listed in Table 1 failed to detect quercetin sulfates in human plasma after feeding subjects onions as a rich source of quercetin glycosides.[14,15] It seems clear to us from metabolic considerations that quercetin sulfates should have been identified under the study conditions employed, and this has been supported by experimental results using direct analytical determination[16] and in many other studies using an indirect approach with sequential enzyme deconjugation.[17] The most likely explanation for the apparent discrepancy in findings relates to the observation that the use of acetone (which would not extract the quercetin sulfates at all efficiently) for extraction was a common feature of those studies failing to observe quercetin sulfates.

5 Storage of Samples

In storage of samples care needs to be taken in respect of (i) possible endogenous enzyme activity and (ii) stability of the polyphenols and their conjugates. Quercetin, for example, is particularly prone to oxidation when the pH exceeds 7.4, which may occur during storage of plasma in a freezer resulting in losses and degradation compounds present as sample artefacts. Quercetin-3-conjugates are more resistant to oxidation, which may also lead to an imbalance in calculation of metabolite concentrations. Quercetin appears to be more stable bound to proteins, as the proteins may afford some protection to oxidation reactions. Antioxidants, such as ascorbic acid, are often added both for storage and during preparation to reduction losses of polyphenols by oxidation.

6 Summary

Efficient extraction of polyphenols from biological matrices, including animal tissues and fluids, is paramount to both qualitative and quantitative analysis, and yet knowledge on the behaviour of polyphenol metabolites in different solvents is less than complete. Consequently researchers are conducting analyses on samples from which metabolites are absent. It is evident from a review of recent literature that more emphasis needs to be placed on characterising and optimising conditions and using appropriate controls. Only when methods are fully validated may controversial data be critically assessed.

References

1. B. Barron, L. Varin, R. K. Ibrahim, J. B. Harborne and C. A. Williams, *Phytochemistry*, 1988, **27**, 2375.
2. J. A. Boutin, F. Meunier, P. H. Lambert, P. Hennig, D. Bertin, B. Serkiz and J-P. Volland, *Drug Metab. Dispos.*, 1993, **21**, 1157.
3. C. Morand, C. Manach, V. Crespy and C. Remesy, *Free Radic. Res.*, 2000, 33, 667.
4. G. Paganga and C.A. Rice-Evans, *FEBS Lett.*, 1997, **401**, 78.
5. A. J. Day, Absorption and metabolism of dietary flavonoid glycosides, Thesis, University of East Anglia, UK, 2000.
6. I. Erlund, G. Alfthan, H. Siren, K. Ariniemi and A. Aro, *J. Chromatogr. B.*, 1999, **727**, 179.
7. I. Erlund, T. Kosonen, G. Alfthan, J. Maenpaa, K. Perttunen, J. Kenraali, Parantainen and A. Aro, *Eur. J. Clin. Pharmacol.*, 2000, 56, 545.
8. R. Gugler, M. Leschik and H. J. Dengler, *Eur. J. Clin . Pharmacol.*, 1975, **9**, 29.

9. C. Manach, C. Morand, C. Demigne, O. Texier, F. Regerat and C. Remesy, *FEBS Lett.*, 1997, **409**, *12*.
10. M. K. Piskula, and J. Terao, *J. Nutr.*, 1998, **128**, 1172.
11. D. Barron, L. D. Colebrook and R. K. Ibrahim, *Phytochemistry*, 1986, **25**, 1719.
12. D. Barron and R. K. Ibrahim, *Phytochemistry*, 1987, **26**, 1181.
13. D. Barron and R. K. Ibrahim, *Phytochemistry*, 1987, **26**, 2085.
14. E. V. Graefe, J. Wittig, S. Mueller, A. K. Riethling, B. Uehleke, B. Drewelow, H. Pfortem G. Jacobasch, H. Derenforf and M. Viet, *J. Clin. Pharmacol.*, 2001, **41**, 492.
15. J. Wittig, M. Herderich, E. U. Graefe and M. Veit, *J. Chromatogr. B*, 2001 **753**, 237.
16. A. J. Day, F. Mellon, D. Barron, G. Sarrazin, M. R. A. Morgan and G. Williamson, *Free Radic. Res.*, 2001, **35**, 941.
17. C. Manach, C. Morand, V, Crespy, C. Démigné, O. Texier, F. Régérat and C. Rémésy, *FEBS Lett.*,1998, **426**, 331.
18. C. Felgines, O. Texier, C. Morand, C. Manach, A. Scalbert, F. Régérat and C. Rémésy, *Am. J. Physiol.*, 2000, **279**, G1148.
19. C. Manach, C. Morand, C. Démigné, O. Texier, F. Régérat and C. Rémésy, *Febs. Lett.*, 1997, **409**, 12.
20. J. H. Moon, T. Tsushida, K. Nakahara and J. Terao, *Free Radic. Biol. Med.*, 2001, **30**, 1274.
21. J. H. Moon, R. Nakata, S. Oshima, T. Inakuma and J. Terao, *Am. J. Physiol.*, 2000, **279**, R461.
22. P. C. H. Holman, M. V. D. Gaag, M. J. B. Mengelers, J. M. P. Van Trijp, J. H. M. De Vries and M. B. Katan, *Free Radic. Biol. Med.*, 1996, **21**, 703.
23. P. C. H. Hollman, J. H. M. De Vries, S. D. Van Leeuwen, M. J. B. Mengelers and M. B. Katan, *Am. J. Clin. Nutr.*, 1995 **62**, 1276.
24. G. Maiani, M. Serafini, M. Salucci, E. Azzini and A. Ferro-Luzzi, *J. Chromatogr. B*, 1997, **692**, 311.
25. A. A. Aziz, C. A. Edwards, M. E. J. Lean and A. Crozier, *Free Radic Res.*, 1998, **29**, 257.
26. L. C. Bourne and C. A. Rice-Evans, *Free Radic Res.*, 1998, **28**, 429.
27. C. Manach, C. Morand, O. Texier, M. L. Favier, G. Agullo, C. Démigné, F. Régérat and C. Rémésy, *J. Nutr.*, 1995, **125**, 1911.
28. D. J. L. Jones, C. K. Lim, D. R. Ferry and A. Gescher, *Biomed. Chromatogr.*, 1998, **12**, 232.
29. P. H. Gamache and I. N. Acworth, *Proc. Soc. Exp. Biol. Med.*, 1998, **217**, 274.
30. T. Walle, Y. Otake, U. K. Walle and F. A. Wilson, *J. Nutr.*, 2000, **130**, *2658*.
31. U. Fuhr and A. L. Kummert, *Clin. Pharmacol. Ther.*, 1995, **58**, 365.
32. D. R. Doerge, H. C. Chang, M. I. Churchwell and C. L. Holder, *Drug Metab. Dispos.*, 2000, **28**, 298.
33. M. J. Lee, Z. Y. Wang, H. Li, L. Chen, Y. Sun, S. Gobbo, D. A. Balentine and C. S. Yang, *Cancer Epidemiol. Biomarkers Prevention*, 1995, **4**, 393.
34. H. W. Peng, F. C. Cheng, Y. T. Huang, C. F. Chenm and T. H. Tsai, *J. Chromatogr. B*, 1998, **714**, 369.

35. A. L. A. Sesink, K. A. O'Leary and P. C. H. Hollman, *J. Nutr.* 2001, **131**, 1938.
36. T. Miyazawa, K. Nakagawa, M. Kudo, K. Muraishi and K. J. Someya, *J. Agric. Food Chem.*, 1999, **47**, 1083.
37. H. Mutsumoto, H. Inaba, M. Kishi, S. Tominaga, M. Hirayama and T. Tsuda, *J. Agric. Food Chem.*, 2001, **49**, 1546.
38. B. Liu, D. Anderson, D. R. Ferry, L. W. Seymour, P. G. de Takats and D. J. Kerr, *J. Chromatogr. B*, 1995, **666**, 149.
39. M. K. Piskula, J. Yamakoshi and Y. Iwai, *FEBS Lett.*, 1999, **447**, 287.
40. J. F. Young, S. E. Nielsen, J. Haroldsdottir, B. Daneshvar, S. Lauridsen, P. Knuthsen, A. Crozier, B. Sandstrom and L.O. Dragsted, *Am. J. Clin. Nutr.*, 1999, **69**, 87.
41. D. G. Watson and A. R. Pitt, *Rapid Commun. Mass Spectrom.*, 1998, **12**, 153.
42. P. G. Pietta, C. Gardana and P. L. Mauri, *J. Chromatogr. B*, 1997, **693**, 249.
43. B. Ameer, R. A. Weintraub, J. V. Johnson, R. A. Yost and R. L. Rouseff, *Clin. Pharmacol. Ther.*, 1996, **60**, 34.
44. K. Shimoi, H. Okada, M. Furugori, T. Goda, S. Takase, M. Suzuki, Y. Hara, H. Yamamoto and N. Kinae, *FEBS Lett.*, 1998, **438**, 220.
45. K. Ishii, T. Furuta and Y. Kasuya, *J. Chromatogr. B*, 1997, **704**, 299.
46. K. Ishii, T. Furuta and Y. Kasuya, *J. Agric. Food Chem.*, 2000, **48**, 56.
47. R. Choudhury, G. Chowrimootoo, K. Srai, E. Debnam and C. A. Rice-Evans, *Biochem. Biophys. Res. Commun.*, 1999, **265**, 410.
48. H. C. Chang, M. I. Churchwell, K. B. Delclos, R. R. Newbold and D. R. Doerge, *J. Nutr.*, 2000, **130**, 1963.

CHAPTER 3

Application of LC–MS for the Identification of Polyphenols

S. DE PASCUAL-TERESA AND J. C. RIVAS-GONZALO

1 Introduction

In general, mass spectrometry has proved to be a very useful technique for the identification of flavonoids, whose main advantage is the small quantities of compound needed for its analysis. Fast atoms bombardment mass spectrometry (FAB) has been widely used for the determination of the molecular weight of flavonoids, most often using xenon atoms and as a matrix glycerol, thioglycerol or dithiothreitol–dithioerythritol mixtures,[1,2] Results have been obtained using both positive mode (molecular species $[M+H]^+$) and negative mode (molecular species $[M-H]^-$), even though negative mode FAB seems to be more sensitive.[3]

The appearance of atmospheric pressure ionisation interfaces (API) enabled great advances to be made in the analysis of flavonoids, since it incorporated the use of mass spectrometry detection with high-performance liquid chromatography (HPLC) separation. This method was more versatile than gas chromatography (GC), the latter needing derivatisation of the compounds with the aim of making them volatile prior to anlysis.[4] Moreover, this kind of interface has permitted the use of different ionisation sources that give information both on the molecular ion and the fragmentation pattern of the target compound.

Atmospheric pressure chemical ionisation (APCI) source has been used for the analysis of flavonoids, especially flavonols, flavones, flavanones and chalcones,[5–8] with the analysis achieving high sensitivity and reproducibility. Nevertheless, until now there have been a limited number of reports describing its use for the analysis of condensed tannins and anthocyanins. In general, this type of ionisation is more suitable to compounds that are partially volatile and have a medium polarity, which is not the case with condensed tannins and anthocyanins. It should be noted that APCI is a gaseous-phase ionisation technique and thus the acidity or alkalinity of the phase formed by the target compound and the solvent when under evaporation plays an important role in the ionisation process.

The advantage of electrospray ionisation (ESI) is the ease with which it can ionise polar and non-volatile compounds, like condensed tannins and anthocyanidins. This kind of ionisation is of great value in the analysis of flavonoids since it permits the detection of the molecular ion but does not cause the fragmentation of the molecule as occurs with other types of ionisation such as APCI.

Some authors have used mass spectrometry with ESI ionisation for the analysis of monomer flavan-3-ols in plant extract samples and in plasma samples with results achieving levels of detection of 20 ng.[9] Furthermore, this technique allows the assignation of structures of highly polymerised compounds by the interpretation of the fragmentation profiles of the molecules and the charge of the formed ions that can either be monocharged or appear with multiple charges.[10] In that sense, the appearance of new, more powerful mass spectrometers has allowed one to obtain MS–MS spectra corresponding to the fragmentation of the molecular ion previously isolated in the ionisation chamber in a selective way, and moreover, the obtention of MS^n spectra, up to $n = 10$. This facilitates the structural elucidation of unknown compounds. Additionally, most of this equipment allows the recording of a desired range of masses in the full scan mode, as well as the recording in select ion monitoring mode (SIM), which is very useful, especially when following the kinetics of some reactions, due to its high sensitivity.

Recently with the introduction of bench-top instrumentation, LC–MS has evolved into a routine technique that enables the collection of significant data on the structures of phenolic compounds. In general, most authors prefer negative ionisation for the analysis of phenolic acids and other derived compounds[11] and positive mode for the analysis of catechins, proanthocyanidins and anthocyanidins,[12–17] with negative ion mode being used in some cases.[18] Flavonols and flavones and their metabolites have been analysed in both positive and negative mode.[19–25]

In this chapter we describe a number of methods that can be of use for the analysis and identification of flavonoids by LC–MS. We illustrate some results obtained on the analysis of anthocyanidins and derived pigments, catechins and proanthocyanidins, and flavonols and flavonol glucuronide conjugates. With this aim we have chosen some samples analysed in our laboratory as examples of the different groups of compounds. The results we demonstrate for the anthocyanins and derived pigments and catechins and proanthocyanidins were obtained using ESI ionisation while APCI was used for the flavonol analysis.

2 Analysis of Anthocyanins and Derived Pigments

Anthocyanins are flavonoid pigments present in a wide range of plant products. They are water and alcohol soluble and they appear in plants mainly in their glycosylated forms. Their free form, the aglycone, known as anthocyanidin, derives from the flavilium ion, and has a structure which possesses a tetravalent oxygen in its molecule, which is stable due to its aromatic characteristics (Figure 1). This positive charge in the anthocyanidin structure makes this kind of

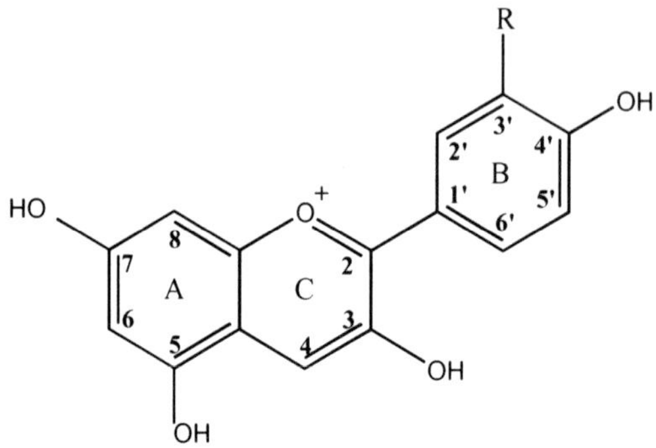

Figure 1 *General structure of an anthocyanidin*

compound more suitable for mass spectrometry analysis in positive mode at low voltages.

Anthocyanins frequently appear in plants acylated with aromatic or aliphatic acids. They are the main flavonoid compounds in some processed foods and beverages, such as wine and fruit juices. They are also used in the food industry as colorant additives. In most of these products they appear as derived pigments formed from the original anthocyanidin. These derived forms of anthocyanins usually are formed during the processing or stocking of the product and in many cases contribute to the organoleptic characteristics of these products. This group of flavonoids, which possess common structural characteristics but different substitution pattern, are a good target for LC–MS and MS–MS analysis.

Methodology

Samples were dissolved in 0.01N HCl and analysed by HPLC using double on-line detection by diode array spectrophotometry and mass spectroscopy (HPLC–DAS–MS). HPLC Hewlett-Packard 1100 equipment was used, provided with a quaternary pump and a photodiode array detector. The column was an Aqua® C_{18}, 5 μm, 150 × 4.6 mm (Phenomenex®) maintained at 35 °C. Solvents were 0.1% trifluoroacetic acid (A) and acetonitrile (B), establishing the following gradient: isocratic 10%B over 5 min, from 10 to 15%B over 15 min, isocratic 15%B over 5 min, from 15 to 18%B over 5 min, and from 18 to 35%B over 20 min, at a flow rate of 0.5 ml min^{-1}.

Mass spectrometry was performed using a Finnigan LCQ equipped with an API source, using an electrospray ionisation (ESI) interface. The HPLC system was connected to the probe of the mass spectrometer *via* the diode array detector cell outlet, using PEEK tubing. Both the auxiliary and the sheath gas were a mixture of nitrogen and helium. The capillary temperature was 180 °C and the capillary voltage 3 V. The MS detector was programmed to do a series of three consecutive scans: a full scan from 150 to 1500 m.u., a zoom scan of the most abundant ion in a ±5 m.u. range and an MS–MS of the most abundant ion in the first scan using a relative collision energy of 20. Spectra were recorded in positive ion mode. In some cases, when the compound is pure the injection can be done directly in the MS using 10 $\mu l\ min^{-1}$ and the same conditions described before but in this case it is possible to do MS^n giving more structural information.

Results

As an example we show the results obtained in the analysis of a purple corncob extract.[45] Figure 2 shows the HPLC anthocyanin profile of the sample registered at 520 nm. Cyanidin-3-glucoside (peak 2), pelargonidin-3-glucoside (peak 3) and peonidin-3-glucoside (peak 4) were identified by LC–MS analysis, with positively monocharged molecular ions at *m/z* 449, 433 and 463 m.u., corresponding to their HPLC peaks respectively. In their MS–MS spectra, the majority of the fragments appeared at *m/z* 287, 271 and 301 m.u. (−162 m.u., loss of a glucose moiety), corresponding to the aglycones cyanidin, pelargonidin and peonidin respectively.

The mass spectrum of peak 5 showed a monocharged molecular ion at *m/z* 535 m.u. In its MS–MS spectrum the major fragments were observed at *m/z* 449 (cyanidin-3-glucoside) and 287 m.u. (cyanidin) (Figure 3). The loss of 86 mass units was consistent with the loss of a malonyl moiety to give the fragment at *m/z* 449. The compound was thus associated to cyanidin 3-(6″-malonylglucoside), previously identified in maize leaves by Harborne and Self in 1987.[26]

Similar observations were made for compounds 6 and 7 (see Table 1) in relation to pelargonidin-3-glucoside (peak 3) and peonidin-3-glucoside (peak 4), which were thus identified as pelargonidin 3-(6″-malonylglucoside) and peonidin 3-(6″-malonylglucoside), respectively.

The UV-vis spectrum of compound 8 showed maximum absorption in the visible region at 517 nm, similar to cyanidin-3-glucoside. The spectra showed a positive molecular ion at *m/z* 563 m.u., whose fragmentation produced two major peaks at *m/z* 449 (cyanidin-3-glucoside) and 287 m.u. (cyanidin). The difference

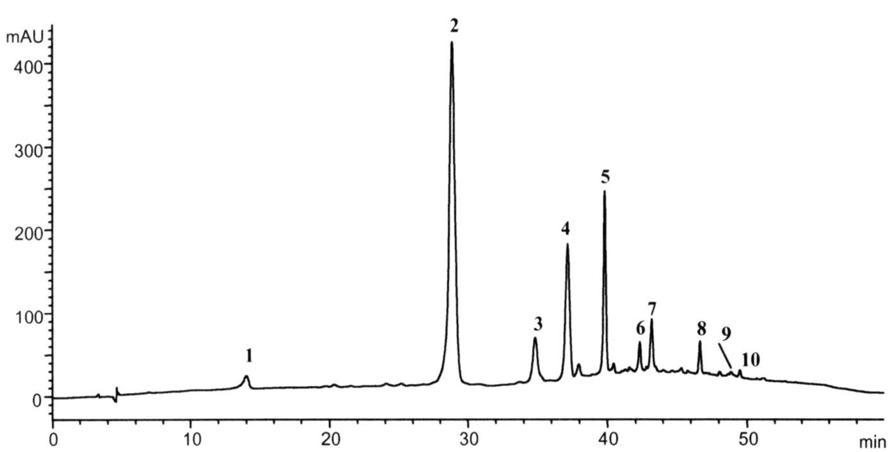

Figure 2 *Chromatogram at 520 nm corresponding to the original cob extract*

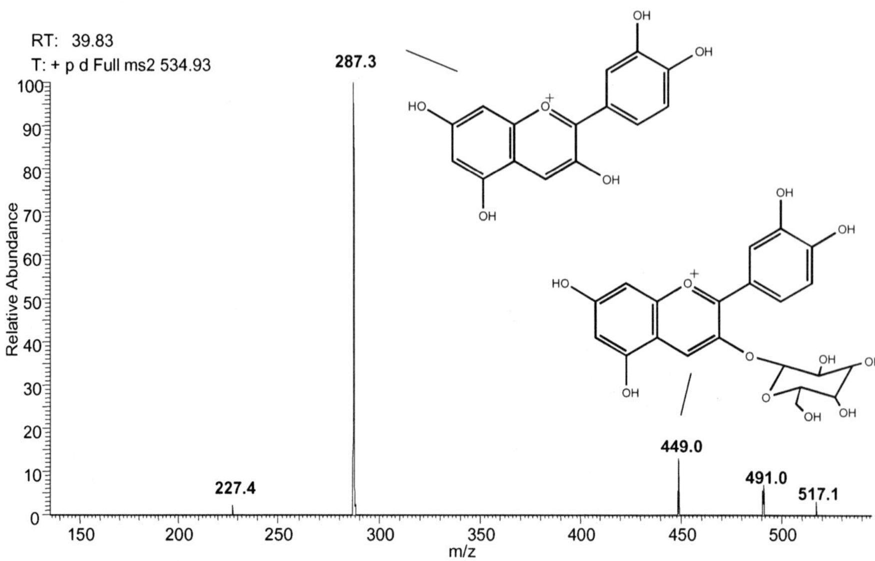

Figure 3 *MS–MS spectra of peak 5 and structure of the fragments*

of mass between compound 8 and cyanidin-3-glucoside (114 m.u.) led us to the conclusion that a malonyl (86 m.u.) and a C_2H_4 (28 m.u.) residue exist. In that case, the fact that both moieties were lost together in the MS–MS spectrum suggested the possibility of a second esterification of the malonyl residue with ethanol. As this kind of acyl derivative had not been described previously, it was thought that it could have been formed during the extraction process to obtain the commercial maize powder. In order to check it, 1 mg of cyanidin 3-(6″-malonylglucoside), previously isolated by semi-preparative HPLC, was dissolved in 60% aqueous ethanol containing 1% citric acid and introduced in a water bath at 60 °C for 3 h. The HPLC analyses carried out at 0 h, 1 h, 2 h and 3 h showed in the chromatograms recorded in the SIM mode at 535 and 563 m.u. the progressive decrease of the peak corresponding to cyanidin 3-(6″-malonylglucoside) from the chromatogram giving rise to the formation of a new product, whose spectral and chromatographic characteristics, molecular ion and MS–MS spectrum were identical to those of compound 8 (Figure 4). It was thus concluded that compound 8 corresponded to cyanidine-3-(6″-ethylmalonylglucoside) and that it was formed during the industrial processing to prepare the maize powder.

It was also observed that peaks 9 and 10 present in the chromatogram of Figure 2 showed molecular ions and MS–MS spectra that followed similar patterns to compound 8 (see Table 1). Therefore, they could correspond to the ethylmalonyl derivatives of pelargonidin 3-glucoside and peonidin 3-glucoside, respectively.

Compound 1 showed a molecular ion at *m*/*z* 899 and fragments at *m*/*z* 737, 575 and 449 m.u. (cyanidin 3-glucoside) were observed in its MS–MS spectrum. The two first fragments derived from the respective loss of two hexose residues (−162 m.u.), indicated the existence of at least three sugar moieties linked to

Table 1 *Chromatographic and spectral characteristics of peaks in Figure 2*

Peak	*Compound*	R_t *(min)*	λ_{max} *(nm)*	$[M + H]^+$ *(m.u.)*	*Fragments* $[M + H]^+$ *(m.u)*
1	Unknown	14.0	530, 454, 330, 286	899	737, 575, 449
2	Cyanidin-3-glucoside	28.8	517, 282	449	287
3	Pelargonidin-3-glucoside	34.8	502, 422, 330, 278	433	271
4	Peonidin-3-glucoside	37.2	517, 282	463	301
5	Cyanidin-3-(6″-malonylglucoside)	39.8	517, 282	535	449, 287
6	Pelargonidin-3-(6″-malonylglucoside)	42.3	502, 422, 330, 278	519	433, 271
7	Peonidin-3-(6″-malonylglucoside)	43.2	517, 282	549	463, 301
8	Cyanidin-3-(6″-ethylmalonylglucoside)	46.7	517, 282	563	449, 287
9	Pelargonidin-3-(6″-ethylmalonylglucoside)	48.9	502, 422, 330, 278	547	433, 271
10	Peonidin-3-(6″-ethylmalonylglucoside)	49.5	517, 282	577	463, 301

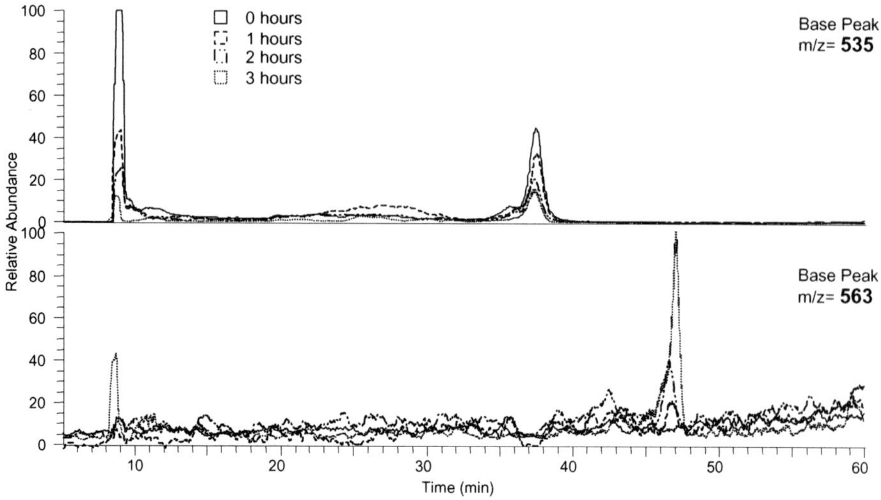

Figure 4 *Chromatograms in SIM mode of the ions corresponding to peak 5 (535 amu) and peak 8 (563 amu) after 0, 1, 2 and 3 hours of reaction of peak 5 in 60% aqueous ethanol at 60 °C*

cyanidin, which is consistent with the increased polarity of the compound (elution at short retention time). As these results show, LC–MS can be very useful for the identification of anthocyanins and derived pigments, especially when analysing glycosides and acylated anthocyanins derived from the same anthocyanidin.

3 Analysis of Catechins and Proanthocyanidins

Proanthocyanidins (PA), also called condensed tannins, are phenolic compounds of a flavonoid nature that are present in a wide range of plant products. They possess a polyflavan-3-ol structure, whose basic units differ in their hydroxylation pattern; the most common units are dihydroxylated (catechins) or trihydroxylated (gallocatechins) in the B-ring of the flavonoid skeleton (Figure 5).

Proanthocyanidins have several biological activities, established from *in vitro* studies. Among their biological activities, their ability to complex proteins and

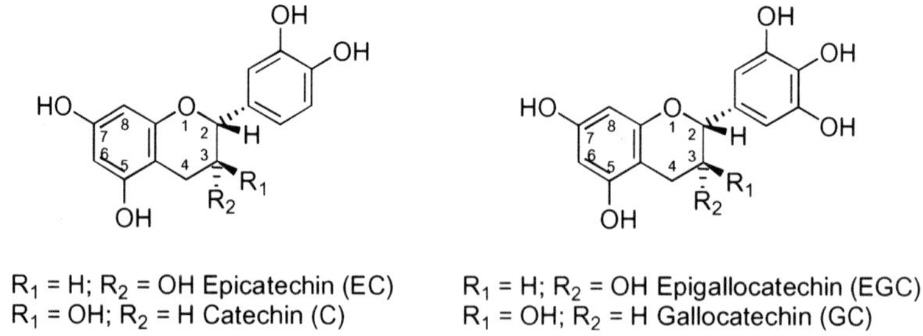

Figure 5 *Structures of catechins and gallocatechins*

metallic ions and to act as antioxidants and radical scavengers has been shown.[27-31] They contribute to astringency in plant foodstuffs[32] and their presence in the diet has been related to protection against some chronic diseases[33-36]. Wine PA have been associated with a cardiovascular protective effect known as the 'French paradox'[37] and they are involved in the definition of the colour in red wine through their interactions with anthocyanins.[38,39]

HPLC coupled to UV detection is the most commonly used technique for the analysis of PA. However, interferences usually exist with co-eluting substances, particularly other phenolics, present in higher amounts and/or with higher UV extinction than PA. These interfere with the adequate observation of their peaks in the chromatograms. Another challenge is the accurate identification of PA, which usually requires their previous isolation from a known source, as no standards are commercially available. LC–MS has proved a powerful tool for the detection and identification of proanthocyanidins in complex samples, without requiring their previous isolation or excessively long elution programs to provide perfect separations of peaks.[14,40,41]

As an example, we have chosen the LC–MS method used for the analysis of proanthocyanidins in wine to show the great utility of this technique when working with very complex samples.[46]

Methodology

Ethanol was eliminated under vacuum and the aqueous phase repeatedly extracted with ethyl acetate. After adding a small volume of water, the ethyl acetate was eliminated by evaporation at reduced pressure and the extract obtained was made up with water to a volume 10 times lower than the starting volume of wine. 100 μl of this solution were then injected into the chromatographic system.

The HPLC system and mass spectrometer were as for the anthocyanin analysis. The column was an Spherisorb S3 ODS2 (4.6×150 mm), 3 μm (Waters, Wexford, Ireland), and solvents were A: formic acid 4.5%:acetonitrile (90:10), and B: acetonitrile. A gradient was established from 0 to 15% of B in A for 30 min, 15 to 50% of B for 5 min, and 50% of B isocratically for 5 min, at a flow rate of 0.5 ml min^{-1}. A first detection was made in the photodiode apparatus selecting 280 nm as a preferred wavelength, followed by a second detection in the mass spectrometer.

For mass spectrometric analysis the capillary temperature was 200 °C and the capillary voltage 31 V. Spectra were recorded in positive mode and the MS was programmed to do a series of three scans: a full mass, a zoom scan of the most abundant ion in the first scan and an MS–MS of the most abundant ion using a collision energy of 20.

Results

In the fragmentation of catechins a loss of 152 mass units (m.u.) is produced due to their retro-Diels–Alder fission (RDA). This same loss is observed for PA containing catechin elementary units (Figure 6). In the case of trihydroxylated units (gallocatechins) the fragment lost in the RDA fission is of 168 m.u. As a consequence, mass spectra show signals at *m/z* 139 in the case of monomeric flavanols (either catechins or gallocatechins), at *m/z* 427 in that of PC dimers containing two (epi)catechin units (molecular ion [M^+] 579 m.u.), and at *m/z* 443, in PD dimers possessing two (epi)gallocatechin units (molecular ion [M^+] 611 u.m.). In the mass spectra of these PA peaks corresponding to the ion [M^+] of their elementary units (291 m.u. for catechins and 307 m.u. for gallocatechins) are also observed.

An important aspect was to establish whether the RDA fission took place on the lower or the upper units of the PA structure and which of these units were released in their free form after MS–MS. Thus, the identification of oligomers could be possible knowing their molecular ion and fragmentation pattern. This would be of the greatest interest in the case of heterogeneous PA containing both (epi)catechin and (epi)gallocatechin units, as well as for those presenting galloyl moieties. With this objective, mass spectra of the dimers gallocatechin-(4,8)-gallocatechin (GC–GC), gallocatechin-(4,8)-catechin (GC–C) and catechin-(4,8)-gallocatechin (C–GC) were obtained. The spectra of the three compounds showed signals corresponding to their positive molecular ions (611 m.u. for GC–GC and 595 m.u., for C–GC and GC–C). In the MS–MS spectra, main signals appeared at *m/z* [M^+] 443 (-152 m.u.) in the case of C–GC, and at *m/z* [M^+] 427

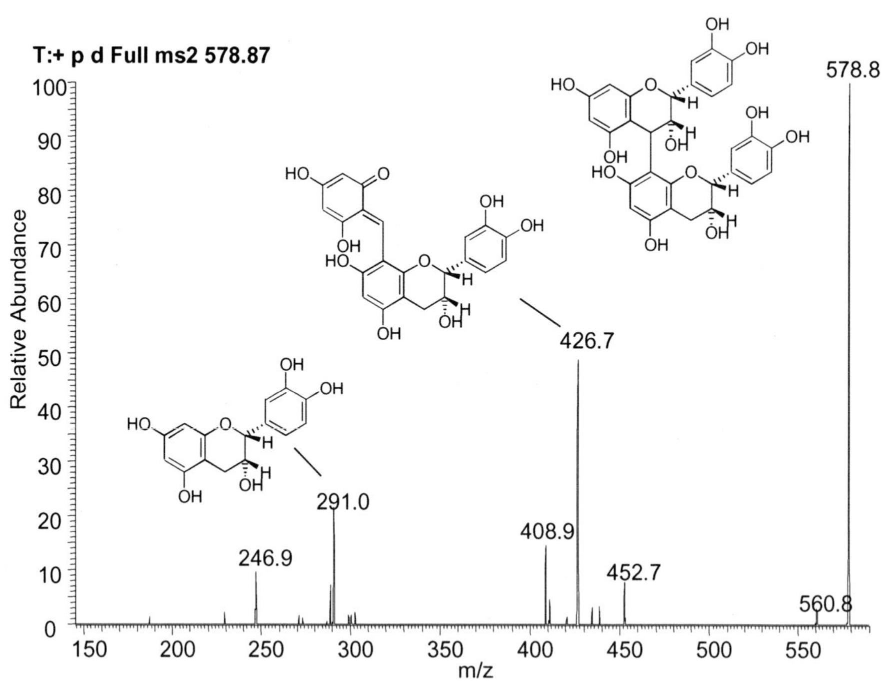

Figure 6 *MS–MS spectra of the procyanidin dimer B2 (epicatechin-4,8-epicatechin)*

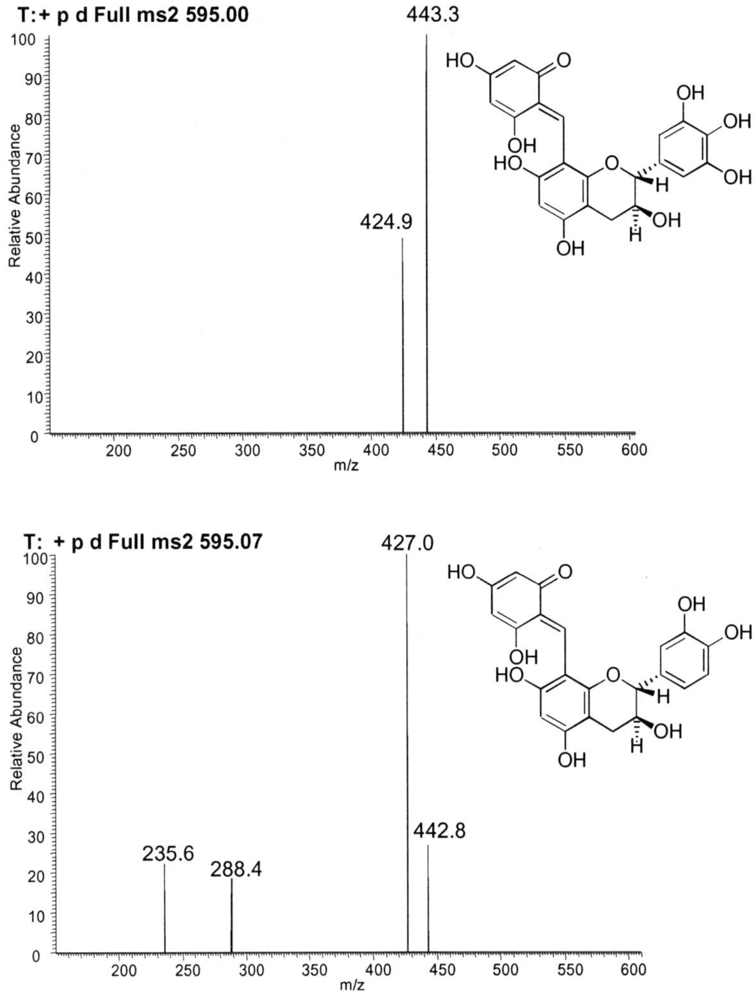

Figure 7 *MS–MS spectra of the dimers catechin-(4,8)-gallocatechin (upper part) and gallocatechin-(4,8)-catechin (lower part) and structure of the fragments obtained after their RDA fission*

(−168 m.u.) in that of GC–C (Figure 7). These results demonstrate that RDA fission is produced fundamentally in the upper subunit of the PA.

4 Analysis of Flavonols and Flavonol Metabolites

Flavonols are secondary metabolites of plants, which are regularly consumed by humans. They are distributed in fruits and vegetables and their daily intake has been estimated at between 3 and 70 mg, essentially quercetin (60–70%). This group of flavonoids are probably the ones that have been most widely studied, since Hertog *et al.* first correlated their consumption to a decrease in cardiovascular disease in 1995.[42]

Quercetin is the major flavonol in the plant kingdom and is found ubiquitously in the diet. It is found in plant food conjugated to various sugars and in human plasma. It has been recently described as its conjugated forms quercetin-3-glucuronide, 3′-methyl-quercetin-3-glucuronide and quercetin-3′-sulfate by Day *et al.* in 2001.[43]

Many authors have used LC–MS for the detection and identification of flavonols in different plant sources and some have identified quercetin metabolites in human or rat samples using this same technique, usually in the selected ion monitoring mode.[25,43]

In this article we describe a LC–MS method valid for the identification of both quercetin glycosides and glucuronides by means of their fragmentation profiles. As an example we show here the results obtained in the analysis of quercetin glucuronides mixture obtained by enzymatic synthesis (see Chapter 5).

Methodology

Samples were suspended in a mixture of methanol:water (80:20) to a final concentration of 50 $\mu g\, ml^{-1}$ and analysed by HPLC using double on-line detection by diode array spectrophotometry and mass spectroscopy (HPLC–DAS–MS).

The HPLC system and mass spectrometer were as before. The column was a Spherisorb S3 ODS2 (4.6 × 150 mm), 3 μm (Waters, Wexford, Ireland), and solvents were A: acetic acid 2.5%, B: acetic acid 2.5%: acetonitrile (90:10) and C: acetonitrile; gradient: from 0 to 100% of B in A for 5 min, 0 to 15% of C in B for 25 min, 15 to 50% of C in B for 5 min and 50% of C isocratically for 5 min; flow: 0.5 ml min^{-1}. The injection volume was 20 μl. A first detection was made in the photodiode apparatus selecting 370 nm as a preferred wavelength, followed by a second detection in the mass spectrometer.

For mass spectrometric analysis using a chemical ionisation (APCI) interface both the auxiliary and the sheath gas were a mixture of nitrogen and helium. APCI conditions were optimised using a quercetin-3-glucoside standard. The capillary temperature was 150 °C and the vaporiser temperature 450 °C, using a capillary voltage of 4 V. The MS was programmed to do a series of three scans: a full mass, a zoom scan of the most abundant ion in the first scan, and an MS–MS of the most abundant ion using a collision energy of 20.

Results

In order to facilitate the study of the fragmentation of the quercetin glucuronides, the MS–MS conditions were first optimised by using a quercetin-3-glucoside standard. The maximum possible intensity for the precursor ion is required for MS–MS analysis, and in this case there is no other possible standard available. Finally the conditions were set up as described in the methodology.

In Table 2 we show the results obtained in the analysis of the compounds present in this sample. In the mixture there were five main peaks. The full scan spectra of those peaks showed molecular ions in the negative mode at 301 m.u. corresponding to free quercetin and four peaks at 477 m.u. corresponding to the different quercetin glucuronides.

The MS–MS spectrum of free quercetin showed majority fragments in the negative mode at 273, 257, 227, 179, 151, 135 and 107 m.u., in accordance with the results previously reported by Ma *et al.* (1997), Wittig *et al.* (2001) and Justesen *et al.* (2001).[25,43,44] (Figure 8).

In the case of the other five compounds present in the prepared mixture the main fragments observed in the MS^3 spectra were the same as for the MS–MS spectra of free quercetin, confirming their identity as quercetin derivatives.

Peaks 1, 2, 3 and 4 showed molecular ions at 477 m.u., and in every case the most abundant ion in their MS–MS spectra was at 301 m.u. corresponding to a loss of 176 mass units consistent with the loss of the glucuronic moiety.

In the full spectra the ratios between the relative abundance of the molecular ion and the main fragments were different for the four different glucuronides, using the same ionisation conditions for all of them (see Table 2). Quercetin-7-glucuronide and quercetin-4′-glucuronide spectra showed much higher proportions of the fragment at 301 m.u., whereas the fragment at 179 m.u. was non-existent in the case of quercetin-7-glucuronide spectra. The MS–MS spectrum corresponding to quercetin-3-glucuronide showed a characteristic fragment at 373 m.u. that did not appear in any of the spectra of the other quercetin glucuronides. Another fragment at 343 m.u. appeared only in the MS–MS spectra

Table 2 *Characteristics and identification of each quercetin derivative*

Peak	R_t *(min)*	$[M-H]^-$ *(m.u.)*	MS^2 *fragments* $[M-H]^-$ *(m.u) (% base peak)*	MS^3 *fragments* $[M-H]^-$ *(m.u)*	*Identification*
1	29.7	477	179 (5), 251 (17), 301 (100), 333 (14), 343 (12), 373 (28)	107, 135, 151, 179, 227, 257, 273	Quercetin-3-glucuronide
2	31.1	477	301 (100)	107, 135, 151, 179, 227, 257, 273	Quercetin-7-glucuronide
3	35.9	477	179 (5), 301 (100), 343 (12)	107, 135, 151, 179, 227, 257, 273	Quercetin-4′-glucuronide
4	36.8	477	151 (10), 179 (25), 301 (100), 477 (10)	107, 135, 151, 179, 227, 257, 273	Quercetin-3′-glucuronide
5	39.0	301	107 (5), 135 (4), 151 (52), 179 (100), 227 (6), 257 (18), 273 (27)		Quercetin

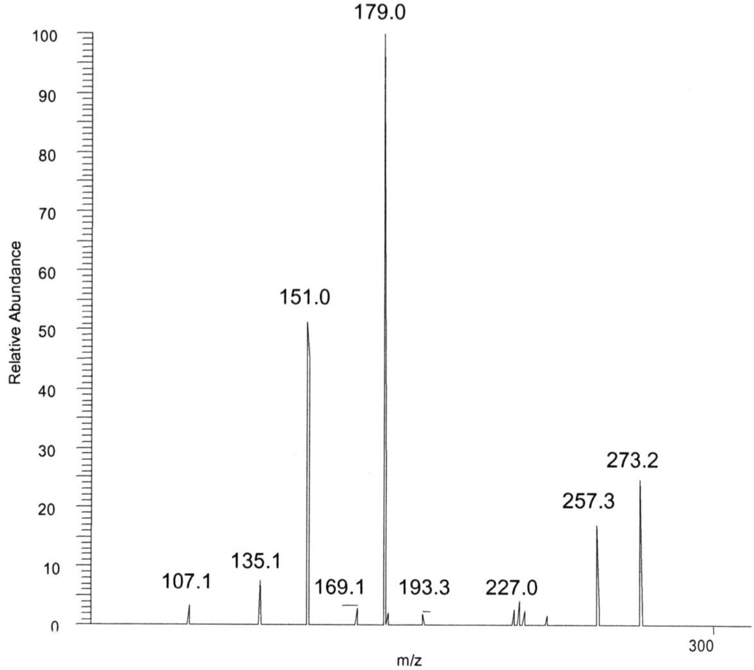

Figure 8 *MS–MS spectra of the ion at 301 m.u. corresponding to free quercetin*

corresponding to quercetin-3-glucuronide and quercetin-4′-glucuronide but not in the others.

These results show that LC/APCI–MS/MS provides an efficient analytical tool for the analysis of quercetin metabolites. The different substitution positions can be identified on the basis of their fragmentation behaviour in their MS/MS spectra.

Acknowledgments

The work in this paper was funded jointly by the Spanish 'Comision interministerial de Ciencia y Tecnologia' (CICYT ALI1997-0773) and the European Union (project FAIR-CT95-0653 and POLYBIND – QLK1 – 1999-00505). Thanks are also due to Ricardo Sarmiento (PRONEX S.A., Peru) for providing of the anthocyanin industrial extract and to Prof. Gary Williamson and Dr. Karen O'Leary (IFR, UK) for providing the flavonol metabolites mixture.

References

1. Y. Cai, F.J. Evans, M. F. Roberts, J. D. Phillipson, M. H. Zenk and Y. Gleba, *Phytochemistry*, 1991, **30**, 2033.

2. L. Foo and J. J. Karchesy, *Phytochemistry*, 1989, **28**, 3185.
3. K. R. Markham, *Techniques of Flavonoid Identification*, Academic Press, London, 1982.
4. G. C. Galletti, V. Modafferi, M. Poiana and P. Bocchini, *J. Agric. Food Chem.*, 1995, **43**, 1859.
5. U. Justesen, P. Knuthsen and T. Leth, *Cancer Lett.*, 1997, **114**, 165.
6. U. Justesen, P. Knuthsen and T. Leth, *J. Chromatogr. A*, 1998, **799**, 101.
7. S. E. Nielsen, M. Kall, U. Justesen, A. Schou and L. O. Dragsted, *Cancer Lett.*, 1997, **114**, 173.
8. M. Yilmazer, J. F. Stevens and D. R. Buhler, *FEBS Lett.* 2001, **491**, 252.
9. J. J. Dalluge, B. C. Nelson, J. B. Thomas, M. Welch and L. C. Sander, *Rapid Commun. Mass Spectrom.*, 1997, **11**, 1753.
10. S. Guyot, T. Doco, J. M. Souquet, M. Moutounet and J. F. Drilleau, *Phytochemistry*, 1997, **44**, 351.
11. G. Vanhoenacker, A. De Villiers, K. Lazou, D. De Keukeleire and P. Sandra, *Chromatographia*, 2001, **54**, 309.
12. A. Chandra, J. Rana and Y. Li, *J. Agric. Food Chem.*, 2001, **49**, 3515.
13. C. Cren-Olivé, S. Deprez, S. Lebrun, B Coddeville and C. Rolando, *Rapid Commun. Mass Spectrom.*, 2000, **14**, 2312.
14. S. de Pascual-Teresa, Y. Gutierrez-Fernandez, J. C. Rivas-Gonzalo and C. Santos-Buelga, *Phytochem. Anal.*, 1998, **9**, 21.
15. D. Favretto and R. Flamini, *Am. J. Enol. Vitic.*, 2000, **51**, 55.
16. M. Murkovic, U. Mülleder, U. Adam and W. Pfannhanser, *J. Sci. Food Agric.*, 2001, **81**, 934.
17. R. L. Prior, S. A. Lazarus, G. Cao, H. Muccitelli and J. F. Hammerstone, *J. Agric. Food Chem.*, **2001**, 49, 1270.
18. E. Hvattum, *Rapid Commun. Mass Spectrom.*, 2002, **16**, 655.
19. U. Justesen and E. Arrigoni, *Rapid Commun. Mass Spectrom.*, 2001, **15**, 477.
20. U. Justesen, E. Arrigoni, B. R. Larsen and R. Amado, *Lebensm.-Wiss. U.-Technol.*, 2000, **33**, 424.
21. P. L. Mauri, L. Iemoli, C. Gardana, P. Riso, P. Simonetti, M. Porrini and P. G. Pietta, *Rapid Commun. Mass Spectrom.*, 1999, **13**, 924.
22. S. E. Nielsen, V. Breinholt, C. Cornett and L. O. Dragsted, *Food Chem. Toxicol.*, 2000, **38**, 739.
23. A. Schieber, P. Keller, P. Streker, I. Klaiber and R. Carle, *Phytochem. Anal.*, 2002, **13**, 87.
24. G. Stecher, C. W. Huck, M. Popp and G. K. Bonn, *Fresenius' J. Anal. Chem.*, 2001, **371**, 73.
25. J. Wittig, M. Herderich, E. U. Graefe and M. Veit , *J. Chromatogr. B*, 2001, **753**, 237.
26. J. B. Harborne and R. Self, *Phytochemistry.*, 1987, **26**, 2417.
27. I. F. F. Benzie and Y. T. Szeto, *J. Agric. Food Chem.*, 1999, **47**, 633.
28. E. Haslam, *J. Nat. Prod.*, 1996, **59**, 205.
29. G. W. Plumb, S. de Pascual-Teresa, C. Santos-Buelga, V. Cheynier and G. Williamson, *Free Radic Res.*, 1998, **29**, 351.

30. N. Saint-Cricq de Gaulejac, C. Provost and N. Vivas, *J. Agric. Food Chem.*, 1999, **47**, 425.
31. P. L. Teissedre, E. N. Frankel, A. L. Waterhouse, H. Peleg and J. B. German, *J. Sci. Food Agric.*, 1996, **70**, 55.
32. E. Haslam, in *The Biochemistry of Plants*, ed. E. Conn, Academic Press, New York, 1981.
33. M. Grønbæk, U. Becker, D. Johansen, H. Tønnensen, G. Jensen, and T. I. A. Sørensen, *Br. Med. J.*, 1998, **317**, 844.
34. M. G. L. Hertog, E. J. M. Feskens, P. C. H. Hollman, M. B. Katan and D. Kromhout, *Lancet*, 1993, **342**, 1007.
35. S. C. Renaud, R. Gueguen, J. Schenker and A. d'Houtaud, *Epidemiology*, 1998, **9**, 184.
36. A. S. St. Leger, A. L. Cochrane and F. Moore, *Lancet*, 1979, **i**, 1294.
37. S. Renaud and de M. Lorgeril, *Lancet*, 1992, **339**, 1523.
38. E. M. Francia-Aricha, M. T. Guerra, J. C. Rivas-Gonzalo and C. Santos-Buelga, *J. Agric. Food Chem.*, 1997, **45**, 2262.
39. C. Saucier, G. Bourgeois, C. Vitry, D. Roux and Y. Glories, *J. Agric. Food Chem.*, 1997, **45**, 1045.
40. V. Cheynier, Y. H. Arellano, J. M. Souquet and M. Moutounet, *Am. J. Enol. Vitic.*, 1997, **48**, 225.
41. H. Fulcrand, S. Remy, J. M. Souquet, V. Cheynier and M. Moutounet, *J. Agri. Food Chem.*, 1999, **47**, 1023.
42. M. G. Hertog, D. Kromhout, C. Aravanis, H. Blackburn, R. Buzina, F. Fidanza, S. Giampaoli, A. Jansen, A. Menotti and S. Nedeljkovic, *Arch. Intern. Med.*, 1995, **155**, 381.
43. A. J. Day, F. Mellon, D. Barron, G. Sarrazin, M. R. Morgan and G. Williamson, *Free Radic. Res.*, 2001, **35**, 941.
44. Y. L. Ma, Q. M. Li, H. Van den Heuvel and M. Claeys, *Rapid Commun. Mass Spectrom.*, 1997, **11**, 1357.
45. S. de Pascual-Teresa, C. Santos-Buelga and J. C. Rivas-Gonzalo, *J. Sci. Food Agric.*, 2002, **82**, 1003.
46. S. de Pascual-Teresa, J. C. Rivas-Gonzalo and C. Santos-Buelga, *Int. J. Food. Sci. Technol.*, 2000, **35**, 33.

CHAPTER 4

The Use of HPLC with Coulometric Array Detection in the Analysis of Flavonoids in Complex Matrices

CLAUDINE MANACH

1 Introduction

Flavonoids are secondary metabolites of plants. More than 4000 molecules exhibiting the typical phenylbenzopyrone structure have been identified, and classified as anthocyanidins, flavonols, flavones, isoflavones, flavanones, flavanols, and other minor classes (Figure 1). Flavonoids are abundant in our diet and a high number of experimental data indicate that they could exhibit some health protective effects towards degenerative processes linked to ageing or oxidative stress.[1]

Flavonoids have been analysed for a long time in plants, using various techniques. HPLC methods have been recently reviewed by Merken and Beecher.[2] Usually the flavonoid content in plant foods is quite high and the difficulties that may be encountered in analysis are mainly due to extraction or stability problems, rather than to a lack of sensitivity of the detection method. Problems of resolution may also occur for food samples containing a complex mixture of flavonoids. A huge amount of work still has to be done on food samples, especially to establish a reference table of the flavonoid content in fruits, vegetables and beverages, and to evaluate the influence of the variety, of the mode of production, and of the domestic or industrial processing on the flavonoid content of plant foodstuffs. For these matters, the technique of HPLC coupled to coulometric array detection may be of specific interest compared with the classic HPLC-UV method, with a higher resolving power and the possibility to use a fingerprint or pattern recognition approach.

In the past 15 years, research on flavonoids has moved towards issues of

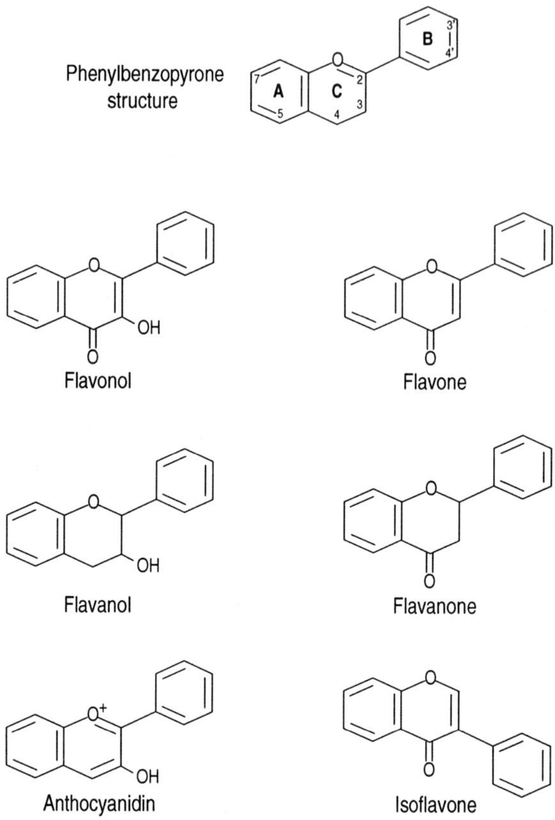

Figure 1 *Chemical structures of flavonoids*

bioavailability and health promoting effects, leading to a requirement of highly sensitive and selective detection methods. Coulometric detection, which was already used in a field requiring high sensitivity, *e.g.* the analysis of neurotransmitters in biological fluids, proved to be a powerful tool for the characterisation and quantification of flavonoids present in trace levels in complex biological matrices. This detection technique is still poorly known and has probably not achieved the place that it actually deserves according to its numerous advantages.

2 Basics of Electrochemistry

Electrochemical detectors measure a compound's chemical properties, not a physical property such as UV absorption. Electrochemistry involves chemical reactions in which electrons are transferred from one compound to another. As such, electrochemistry is a destructive method, with relatively few compounds able to be reduced to the original structure once oxidised. Typical half-reactions are shown below for the redox couple Hydroquinone (HQ)/Quinone (Q):

$$HQ \rightarrow Q + 2e^- + 2H^+ \qquad \text{(oxidation)}$$

$$Q + 2e^- + 2H^+ \rightarrow HQ \qquad \text{(reduction)}$$

The analyst is concerned with only one half-reaction at one time, usually the oxidation. In order for an oxidation or a reduction to occur, energy in the form of an electric potential is required. This potential is applied on a working electrode and when compounds eluting from the column enter in contact with the electrode and are subjected to a potential sufficient to drive the reaction, they undergo molecular rearrangement at the working electrode's surface with a loss (oxidation) or gain (reduction) of electrons. The electrons are detected as a current generated in real time. Depending on the design of the working electrode, the electrochemical detector can be amperometric or coulometric (Figure 2). In amperometric detectors, compounds eluting from the HPLC column flow by the working electrode and only a small fraction, typically 5–15%, of the electroactive compounds can actually react on its surface. In a coulometric detector, compounds flow through a working electrode constituted of porous graphite. Due to the high surface of contact offered by the structure of porous electrodes, 100% of the species of interest is oxidised (or reduce) so that the magnitude of the current is directly proportional to the injected quantity of compound. On the chromatogram, peak areas represent the total current as a function of time, that is 'charge', measured in coulombs. If the mass injected in the column and the molecular weight of the compound of interest are known, it is possible to predict the peak area (in nanocoulombs) that will be obtained for this compound *via* Faraday's law:

$$\text{Peak area} = Q = mFn/MW$$

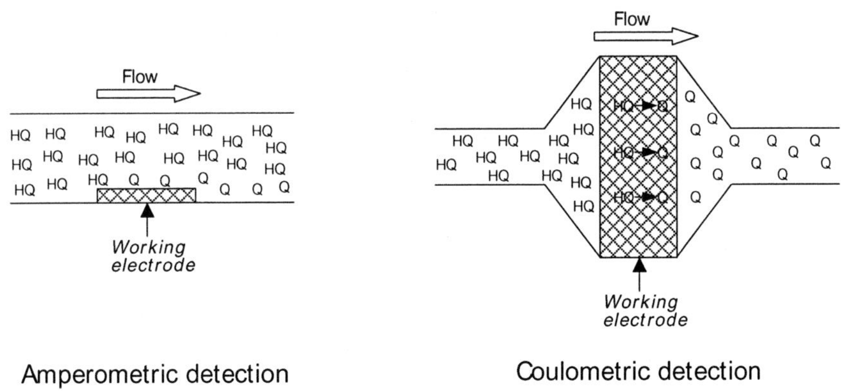

Figure 2 *Electrode design in electrochemical detection*

where:

m is the injected mass of compound (ng)
F is Faraday's constant (96500 coulombs/equivalent)
n is the number of electrons transferred per molecule (equivalent mol^{-1})
MW is the molecular weight (g mol^{-1})

It should be noted that chromatographic variables are not required for this calculation.

A wide variety of compounds can be analysed by electrochemical methods, including antioxidants (*e.g.* BHT), polyphenols, catecholamines, carbohydrates, pesticides, vitamins, pharmaceuticals ... In general, electroactivity is dependent on the presence of an electroactive functional group such as the hydroxyl group of a phenol or the amino group of an aniline, and differences in optimal potential between compounds depend on their relative ability to adapt to the loss of electrons, by delocalising charge *via* resonance or inductive charge stabilisation.[3]

The potential used for an analyte is an important issue. Each compound is characterised by an optimal potential of oxidation or reduction, as it has a specific wavelength at which it exhibits maximum absorption. This potential is dependent on several factors including the nature of the electrode surface, the pH, the composition of the mobile phase, especially its ionic strength, and of course the chemical structure of the species of interest. The optimal potential can be determined by measuring the oxidative (or reductive) current obtained for a constant concentration of the analyte over a range of working electrode potentials. A plot of current generated (peak height or area) *vs.* applied potential is commonly called a hydrodynamic voltammogram (HDV) or a current/voltage (C/V) curve. A typical HDV is shown in Figure 3. The curve indicates the potential at which the oxidative current is maximal. In the example shown, E_{max} = 300 mV. If the applied potential is lower than 100 mV, the compound will not oxidise and will be invisible to the detector. If the applied potential is between 100 and 300 mV, the sensitivity will not be maximal. If the applied potential is higher than 300 mV, no gain in the current intensity will be obtained, but the risk that other species may also be oxidised and interfere with the analyte of interest will increase. It is always better to use a potential as low as possible to avoid contaminant peaks, to minimise background current due to mobile phase

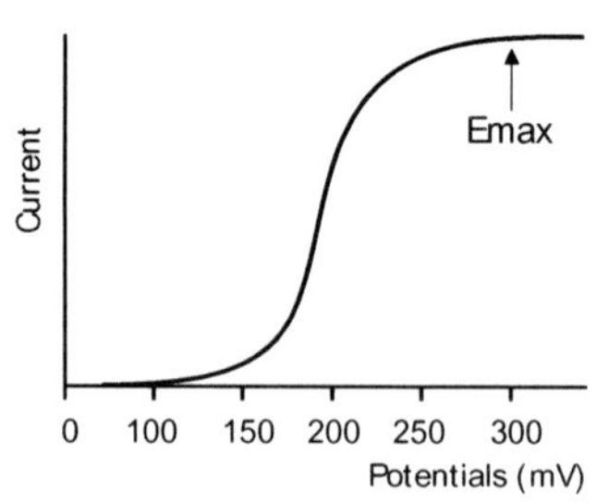

Figure 3 *Typical hydrodynamic voltammogram*

oxidation, and to extend electrode-life. Thus the applied potential is typically chosen about 20 mV above E_{max}.

If more than one compound in the sample is of interest, the potential should be set to maximise the sensitivity of the compound with the largest oxidation potential. For example, if two compounds, A with $E_{max} = 250$ mV and B with $E_{max} = 600$ mV, are to be quantified, a working potential of 600 mV is required. But in this case the risk of interferences with co-eluting compounds present in the samples will increase for compound A which could be detected at a lower potential. Coulometric detectors are often designed with two or more electrodes, so that the first electrode can be set at 250 mV and oxidise only compound A, and the second electrode can be set at 600 mV to oxidise compound B (Figure 4). This mode is particularly interesting when compounds A and B have the same retention time. Compound A will then be quantified on the chromatogram recorded at 250 mV and compound B on the chromatogram recorded at 600 mV, with complete separation despite the co-elution. For this purpose, it is crucial that electrolysis of compound A does not result in a product that will react on the second electrode, since some compounds, especially flavonoids, can bear several sites of oxidation and thus react successively at increasing potentials.

The two electrodes can also be used in a screen mode, to eliminate interferences. In this case, the potential set at the first electrode will be lower than that required to oxidise the compounds of interest so that only interfering compounds will be oxidised on that electrode, and the compounds of interest will be quantified on the second electrode. Another possibility is that mobile phases contain impurities which oxidise at a higher potential than that of one of the compounds of interest. In this case, a guard cell can be put between the pump and the injector, with a potential higher than those of the working electrodes. All interfering species present in the mobile phase will then be invisible to the working electrodes.

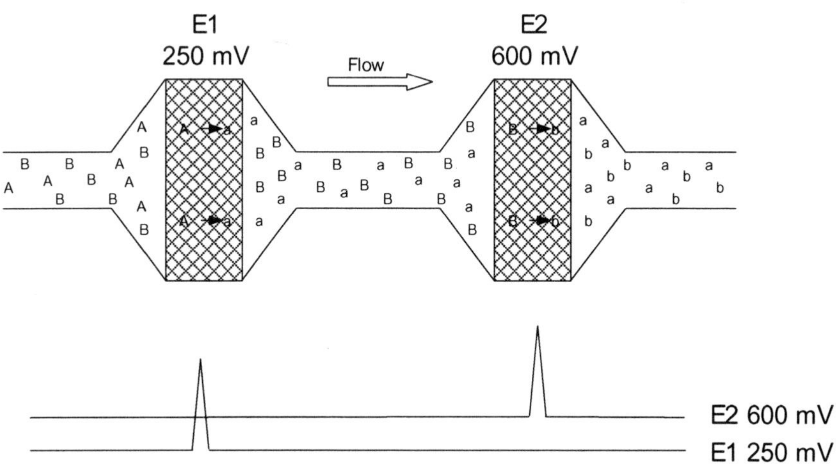

Figure 4 *Use of coulometric detector with two electrodes for detection of compounds A and B, oxidising at 250 and 600 mV respectively*

3 Principle of Coulometric Array Detection

The coulometric array technology developed by Matson *et al.*[4] consists of two, three or four cell packs in series, each pack containing four porous graphite working electrodes with associated palladium reference electrodes and platinum counter-electrodes. The reference electrode provides a stable potential to monitor the potential of the working electrode. The role of the counter-electrode is to maintain the cell potential constant: when an analyte is oxidised or reduced on the working electrode, a complementary electrolytic reaction takes place on the counter-electrode, providing or removing the same number of electrons as are involved in the oxidation or reduction of the analyte.[3] Different potentials are applied on the working electrodes and eight, twelve or sixteen chromatograms are recorded simultaneously. The design of the cells is such that the chromatographic separation is the same on the last electrode as on the first one, with no significant differences in the retention times of the analytes. A system called Coulochem® Electrode Array System (CEAS) was first commercialised by ESA (Chelmsford, USA), and is now replaced by an improved version called CoulArray™. In most cases, analytes eluting from the column oxidise on three (or more) successive electrodes, termed the preceding, dominant, and following electrodes. Although each electrode is 100% (*i.e.* coulometrically) efficient, the analyte reacts across three adjacent electrodes because none of them are set to its maximal oxidation potential, but rather to voltages along its hydrodynamic voltammogram (Figure 5). Thus the first electrode will oxidise a small portion of the compound, the second or dominant electrode oxidises a large portion of the compound, and the third electrode oxidises the remainder of the compound. The response ratios obtained between adjacent electrodes (preceding/dominant and following/dominant) are characteristic of the voltammetric behaviour of each analyte in the specified chromatographic conditions used, and are independent of the analyte concentration. Unknown compounds present in samples can thus be identified by comparison with pure standards on the basis of their retention time

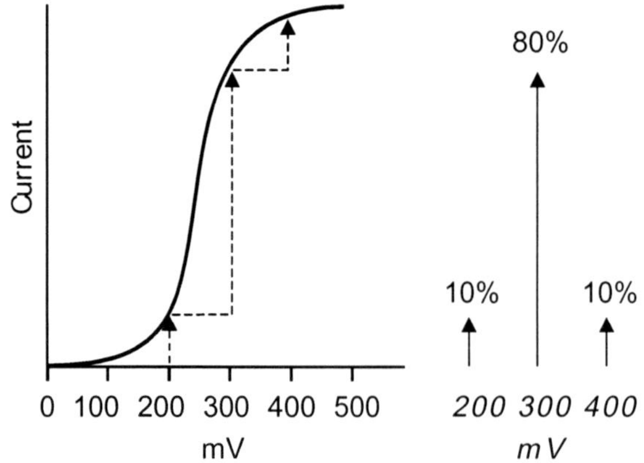

Figure 5 *Repartition of oxidation on successive electrodes*
(Reproduced with permission from ESA, Inc.)

and their electrochemical behaviour on the coulometric array. Any difference in the ratios observed for the unknown compound and the pure standard signifies that either there is some contamination, co-elution of this compound with another unknown compound, or that this compound is not the same as the standard compound. The software calculates 'ratio accuracy':

Ratio accuracy (%) = [ratio of the preceding electrode to the dominant electrode for the analyte in the sample]/[ratio of the same electrodes for the standard] × 100

A second ratio accuracy is obtained by comparing the response ratio for the following and dominant electrodes. By default in the ESA software, each ratio accuracy must be higher than 70% for it to be considered that the compound is pure and identical to the standard. This setting limit can be modified by the user to a value approaching 100%, but it must be noted that a small contamination generally leads to a dramatic change of the response ratios.

Some compounds, such as flavonoids, present several waves of oxidation across the array, corresponding to several moieties capable of undergoing oxidation. Only one wave, usually the lowest wave of oxidation, can be used to quantify the compound, but the whole electrochemical behaviour across the array of the compound can also help to identify it.

4 Determination of Analysis Conditions

The instrumental design is not fundamentally different from a traditional HPLC apparatus except that the usual UV detector is replaced by a coulometric array detector. The first step in the development of an HPLC–CoulArray assay is the determination of the chromatographic conditions adapted to the separation of the compounds of interest. They do not differ from those used with UV detection, except that mobile phases must contain an electrolyte to facilitate the current flux, typically 30–100 mM sodium or lithium phosphate. The pH must be carefully controlled since it influences the current/voltage curve of the compounds, and thus their optimal oxidation potential. Gradient elution is generally considered not suitable for electrochemical detection, because the mobile phase changes during the gradient cause a great baseline drift that impairs peak measurement. However, the CoulArray system is equipped with software containing a gradient-correcting algorithm that removes the baseline drift and allows the use of gradient elution without restriction.

With a single electrode system, the working potential must be chosen 20 mV above the optimal potential of oxidation of the analyte of interest to provide the largest signal while minimising interference from contaminants. In some cases, an estimation of the optimal potential of oxidation is reported in the literature. It must be kept in mind that this value depends on the chromatographic conditions, such as the composition of the mobile phase, and thus should be checked and maybe modified in other conditions. Furthermore, the reference type must be considered. In the CoulArray detector, the indicated potentials are referred to a solid-state palladium reference electrode; their absolute value is 300 mV lower than the corresponding potential measured using an Ag/AgCl reference electrode. It means that if the working electrode was set to 600 mV against an Ag/AgCl reference then it would be equivalent to 300 mV against a palladium reference electrode.

Even if an approximate potential obtained from the literature is useful, it is better to use hydrodynamic voltammetry to determine the optimal potential for each compound to be analysed. A constant concentration of a standard is injected into the HPLC, with chromatographic conditions as close as possible to the conditions of the analysis to be performed. The peak height or area is measured at a series of fixed potentials, after equilibration of the electrode at each change of potential. A hydrodynamic voltammogram is manually generated (peak height or area *versus* potential) to determine the maximum potential of oxidation for the compound of interest, that is the potential at which the peak height or area has reached a plateau.

With a multielectrode detection system, the choice of the applied potentials depends on the objective of the analyst.

In most cases, the electrochemical behaviour of the compounds is used qualitatively to examine peak purity and help in structural identification of unknowns. To allow the calculation of ratio accuracy, the analytes must react on several consecutive electrodes. In general, hydrodynamic voltammograms show an oxidation wave of the analytes over a range of 200–300 mV, so that application of increasing potentials with increments of 60–100 mV will result in a response on three consecutive electrodes for most compounds. The potentials can be more precisely adjusted to give an optimal repartition of the responses on the three adjacent electrodes, that is about 80% of the response on the dominant electrode, and 10% on the preceding and following electrodes. This repartition allows the calculation of ratio accuracy in good conditions, with maximum sensitivity.

In other cases, a peak may be well identified, and the objective of the analyst may be to quantify low levels of the compound in routine analysis. In this situation, a potential 20 mV above the maximum potential of oxidation of the analyte will be applied, to provide the maximum signal. A potential too low to oxidise the compound may be applied on the preceding electrode to serve as a screen channel for minimising contaminants. The totality of the signal for the analyte will then be concentrated on the second electrode. There will be a gain in sensitivity but a loss in electrochemical information, with no possibility of calculating ratio accuracy.

Ideally, chromatographic conditions provide well-separated analytes to the detector. However, in practice, some co-elutions can be very difficult to resolve, even after modification of the chromatographic conditions (mobile phases, column, flow rate) or optimisation of the sample preparation procedure. Co-elutions with contaminants can also occur, especially when trace levels of compounds must be analysed in complex matrices such as urine. The magnitude of this problem will increase with the potential needed to oxidise the compounds of interest. Under these conditions, further resolution can be obtained with coulometric array detection. Provided that the co-eluting compounds have different oxidation potentials, they can be separated across the array. Each compound of interest is then quantified on one or several electrodes at which the interfering compound is not oxidised. Since the properties that influence the voltammetric behaviour of a compound are fundamentally different from those determining its chromatographic behaviour, a powerful multidimensional resolution technique is obtained. This selectivity allows a simplification of the clean-up procedures of samples before HPLC analysis, which can be time-consuming and present a risk of changing the nature or the concentration of the compounds of interest.

In conclusion, when several compounds of interest are present in the sample, potentials should be chosen to optimise the response for each of them, but also to separate electrochemically the co-eluting compounds across the electrode array. Some compromises may be made, especially when a high number of analytes are to be quantified, when they are not well-resolved chromatographically, and when complex matrices are analysed.

5 Sensitivity in Coulometric Detection

The sensitivity of an analytical method is a function of the signal/noise ratio. We have seen that in coulometry the total peak area, that is the signal, obtained with the oxidation of a compound is governed by the Faraday law, and depends on parameters that cannot be changed by the analyst (*e.g.* mass of compound and number of electrons involved in the oxidation reaction/mol). The analyst can only improve the signal by optimising the chromatographic conditions to obtain thin peaks, easier to measure than tailing peaks, and in the case of a coulometric array by judiciously choosing the applied potentials to concentrate the response on the dominant electrode as described above. However, the best way to improve the sensitivity of the detection is to minimise the noise.

The background current is highly dependent on the composition and the purity of the mobile phases. Water must be obtained from a reverse osmosis system (milli-Q or equivalent) and have a resistance of 18.2 MΩ cm^3. It may be further filtered through a 0.2 μm filter. It is also necessary to use the highest purity organic solvents (HPLC grade). Chemicals must be selected based on purity with low ppm (or better) of divalent metal ions capable of undergoing changes in oxidation state, *e.g.* iron, chromium. To adjust the pH of the mobile phase, concentrated bases or acids of the highest purity should be used. The mobile phase must be sparged with an inert gas such as helium to remove oxygen, which is electrogenic, and to limit the formation of small bubbles which can degrade the quality of the baseline.

The quality of the HPLC system is also important. The pump must be equipped with pulse dampers to minimise the noise of the baseline. Tubings and other HPLC components should be made of inert polymer material such as PEEK rather than of stainless steel, to lower as much as possible the metal ion concentration in the system. In-line filters (made of PEEK preferably) can be put between the pump system and the injector and between the analytical column and the detector.

To increase the stability of the system, the ideal situation is to maintain the mobile phase flow and the applied potentials at the level used for analysis when the detector is not being used. For this purpose, a recirculation of the mobile phase can be done between analyses by placing the outlet tube from the detector into the solvent bottle. This must be performed in isocratic conditions, with 100% phase A. When it is not possible, it is crucial to establish a stable baseline before a batch of analyses. One or two runs must be performed to stabilise the system. In theory, minor day-to-day changes in flow rate, temperature, and mobile phase composition will not affect the total peak area with a coulometric detection. However, we have observed that the response ratios between consecutive electrodes may be slightly modified. Thus, it may be useful to analyse regularly some standards between samples in each batch of analyses, to evaluate the electrochemical changes and facilitate the analyte identification by the use of ratio accuracy.

The palladium reference electrode and the flow-through working electrodes are both maintenance-free and long lived. The expected life of a cell is highly dependent on the potential and the current, and the use of potentials above 900 mV may shorten its lifetime. The only maintenance operation advised is to electrochemically clean the working electrode surface at the end of each run, by applying an elevated potential (1000 mV) for a short time, typically 15 seconds, before restoring the initial conditions of analysis.

The same care as with HPLC–UV systems must be taken to maintain the best chromatographic conditions. The use of a precolumn will protect the analytical column and prolong its lifetime. A high concentration of organic solvent at the end of a gradient procedure will clean the column and the electrodes. Holding the column at a constant temperature in a oven will provide more constant retention times from analysis to analysis, facilitating the identification and quantification of the peaks.

When performed with care, coulometry is an extremely sensitive technique, with typical detection limits about 10–100-fold higher than with UV detection and 10-fold higher than with fluorescence detection, depending on the compound studied.

The linear response range of an electrochemical detector is approximately 10^5 (*versus* 10^4 for UV absorption). The CoulArray software possess an automatic gain function which encompasses eight orders of magnitude in response (full screen current from pA to 100 μA) and thus enables detection of low pg to high μg levels of different compounds in the same injection.

6 Analysis of Flavonoids Using HPLC with Coulometric Array Detection

HPLC with coulometric detection is particularly suited to the analysis of flavonoids since they often have a low potential of oxidation. Their detection can thus be performed at a potential at which only few interfering compounds present in the sample will be oxidised. Thus, high quality chromatograms will be obtained. Furthermore, coulometric detection is one of the most sensitive detections available. This is a crucial point, especially for bioavailability studies in which trace levels of metabolites must be identified and quantified in plasma, urine or animal tissue samples. The coulometric array detection offers further advantages, with the possibility to selectively use the electrochemical behaviour of a compound across the array.

7 Voltammetric Characteristics of Flavonoids

UV detection, with a single wavelength or a photodiode array, has been widely used in flavonoid analysis and is suitable for number of applications. However, it may lack sensitivity and selectivity. The UV spectrum is characteristic of the class of flavonoid considered and anthocyanins, flavonols, flavones, flavanones, iso-

Table 1 *UV absorption range for flavonoids*[a]

Flavonoid type	*Band II (nm)*	*Band I (nm)*
Flavones	250–280	310–350
Flavonols (3-OH substituted)	250–280	330–360
Flavonols (3-OH free)	250–280	350–385
Isoflavones	245–275	310–330
Flavanones	230–270	300–330
Anthocyanidins	270–280	465–560

[a] K. R. Markham, *Techniques of Flavonoid Identification*, Academic Press, NY, 1982.

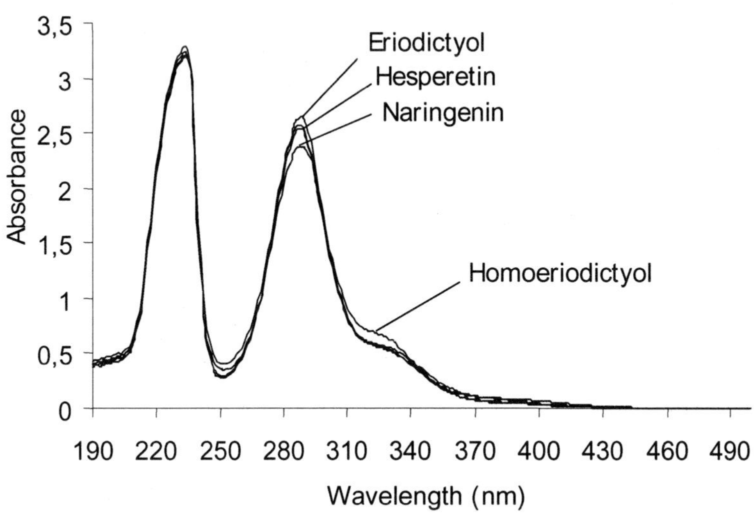

Figure 6 *UV spectra of flavanones in methanol*

flavones or flavanols have different maximum wavelength of absorption[5] (Table 1). However, the UV spectrum is only weakly affected by the substitution pattern of the flavonoid. Compounds within the same class have very similar spectra, as can be seen for the flavanones eriodictyol (5,7,3′,4′-tetrahydroxyflavanone), naringenin (5,7,4′-trihydroxyflavanone), hesperetin (4′-methoxy-3′,5,7-trihydroxy-flavanone), and homoeriodictyol (3′-methoxy-4′,5,7-trihydroxyflavanone) in Figure 6.

In contrast, the hydrodynamic voltammograms presented in Figure 7 clearly

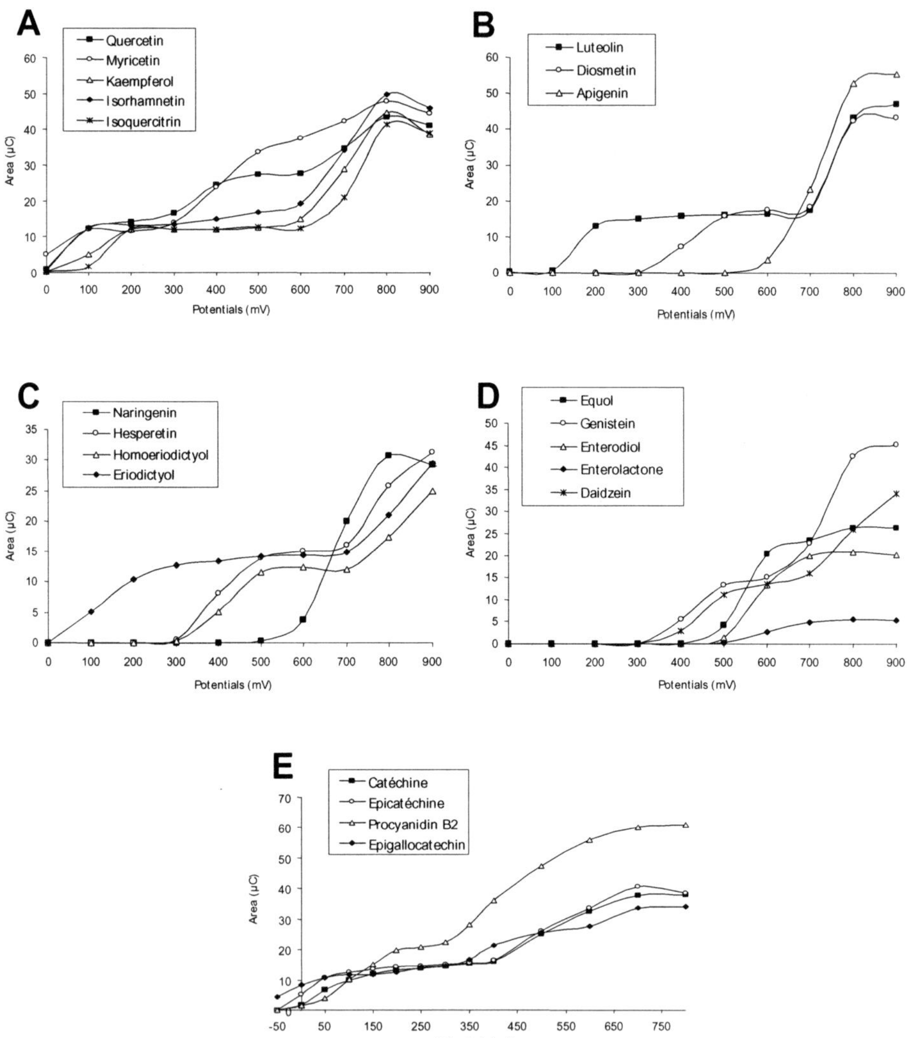

Figure 7 *Voltammograms of flavonoids. Analyses were performed in isocratic conditions (0.8 ml min^{-1}) with a 30 mM NaH_3PO_4 buffer, pH 3 containing, 35% acetonitrile (or 15% acetonitrile for flavanols)*

show that the electrochemical behaviour of flavonoids is directly affected by small changes in their structure. The oxidation potentials of the flavonoids increase on a scale roughly based on the nature of the hydroxyl substitutions of the B ring, as was also noticed by Milbury.[6] The presence of three hydroxyl groups on the B ring leads to a very low potential of oxidation, as observed for myricetin and epigallocatechin. Compounds with a catechol moiety (catechin, epicatechin, procyanidin B_2, quercetin, luteolin, eriodictyol) are oxidised at slightly higher, but still very low potentials (100–200 mV). The potential of oxidation further increases for compounds with a methoxylated catechol (diosmetin, isorhamnetin, hesperetin, homoeriodictyol). Monohydroxylated flavonoids (kaempferol, apigenin, naringenin, genistein, daidzein) come later in the potential scale. Other features of the flavonoid structure influence the potential of oxidation: the presence of unsaturation in the heterocycle (absent in flavanones) and the presence of a 3-hydroxyl (in flavonols) facilitate the delocalisation of electrons and lead to a lower oxidation potential of the compound. The only shift of a methoxyl group from one position to another (3′ in hesperetin and 4′ in homoeriodictyol) changes the hydrodynamic voltammogram. The use of voltamograms is thus an excellent identification method for electroactive compounds and it may be possible to distinguish structurally closely related compounds such as isomers. Coulometric detection, therefore, can achieve greater resolution of flavonoids from the same class than photodiode array detectors.

8 Conditions of Analysis for Each Class of Flavonoid

Coulometric detection is just beginning to be used in the field of flavonoid analysis and only a few methods have been described. In this section will be given the analysis conditions developed in our laboratory for each class of flavonoids and the literature references from other researchers when available.

All our analyses have been performed using a system consisting of two HPLC pumps (Model 580, Eurosep, France or ESA, USA) for high pressure gradient, an eight-channel CoulArray detector (Model 5600, Eurosep, France or ESA, USA) with a 150 × 4.6 mm Hypersil BDS C_{18}-5 μm column thermostated at 35 °C. All the mobile phases are 30 mmol l^{-1} NaH_3PO_4 solutions containing a variable percentage of acetonitrile and adjusted to pH 3 with ortho-phosphoric acid. Thus only the percentage of acetonitrile in mobile phases, the conditions of elution and the potentials applied (*vs.* palladium) will be specified for each method. The quantification limits are in the range 5–20 nmol l^{-1} for all the compounds.

We have developed a method to detect all the flavonoids in the same run. This method is useful for the analysis of complex mixtures of flavonoids such as plant extracts (see Section 10). The mobile phases contain 5% acetonitrile (A) and 50% acetonitrile (B). The elution conditions, with a flow rate of 0.8 ml min^{-1}, are: 0–3 min: 100% A; 3–70 min: linear gradient from 100% A to 100% B; 70–73 min: 100% B; 73–95 min: re-equilibration with 100% A. The order of elution is benzoic acids, hydroxycinnamic acids and flavanols (10–20 min), glycosides of isoflavones, flavonols and flavones (20–25 min), glycosides of flavanones

(25–30 min), then aglycones (30–45 min), with methoxylated compounds eluting at the end. The potentials are set at 0, 120, 240, 360, 480, 600, 720 and 840 mV, but can be modified according to the compounds to be analysed, especially in case of co-elutions. Similar methods were described by Milbury[6] and Gamache *et al.*[7] to detect phenolic acids and flavonoids compounds in juice beverages using CoulArray.

Phytoestrogens are the polyphenols that have been the most studied by coulometric detection. For isoflavone analysis, we use mobile phases containing 20% acetonitrile (A) and 40% acetonitrile (B). Separation is achieved at 0.8 ml min^{-1} using a gradient elution: 0–5 min: 100% A; 5–15 min: linear gradient from 100% A to 100% B; 15–23 min: 100% B then re-equilibration in 100% A. Potentials are set at 200, 280, 430, 520, 560, 650, 700 and 750 mV. A typical chromatogram obtained in these conditions for a 24 h urine of a volunteer who consumed a soy supplement is presented in Figure 8. As usual, urine has been hydrolysed with a β-glucuronidase/sulfatase to allow the quantification of the released aglycones daidzein, genistein and the metabolite equol.

Lignans are not flavonoids but they are polyphenols exhibiting phytoestrogenic effects like isoflavones. For the analysis of the lignans enterodiol and enterolactone in plasma samples (with genistein as internal standard), we use mobile phases containing 23% acetonitrile (A) or 40% acetonitrile (B). The conditions of elution are: 0–10 min: linear gradient from 100% A to 100% B; 10–15 min: 100% B; 15–17 min: 100% B to 100% A; 17–27 min: 100% A; with a flow rate of 0.8 ml min^{-1}. Potentials are set at 300, 460, 500, 560 and 700 mV. An example of a chromatogram obtained in these conditions for a plasma of rat fed a diet supplemented with 1.25% flax seed is presented in Figure 9. For analysis of the same compounds in urine samples, we use mobile phases containing 25%

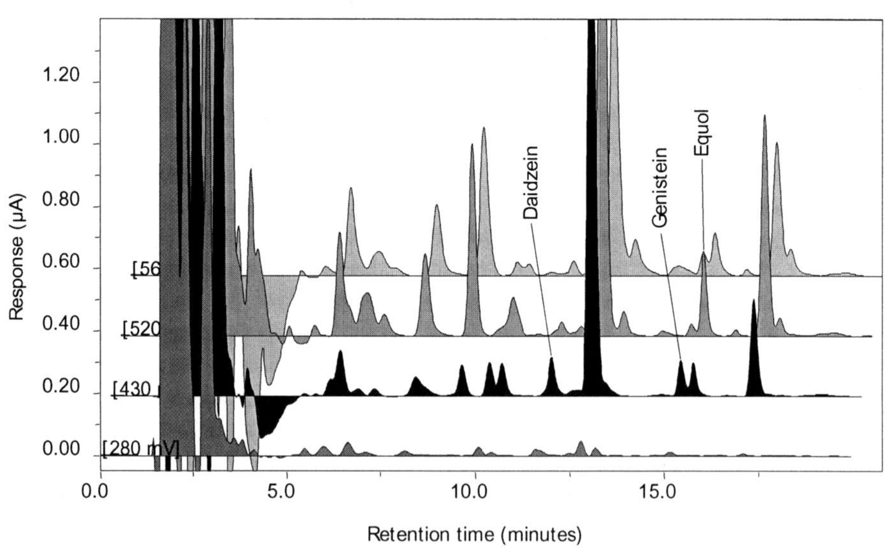

Figure 8 *CoulArray chromatogram of human urine after ingestion of a soy supplement*

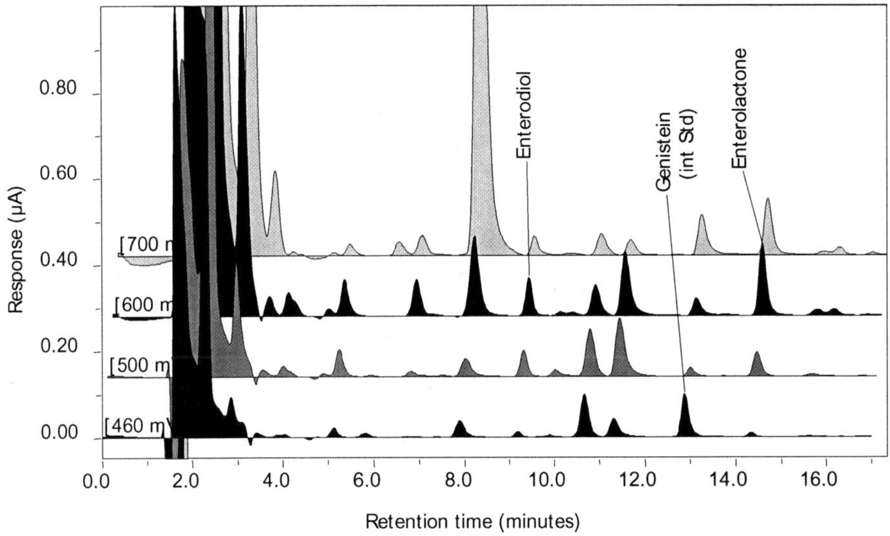

Figure 9 *CoulArray chromatogram of urine of rat fed a diet containing 1.25% flax seed*

acetonitrile (A) and 50% acetonitrile (B), a flow rate of 0.8 ml min^{-1}, and the following gradient conditions: 0–5 min: 100% A; 5–25 min: from 100% A to 100% B; 25–35 min: 100% A. Potentials are the same than for analysis in plasma samples.

Important work has been published by Nurmi and Adlercreutz[8] on the determination of phytoestrogens in plasma using CoulArray. Thirteen molecules can be quantified in the same run: daidzein 7-glucoside, genistein 7-glucoside, daidzein, genistein, dihydrodaidzein, dihydrogenistein, equol, O-desmethylangolensine, secoisolariciresinol, matairesinol, enterodiol, enterolactone and anhydrosecoisolariciresinol. Gamache and Acworth[9] also described a complete CoulArray method to quantify eleven isoflavones, lignans and estrogens in human urine and in rat uterine tissue. The same authors published other conditions for the analysis of these compounds in human serum.[10] Franke and Custer[11] have published a method to quantify isoflavones in soybeans and in human milk and urine with a two-electrode coulometric detector.

The flavanones hesperetin and naringenin can be detected in human plasma or urine with the following mobile phases: phase A: 20% acetonitrile, phase B: 40% acetonitrile. Separation is achieved at 0.8 ml min^{-1}: 0–3 min: 100% A; 3–30 min: linear gradient from 100% A to 100% B; 30–35 min: 100% B; 35–45 min: 100% A. Potentials are set at 25, 320, 480, 550, 650, 750 and 850 mV. A typical chromatogram obtained for a plasma of a volunteer who ingested 500 ml of orange juice (from the study described in Manach *et al.*[12]) is presented in Figure 10. Flavanone glycosides can be quantified in food samples with the same gradient of elution, but with mobile phases containing 5% acetonitrile (A) and 40% acetonitrile (B) and potentials set at 350, 480, 550, 700, 760, 860 and 900 mV.

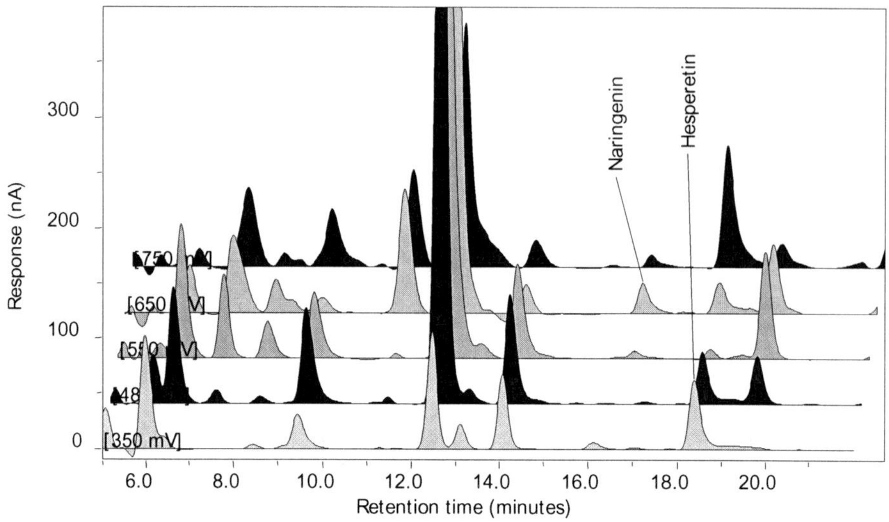

Figure 10 *CoulArray chromatogram of plasma of a volunteer who consumed orange juice*

The flavanols catechin, epicatechin, their 3′- and 4′-methoxylated derivatives and the procyanidins B1, B2 and B3 are separated with phase A containing 5% acetonitrile, phase B containing 50% acetonitrile and a gradient elution at 1 ml min^{-1}: 0–15 min: linear gradient from 100% A to 50% A and 50% B; 15–20% 50% B; 20–25 min 100% B; 25–40 min: 100% A. Potentials are set at 25, 200, 250, 350, 600 and 700 mV. Typical chromatograms of standards and urine from a rat fed a diet supplemented with a grapeseed extract containing catechin and epicatechin (from the study described in Donovan *et al.*[13]) are presented in Figure 11.

We have published two studies on the bioavailability of catechin in rat, using CoulArray detection.[14,15] Several methods have been developed to determine tea or chocolate catechin using coulometric detection. Lee *et al.*[16] described an HPLC-coulometric array detection method to measure epigallocatechin, epigallocatechin gallate and epicatechin in tea, plasma and urine samples. This method has also been used by Sherry Chow *et al.*[17] for the analysis of tea catechins in human plasma after supplementation. Lee *et al.*[18] improved their own method in 2000 to allow the detection of epigallocatechin gallate and the therarubigins from black tea in addition to the catechins determined with the previous method. Yang *et al.*[19] have described a method for the determination of tea catechins in human urine using a single electrode detector. Rein *et al.*[20] also developed a method to quantify catechin and epicatechin using a single-electrode or a coulometric array detector. These authors and Wang *et al.*[21] have applied this method to determine the increase in flavanol concentrations in human plasma after consumption of chocolate.

For the detection of flavonols, especially quercetin and its glycosides, glucuronides and methylated metabolites, we use mobile phases containing 15% acetonitrile (phase A) and 40% acetonitrile (phase B). The elution is performed in gradient conditions at 0.8 ml min^{-1}: 0–13 min: linear gradient from 100% A to

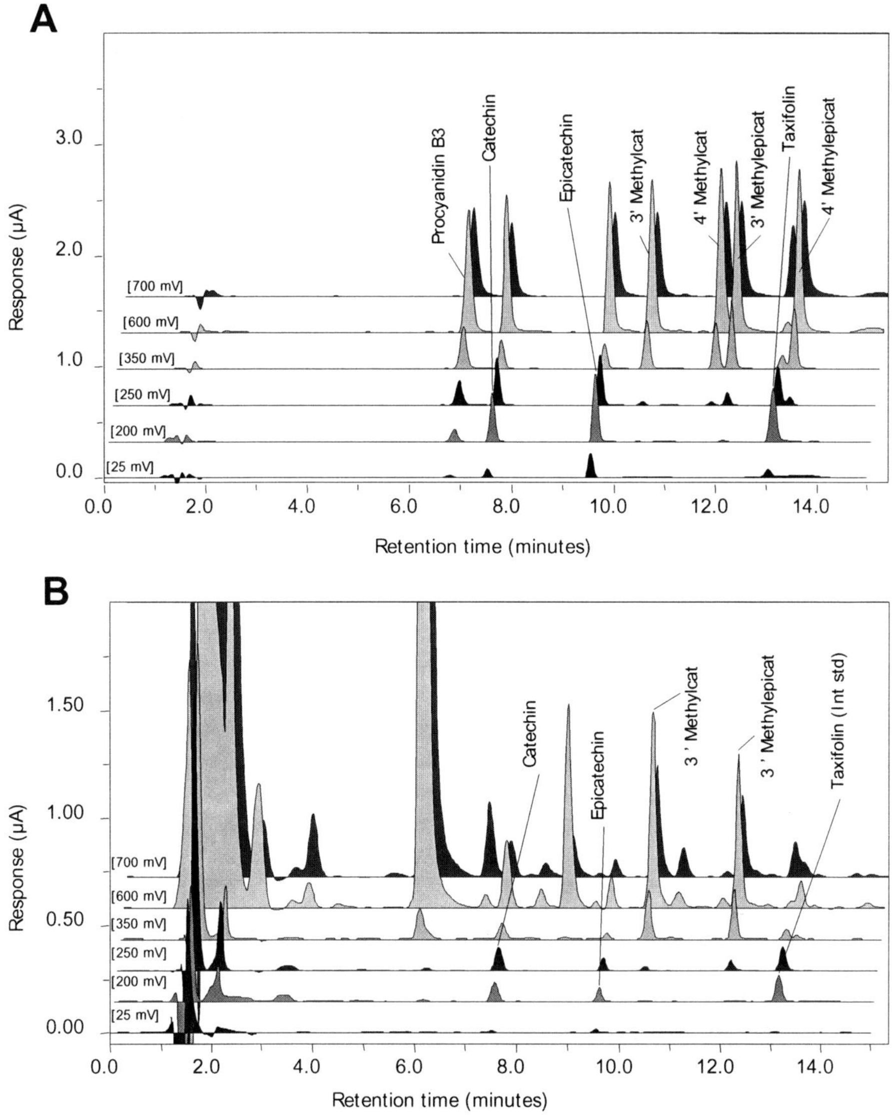

Figure 11 *CoulArray chromatograms of flavanol standards* (A) *and of plasma of rat fed a diet containing a grapeseed extract* (B)

100% B; 13–20 min: 100% B; 20–22 min: 100% B to 100% A; 22–32 min: 100% A. Potentials are set at 0, 50, 100, 150, 200, 250, 300 and 350 mV. A chromatogram of standards is presented in Figure 12. It should be noted that the analysis of quercetin required specific care because this compound may adsorb on various supports. The use of filters in the HPLC system should be avoided. Moon *et al.*[22] have used an amperometric detector to quantify quercetin in human plasma and LDL. Erlund *et al.*[23] have described a coulometric method with a

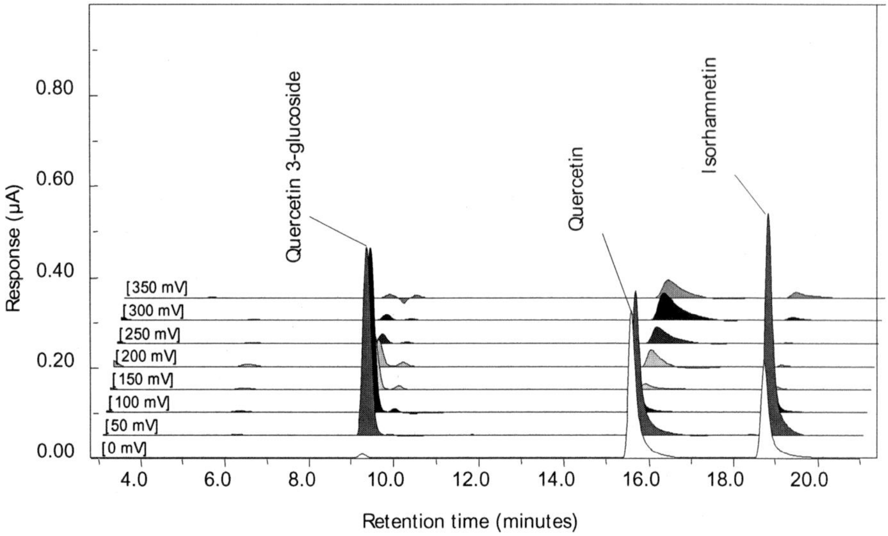

Figure 12 *CoulArray chromatogram of flavonol standards*

single electrode to analyse quercetin in human plasma. As they did not have an electrode array, the authors have generated hydrodynamic voltammograms to check the identity of quercetin by comparison with pure standard. The analysis of flavonols and resveratrol has been also described on an application note of ESA (Chelmsford, MA, USA) entitled 'The measurement of resveratrol, kaempferol, quercetin and isorhamnetin in human plasma'.

We have also developed analysis conditions for phenolic acids. The mobile

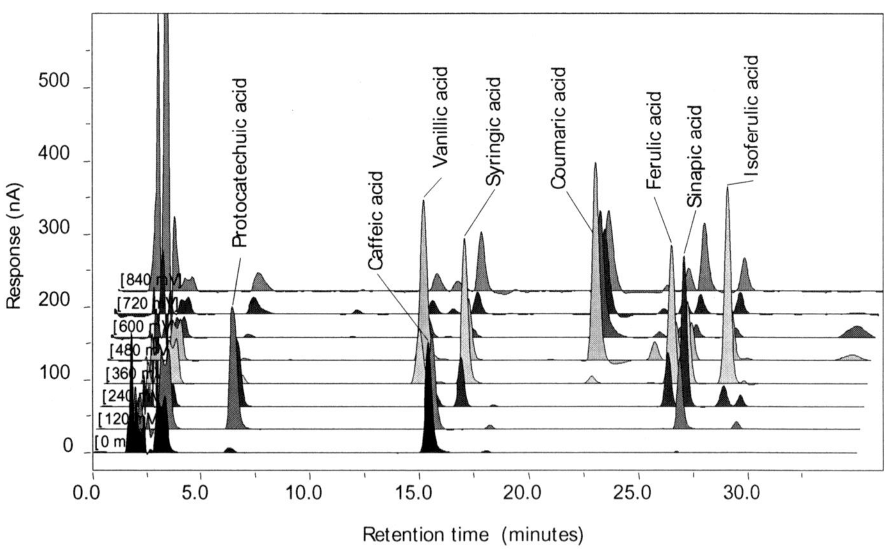

Figure 13 *CoulArray chromatogram of phenolic acid standards*

phases contain 5% acetonitrile (A) and 30% acetonitrile (B). The gradient procedure is: 0–45 min: linear gradient from 100% A to 100% B; 45–50 min: 100% B; 50–60 min: 100% A, with a flow rate of 0.8 ml min^{-1}. Potentials are set at 0, 120, 240, 360, 480, 600, 702 and 840 mV. A chromatogram of standards is shown in Figure 13.

Phenolic acids have been detected in vegetable oils using CoulArray.[24] Another study which must be mentioned is that of Brenes *et al.*[25] for the analysis of polyphenols in olive oil using CoulArray.

9 Specific Applications of Coulometric Array Detection in the Field of Flavonoid Bioavailability

The application of coulometric array detection will be further illustrated by three examples.

Example 1: Use of the selectivity of CoulArray detection to demonstrate the absence of quercetin glucoside in plasma

In the past five years, there has been a debate in the scientific community working on flavonoid bioavailability as to whether quercetin glycosides present in food could be found intact in the plasma. The problem arose from the fact that glucosides are difficult to distinguish from glucuronides, which are the main flavonoid metabolites recovered in plasma. These compounds are not easily separated on reverse-phase columns, and in the studies available analyses had been performed with diode array or fluorimetric detection which are not specific enough to distinguish glucuronides and glucosides. In 2000, we used the selectivity of the CoulArray detection to demonstrate that quercetin 3-glucoside was absent from plasma of rats fed with this flavonoid.[26] The conditions used, with increments of only 50 mV across the electrode array, allowed the separation of the glucoside added to the sample and a co-eluting glucuronide on the basis of their electrochemical behaviour (Figure 14). In 2001 Sesink *et al.*[27] confirmed the absence of circulating quercetin glucosides in humans, using CoulArray. The authors identified the peaks present in the plasma of volunteers who consumed oral solutions of quercetin glucosides as quercetin glucuronides by comparison with standards and using ratio accuracy. Graefe *et al.*[28] also analysed the glucuronides of quercetin in human plasma using CoulArray and showed that they were unchanged, whatever the source of quercetin glucoside that was administered to the volunteers.

Example 2: Use of the electrochemical behaviour to predict some elements of the structure of an unkown metabolite

The combination of retention time with the electrochemical behaviour across the array can be used to elucidate some aspects of the structure of an unknown

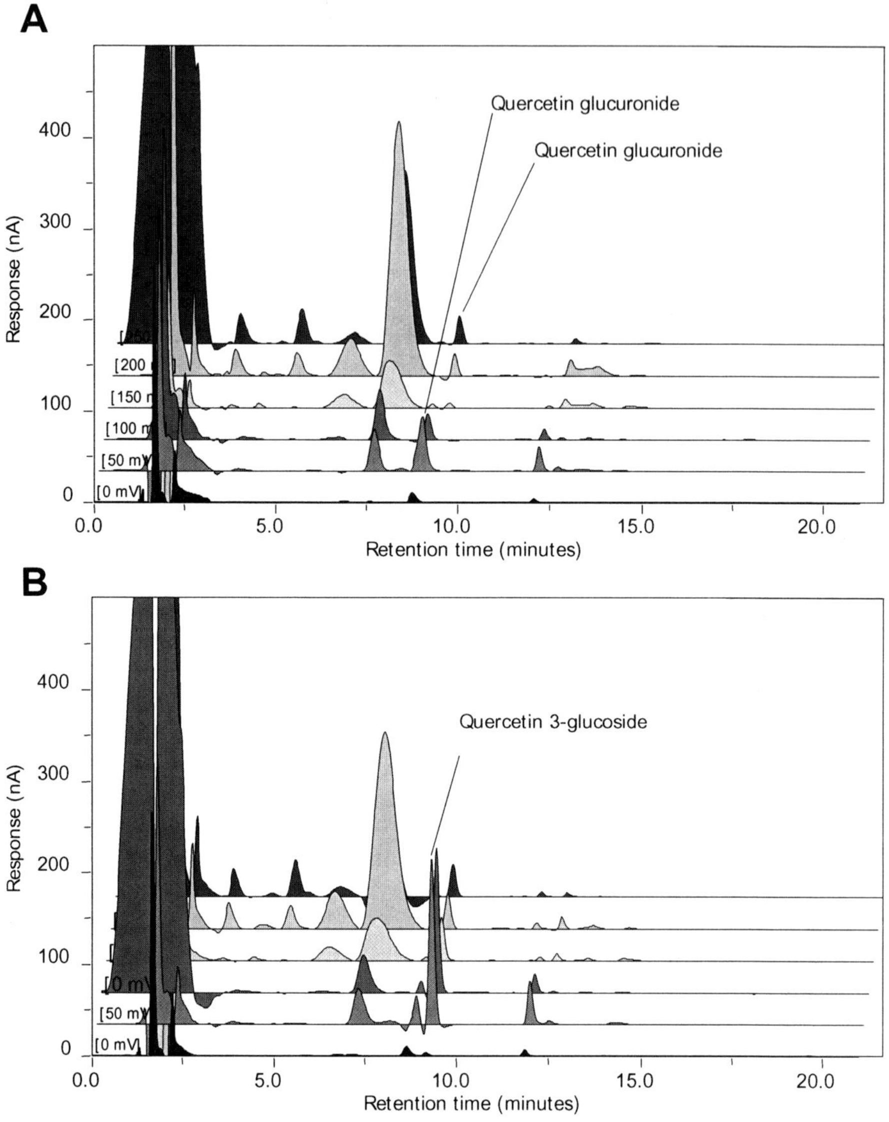

Figure 14 *CoulArray chromatograms of a non-hydrolysed plasma of rat fed a diet containing quercetin 3-glucoside* (A) *and of the same plasma spiked with pure quercetin 3-glucoside* (B). *Conditions of analysis were those described for flavonols in Section 8*

compound. As an example, chromatograms of glucuronides of quercetin and isorhamnetin (the 3′-methylated metabolite of quercetin) are shown in Figures 15A and 15B. The glucuronides have been obtained by *in vitro* glucuronidation with a rabbit liver UDP-glucuronyl transferase. The positions of glucuronidation

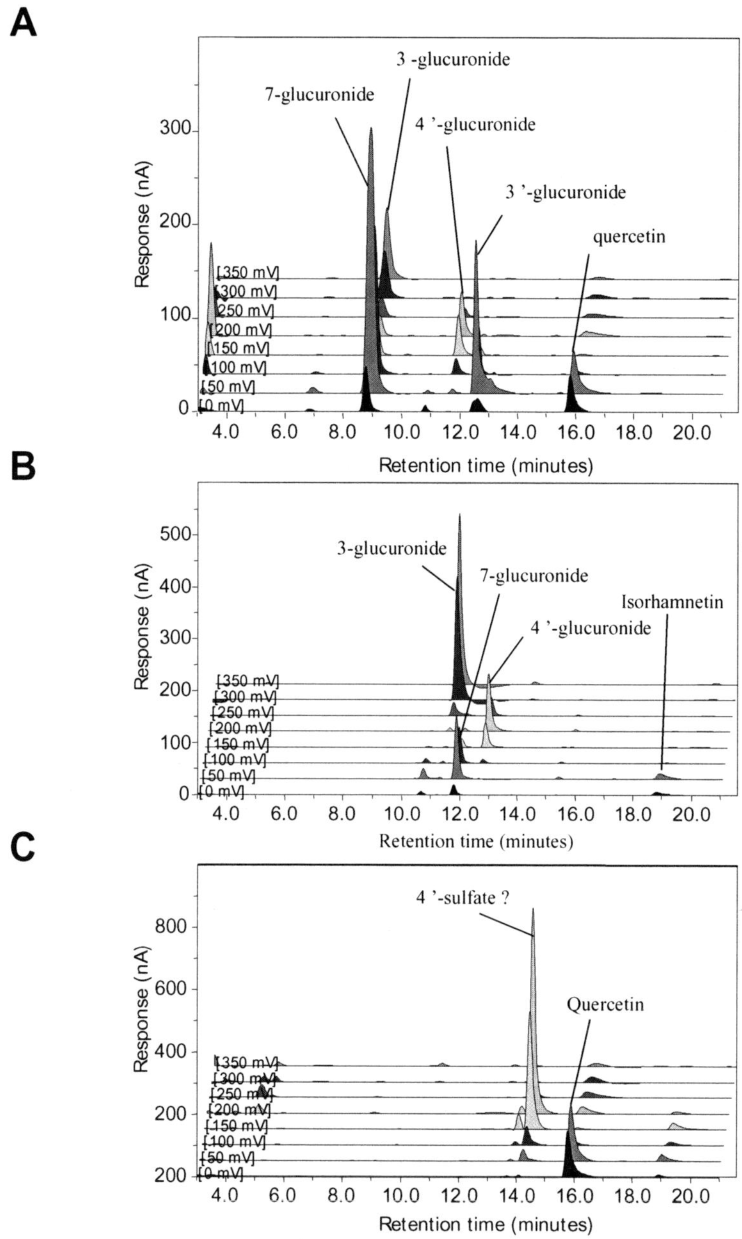

Figure 15 *CoulArray chromatograms of quercetin glucuronides* (A), *isorhamnetin glucuronides* (B) *obtained by in vitro enzymatic synthesis, and of plasma of rat fed a diet containing quercetin 3-glucoside, after hydrolysis by a bovine β-glucuronidase. Conditions of analysis were those described for flavonols in Section 8*

have been assigned using the work of Day *et al.*[29] The authors have identified the positions of glucuronidation of quercetin and isorhamnetin obtained by *in vitro* enzymatic synthesis, and we have identified our glucuronides on the basis of the order of elution of the glucuronides in the HPLC chromatograms. What can be seen in Figure 15 is that the position of glucuronidation strikingly determines the electrochemical response across the array. 7-Glucuronides of quercetin and isorhamnetin are both oxidised on the first three electrodes, like the 3′-glucuronide of quercetin. The 4′-glucuronides are oxidised on the fourth to sixth electrodes, and the 3-glucuronides on the seventh and eighth electrodes. There is a relation between the electrochemical response across the array and the capacity of electron delocalisation, thus the reduction capacity of a compound. It probably means that the 7- and 3′-glucosides keep most of the antioxidant capacity of the corresponding aglycone, whereas 3-glucosides do not. After hydrolysis by a β-glucuronidase of a plasma of rat fed a diet supplemented with quercetin 3-glucoside, we have obtained the chromatogram presented in Figure 15C. The peak eluting at about 14 min corresponds to a sulfate derivative of quercetin, as it released quercetin after further hydrolysis by a sulfatase. As the oxidation of this sulfate occurs on the fourth to sixth electrodes, it may be suggested that it is probably substituted in the 4′ position, by analogy to what was observed for the glucuronides. Of course, the identification is not definitive, but the CoulArray detection may give such useful indications of the nature of unknown compounds. This possibility would deserve to be further exploited.

Example 3: Research of unknown metabolites in a complex matrix, use of pattern recognition

Chromatograms of urine samples obtained from control rats or rats fed a diet supplemented with 15% lyophilised apple are presented in Figure 16. The chromatograms are so complex that it is difficult to spot the peaks that are recovered only in the urine from rats of the apple group and may correspond to metabolites of flavonoids present in apple. The CoulArray software offers the possibility to compare two (or more) samples by considering each electrode separately. Simple chromatograms are obtained, as shown in Figure 16C for the comparison of urine samples from the control or apple groups on the first electrode, set at 0 mV. Among the peaks that are only present on the chromatogram corresponding to the apple group, some can be assigned to flavonoids present in apple or to expected metabolites of these flavonoids, on the basis of their retention time and their oxidation potential. This is the case in Figure 16C for catechin, epicatechin, caffeic acid, ferulic acid, quercetin and isorhamnetin. A chromatogram of standards can be added in the comparison mode to facilitate the identification of peaks.

A simultaneous comparison of all the samples of the apple group can be carried out to find the non-identified peaks that are systematically present in the experimental group and absent from the control group, and thus which are really characteristics of the apple diet. Some information will be obtained on these

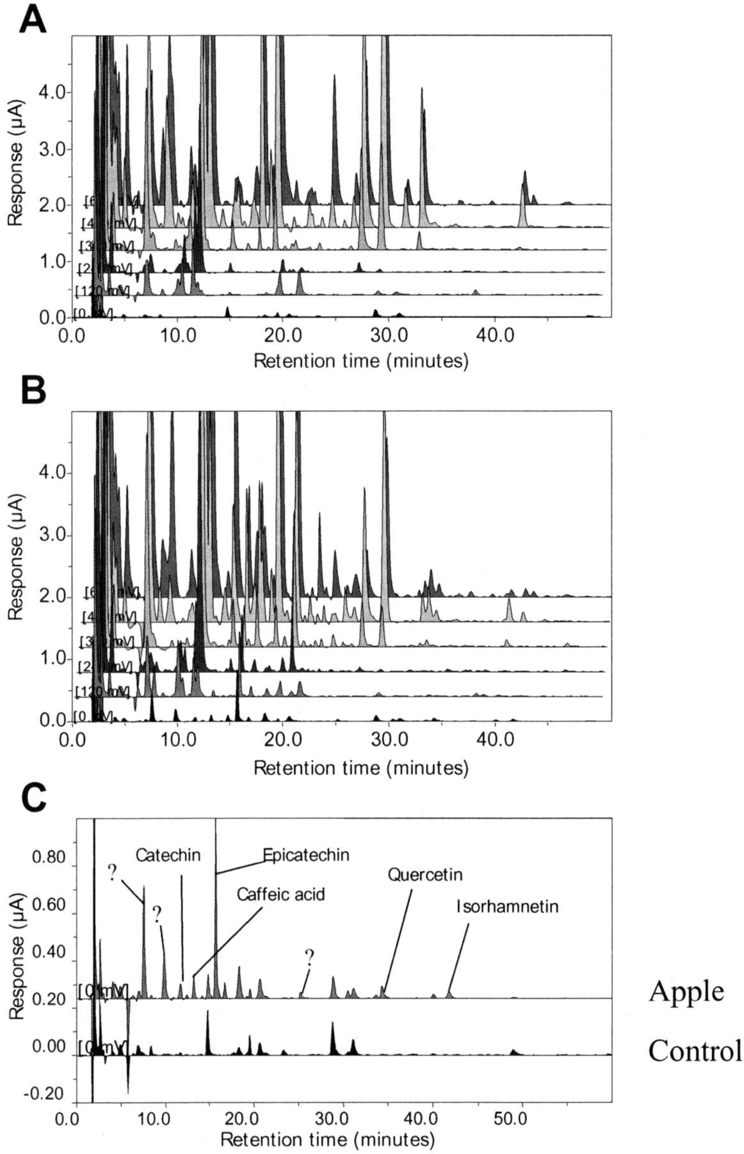

Figure 16 *CoulArray chromatograms of urine of rat fed a control diet* (A)*, or a diet containing 15% lyophilised apple* (B)*. Comparison of chromatograms for control and apple groups on electrode 1, set at 0 mV* (C)*. Conditions of analysis were those described to detect all the flavonoids in Section 8*

compounds on the basis of their retention time and their electrochemical behaviour, although their definitive identification will require additional analyses with other techniques such as diode array detection, NMR or mass spectrometry. It can be noted that a photodiode array detector may be inserted between the HPLC column and the multielectrode coulometric detector, so that a UV spectrum will be obtained for every resolved analyte present in sufficient concentration.

Coulometric array technology, combined with advanced pattern recognition software, could offer the possibility to determine the features in the pattern of metabolites which distinguish control and experimental groups. This would have interesting applications in the field of flavonoid bioavailability, but the objective of the CoulArray conceptors is to extend that to the comparison of samples from individuals in good health or suffering from a particular disease, with the hope that a predisposition to this particular disease may be reflected in the complex pattern of electrogenic compounds found in the blood (or other tissue) and diagnosed using the CoulArray.[3,30]

10 Specific Applications of the Coulometric Array Detection in the Field of Food Analysis

As in biological fluids, the analyst may have to identify and quantify a single or a few flavonoids in food samples. The methods are similar to those described for biological samples, except that the gradient elution conditions may have to be modified if some contaminating peaks are present in the plant matrices. Usually the detection is easier in food samples than in the biological samples due to the higher flavonoid concentrations. The quantification of all individual flavonoids present in each variety of fruit, vegetables and beverages represent a huge work, and most data are still lacking. In general, the main compounds have been quantified, but it may occur that even some flavonoids present in high amount have been ignored in some sources, because these sources have been poorly studied. For some plants or food products, data on the flavonoid composition are still totally lacking. Some fruits contain only one or two families of flavonoids, for example citrus fruits contain flavanones and polymethoxylated flavones, but it must be kept in mind that most of the time the flavonoid composition of plant foods is very complex. Thus for plant products, the interest of the analyst may no longer lie in the absolute level of a few compounds for which standards are available, but in the complex pattern of flavonoids present in the sample.

Use of Fingerprints

The chromatograms obtained with HPLC coupled to coulometric array detection in standardised and reproducible conditions can provide such patterns. The gradient elution must be as wide as possible to allow the detection of polar compounds such as phenolic acids and flavanols and of more hydrophobic compounds such as methoxylated flavonoids, and a wide range of potentials must also be used. In such conditions, complex flavonoid patterns, which can also be referred as 'fingerprints' can be easily obtained for all fruits and vegetables. Even if some of the flavonoid present can be quantified by comparison with standards, most of them are unidentified and the profile is considered as a whole pattern.

The use of 'fingerprints' can have various applications. One of them is the control of some products, such as fruit juices, wine, or herbal supplements, to detect adulterations.[7] For example the addition of grapefruit juice to the orange juice will be easily revealed with a CoulArray analysis, because some flavonoids characteristic of grapefruit juice are normally absent from the orange juice. In some cases the discriminant flavonoids may be non-identified but the comparison of the whole pattern will be sufficient to prove any adulteration. This could be the case for wine control. Another application of fingerprints is the creation of a database library for various plant materials. We have observed that the flavonoid profile is characteristic of the plant species, with only differences in concentration among the varieties of the same species. An example is given with apple in Figure 17. The chromatograms have been obtained in the same conditions for the three apple varieties, Golden, Canada and Douce Couët. The profiles clearly show that the Douce Couët apple, which is used for cider production, is markedly richer in flavonoids than the other varieties. Nevertheless the nature of the flavonoid present is the same for the three varieties. A fingerprint library would greatly help in the determination of the species and varieties containing the highest amount of flavonoids, especially antioxidant flavonoids. As an example, the flavonoid patterns of pumpkin, spinach and eggplant are presented in the Figure 18. They clearly show that eggplant contains a lot of polar compounds oxidising at low potential and that spinach is rich in flavonoids whereas pumpkins are devoid of such compounds. Another interest of fingerprints is the possibility to study the effects of domestic treatments (peeling, cooking...) or industrial processes on the flavonoid composition of plant foodstuffs. A comparison of the flavonoid patterns along the transformation process would give more information than the analysis of only one or a few compounds that may not reflect the behaviour of the other flavonoids. It must be noted that the use of pattern recognition or fingerprints requires an excellent stability of the system, with minimal variations of retention times and voltammetric response.

Estimation of the Overall Antioxidant Capacity

Since coulometric array detection relies on redox chemistry, Guo *et al.*[31] have proposed that there should be a relationship between the sum of peak areas (or heights) of a chromatogram of a plant extract and the antioxidant capacity of this extract. They found a significant positive linear correlation ($R = 0.91$–0.96) in aqueous extracts of various fruits and vegetables between their $ORAC_{ROO^{\circ}}$ activities and their electrochemical data – total peak area or height summed across all peaks from the dominant channels or across all channels – obtained with HPLC coupled with coulometric array detection. The HPLC procedure allowed the detection of various antioxidants including vitamin C, glutathione, phenolic acids and flavonoids. This result should be confirmed, especially with conditions of plant extraction more suitable to recover flavonoids. We also suggest that only peaks obtained at low potentials are to be considered because, at potentials higher than 750 mV, compounds without antioxidant capacity can be

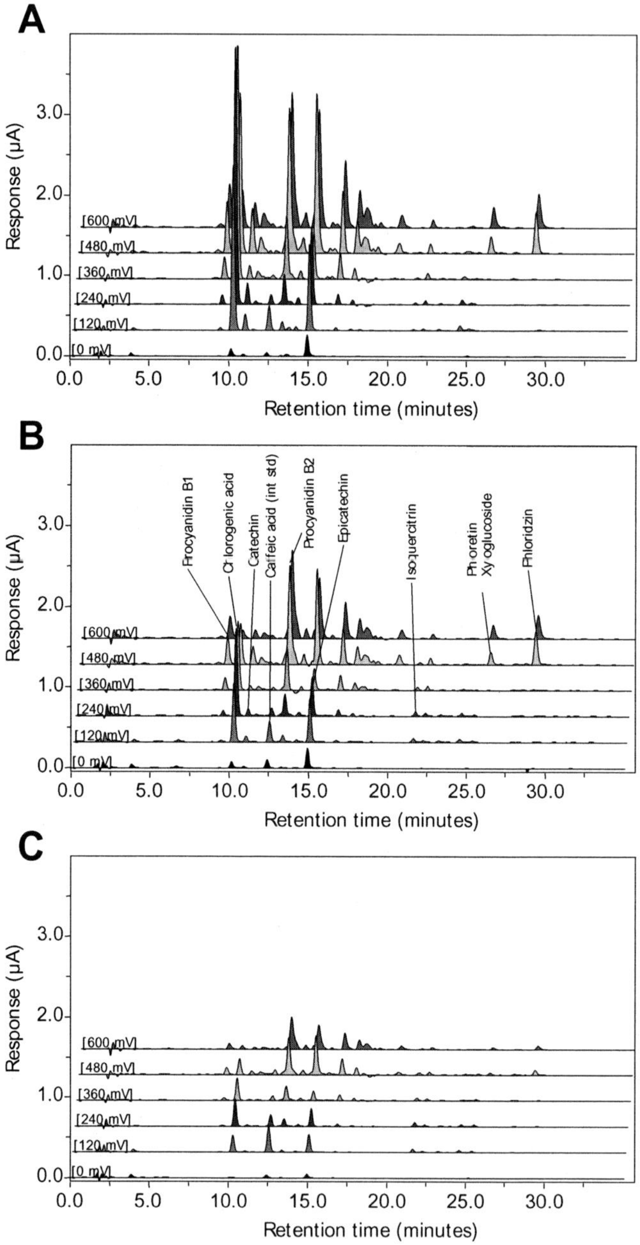

Figure 17 *Comparison of the polyphenol profiles of three apple varieties: Douce Couët* (A)*, Canada* (B)*, Golden* (C)*. Conditions of analysis were those described to detect all the flavonoids in the Section 8*

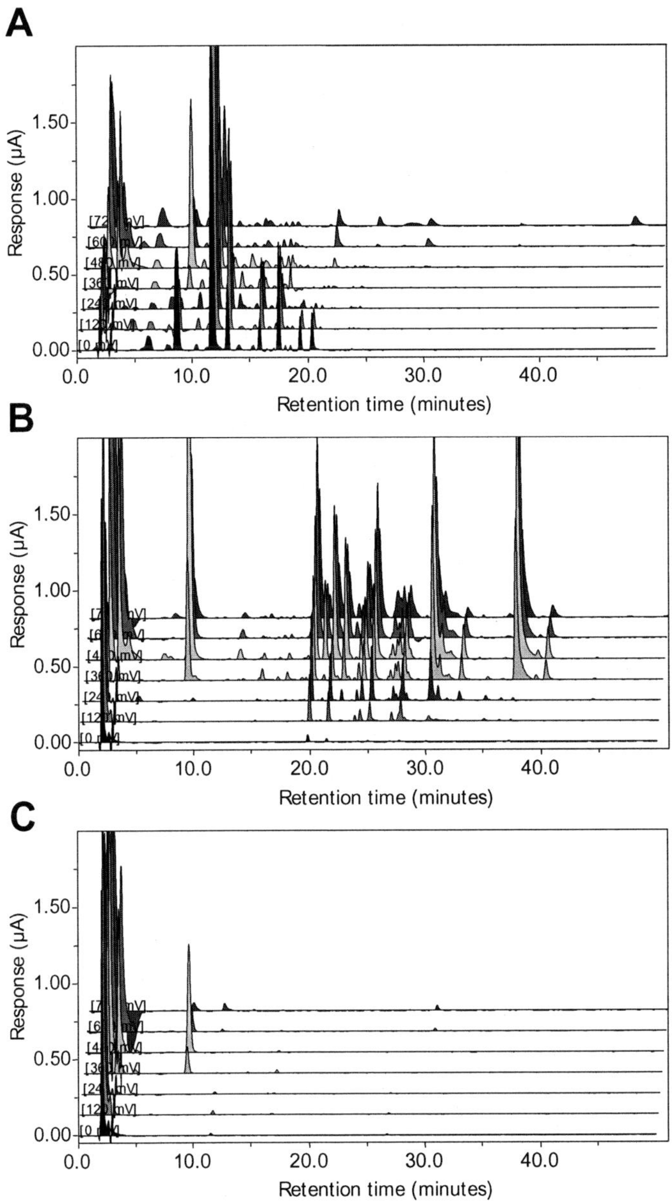

Figure 18 *CoulArray chromatograms of vegetable extracts: eggplant* (A), *spinach* (B), *pumpkin* (C). *Conditions of analysis were those described to detect all the flavonoids in the Section 8*

oxidised and give large peaks. Additional appraisal should be done on the subject because it would be of great interest to have a standardised method using CoulArray detection to roughly evaluate the overall antioxidant capacity of any plant extract.

11 Conclusion

HPLC with coulometric array detection is a high quality technique, with numerous advantages:

- high sensitivity
- high selectivity, even to small changes in the structure
- possibility to check the peak purity by using ratio accuracy
- high confidence in the assignment of peaks
- ability to resolve co-eluting peaks based on differences in oxidation potentials
- simultaneous analysis of many analytes due to the high resolving power of the array
- simplified sample preparation due to the selectivity of detection
- high stability of the system, comparable to that of UV detectors
- ability to make some prediction about the chemical structure of an unknown
- possibility to use a pattern recognition or a fingerprint approach.

Thus this is a technique of choice that should be more widely used for the analysis of flavonoids in biological and food samples.

Acknowledgements

Special thanks to L. Martinez and X. Allain from Eurosep Instruments (France) for their documentation and advice.

References

1. E. Middleton Jr., C. Kandaswami and T. C. Theoharides, *Pharmacol. Rev.*, 2000, **52**, 673.
2. H. M. Merken and G. R. Beecher, *J. Agric. Food Chem.*, 2000, **48**, 577.
3. I. N. Acworth, and M. Bowers, in *Coulometric Array Detectors for HPLC*, eds I. N. Acworth, M. Naoi, H. Parvez and S. Parvez, VS Press, The Netherlands, 1997, p. 1.
4. W. R. Matson, P. Langlais, L. Volicer, P. H. Gamache, E. Bird and K. A. Mark, *Clin. Chem.*, 1984, **30**, 1477.
5. K. R. Markham, *Techniques of Flavonoid Identification*, Academic press, NY, 1982.
6. P. E. Milbury, in *Methods in Enzymology. Flavonoids and Other Polyphenols*, ed. L. Packer, Academic Press, London, 2001, Vol. 335, p. 15.
7. P. Gamache, E. Ryan and I. N. Acworth, *J. Chromatogr.*, 1993, **635**, 143.
8. T. Nurmi and H. Adlercreutz, *Anal. Biochem.*, 1999, **274**, 110.
9. P. Gamache and I. N. Ackworth, *Proc. Soc. Exp. Biol. Mol.*, 1998, **217**, 274.
10. P. H. Gamache, S. M. Freeto and I. N. Acworth, *J. Med. Food*, 1999, **2**, 125.
11. A. A. Franke and L. J. Custer, *Clin. Chem.*, 1996, **42**, 955.

12. C. Manach, C. Morand, A. Gil-Izquierdo, C. Bouteloup-Demange and C. Rémésy, *Eur. J. Clin. Nutr.*, submitted.
13. J. Donovan, C. Manach, L. Rios, C. Morand, A. Scalbert, and C. Rémésy, *Br. J. Nutr.*, 2002, **87**, 299.
14. C. Manach, O. Texier, C. Morand, V. Crespy, F. Régérat, C. Démigné and C. Rémésy, *Free Radic. Biol. Med.*, 1999, **27**, 1259.
15. J. L. Donovan, V. Crespy, C. Manach, C. Morand, C. Besson, A. Scalbert and C. Rémésy, *J. Nutr.*, 2001, **131**, 1753.
16. M. J. Lee, Z. Y. Wang, H. Li, L. Chen, Y. Sun, S. Gobbo, A. A. Balentine and C. S. Yang, *Cancer Epidemiol. Biomarkers Prev.*, 1995, **4**, 393.
17. H. H. Sherry Chow, Y. Cai, D. S. Alberts, I. Hakim, R. Dorr, F. Shadi, J. A. Crowell, C. S. Yang and Y. Hara, *Cancer Epidemiol., Biomarkers Prev.*, 2001, **10**, 53.
18. M. J. Lee, S. Prabbu, X. Meng, C. Li and C. S. Yang, *Anal. Biochem.*, 2000, **279**, 164.
19. B. Yang, K. Arai and F. Kusu, *Anal. Biochem.*, 2000, **283**, 77.
20. D. Rein, S. Lolito, R. R. Holt, C. L. Keen, H. H. Schmitz and C. G. Fraga, *J. Nutr.*, 2000, **130,** 2109S.
21. J. F. Wang, D. D. Schramm, R. R. Holt, J. L. Ensunsa, C. G. Fraga, H. H. Schmitz and C. L. Keen, *J. Nutr.*, 2000, **130**, 2115S.
22. J. H. Moon, R. Nakata, S. Oshima, T. Inakuma and J. Terao, *Am. J. Physiol. Regulatory Integrative Comp. Physiol.*, 2000, **279**, R461.
23. I. Erlund, G. Alfthan, H. Siren, K. Ariniemi and A. Aro, *J. Chromatogr. B*, 1999, **727**, 179.
24. I. Bernwieser and G. Sontag, *Z. Lebensm. Unters. Forsch.*, 1992, **195**, 559.
25. M. Brenes, A. Garcia, P. Garcia and A. Garrido, *J. Agric. Food Chem.*, 2000, **48**, 5178.
26. C. Morand, C. Manach, V. Crespy and C. Rémésy, *Free Rad. Res.*, 2000, **33**, 667.
27. A. L. A. Sesink, K. A. O'Leary and P. C. H. Hollman, *J. Nutr.*, 2001, **131**, 1938.
28. E. U. Graefe, J. Wittig, S. Mueller, A. K. Riethling, B. Uehleke, B. Drewelow, H. Pforte, G. Jacobasch, H. Derendorf and M. Veit, *J. Clin. Pharmacol.*, 2001, **41**, 492.
29. A. J. Day, Y. Bao, M. R. A. Morgan and G. Williamson, *Free Radic. Biol. Med.*, 2000, **29**, 1234.
30. K. Hensley, K. S. Williamson, M. L. Maidt, S. P. Gabbita, P. Grammas and R. A. Floyd, *J. High Resol. Chromatogr.*, 1999, **22**, 429.
31. C. Guo, G. Cao, E. Sofic and R. L. Prior, *J. Agric. Food Chem.*, 1997, **45**, 1787.

CHAPTER 5

On-Line Identification of Flavonoids by HPLC Coupled to Diode Array Detection

CELESTINO SANTOS-BUELGA; CRISTINA GARCÍA-VIGUERA AND FRANCISCO A. TOMÁS-BARBERÁN

1 Introduction

Flavonoids comprise one of the largest groups of secondary plant metabolites occurring in plants. They share a common $C_6-C_3-C_6$ skeleton, with two benzene rings (A and B) linked through a heterocyclic pyrane or a pyrone ring (C) (Figure 1). Several thousands of flavonoids have been described, the larger number corresponding to flavones, flavonols and their *O*-glycosides. Comprehensive revisions on flavonoid types (structures, characteristics, biological properties and analysis) can be found in various regularly updated reviews.[1-3] Flavonoids are found in nearly all plants. Major subclasses present in food correspond to

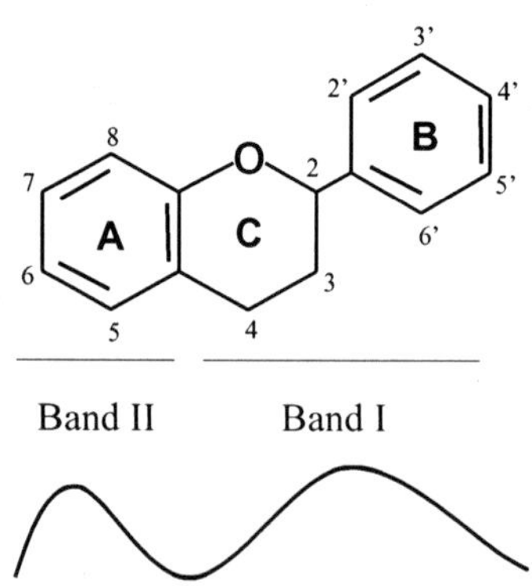

Figure 1 *Basic structure of flavonoids and spectrum. Numbering scheme for flavonoids other than chalcones and aurones*

anthocyanins, flavanols, flavones, flavanones, flavonols and isoflavones. Flavanones and flavones are often present in the same plant (*e.g.* citrus); however, flavones and flavonols are usually not found together, nor are flavanones and anthocyanins.[4] With the exception of flavanols, dietary flavonoids mostly occur in *O*-glycosidic forms.

Since its introduction in the 1970s for flavonoid analysis, HPLC has become the technique most commonly used for their separation, screening, chemotaxonomical surveys or quantitative determination in plants and foodstuffs. Further introduction in the 1980s of diode array spectrophotometers as detectors associated to HPLC (HPLC–DAD) has led to great improvement for flavonoid identification, as not only the retention characteristics but also the UV-visible spectra can be used for this purpose. In HPLC–DAD the chromatographic eluent is scanned for UV-visible spectral data, which are stored and can be later compared with a library for peak identification. This type of detector allows the simultaneous recording of chromatograms at different wavelengths, thus facilitating the selective detection of flavonoids of different groups at their maximum wavelength. This improves the possibilities of quantification, as the sensitivity can be increased by an adequate selection of the recording wavelength, which may make possible the quantification of a compound even if the resolution is poor. Baseline drift and noise may be also reduced by subtracting from the wavelength of recording a reference wavelength at which the compound of interest shows little absorption. Furthermore, conclusions on the purity of the peaks can be obtained through the examination of the different spectra provided for each of them and the calculation of the ratio of absorbances at different wavelengths; if this ratio is constant over the whole width of the peak it can be assumed that it is pure, unless two peaks with the same spectrum shape are overlapped.

There is not an HPLC method that can solve all flavonoid separations; even for a given family of compounds (*i.e.* flavones, flavonols, anthocyanins, isoflavones, *etc.*) the stationary phase, solvents and gradients should be optimised in each case. Performance of the separations will depend on the column characteristics and capacity, which will be also influenced by the solvent composition and gradient and the flow rates used. A wide range of stationary and mobile phase combinations have been reported in the literature either for flavonoids in general or for different flavonoids groups, which it is not intended to discuss here, since excellent reviews exist in the recent literature.[5–10]

Normal phases (*i.e.* silica gel columns) have not been employed often, but they can be useful for the separation of non-polar or weakly polar flavonoid aglycones, such as polymethoxylated flavones, flavanones or isoflavones. A drawback of these phases is that no gradients can usually be employed.[11] Derivatization of flavonoids to their acetates has been used by Galensa and Herrmann[12] before the separation of flavonoid glycosides on silica gel columns (LiChrosorb Si 60), obtaining good resolution, but this separation does not improve on those obtained with reversed-phase columns, which also avoid derivatization. Normal phases (*i.e.* silica gel columns) have been also employed for separation of flavanols (*i.e.* catechins and proanthocyanidins) according to their degree of polymerization.[13–15]

The most commonly stationary phases used for analytical separations of flavonoids are of the reversed-phase (RP) type, usually C_8 or C_{18} alkyl bonded phases. In RP-HPLC the stationary phase is less polar than the mobile phase and thus compounds are eluted in order of decreasing polarity. Typically, for a given group of flavonoids, glycosides will be eluted before their aglycones and, among these, those possessing more hydroxyl groups will be eluted before those with less. Thermostatic control of the column temperature (*e.g.* 30–35 °C) is usually employed to improve separations and keep retention times stable.

Elution systems used in RP are usually binary with an aqueous acidified polar solvent and a less polar organic solvent, most often acetonitrile or methanol; occasionally, other solvents, such as tetrahydrofuran, isopropanol or propan-1-ol, have also been used. The acid modifier is necessary to suppress the ionization of the phenolic hydroxyl groups, thus obtaining sharper peaks and minimizing peak tailing. Usually, the greatest difference observed among the mobile phases used by different authors is the type of acid. Acetic or formic acids are the most usually employed, although other acids (such as phosphoric, citric, trifluoroacetic or perchloric acids), phosphate buffers or ammonium acetate are also used. Less frequently isocratic, tertiary or quaternary systems have been used. Typical flow rates are in the range 1–1.5 $ml\,min^{-1}$, although there is a tendency to decrease these rates, in order to save solvents and facilitate the coupling to novel detection systems, such as mass or NMR spectroscopy. This is facilitated by the recent introduction of short columns and microcolumns containing stationary phases with smaller pore sizes, which allow narrower peaks to be obtained using lower flow rates. The use of these columns also permits reduction of the time of the HPLC runs and the amount of sample injected.

UV-visible spectra have been long used for structural analysis of flavonoids. The typical flavonoid spectrum consists of two maxima in the ranges 240–285 (Band II), presumably determined by the A-ring, and 300–550 nm (Band I), arising from B-ring (Figure 1). The position and relative intensities of these maxima yield information on the nature of the flavonoid and its oxygenation pattern; variation within these ranges will depend on the hydroxylation pattern and on the degree of substitution of the hydroxyls. All flavonoids show wavelength of maximum absorption in Band II, which is thus more unspecific and less useful than Band I for obtaining information regarding identification. Due to the lack of conjugation between their A and B rings, flavanols, flavanones and isoflavones hardly show absorption in Band I, whereas anthocyanins have a peak in Band I around 460–560 nm, and flavones and flavonols about 300–380 nm. Additional structural information may be gained by use of some reagents which induce changes in the spectrum shape or wavelengths of maximum absorption. Excellent revisions of the use of absorption spectroscopy for the identification of flavonoids are available in the literature.[16–18] Spectral data collected in these sources are referred to measurements made in methanol (acidified with HCl in the case of the anthocyanins); however, the spectral characteristics of flavonoids obtained on-line after HPLC separation may differ from those collected in the literature, as they may be influenced by the nature of the solvent, which varies according to the composition of the mobile phase. In the following pages chromatographic and on-line spectral characteristics for selected groups of flavonoids are revised with the objective of contributing data that can be useful for flavonoid identification through the HPLC–DAD technique.

2 Anthocyanins

The anthocyanins are glycosylated derivatives of different naturally occurring anthocyanidins, from which only six are widespread and usually observed in fruits and vegetables. They are: pelargonidin, cyanidin, peonidin, delphinidin, petunidin and malvidin. These are based on the same 2-phenylbenzopyrylium (flavylium) skeleton hydroxylated in the 3, 5 and 7 positions, and differing in the number and position of hydroxyl and methoxyl groups in the B-ring. Derived from these anthocyanidins a wide variety of anthocyanins exist which differ in the nature and number of sugars attached to the aglycone and the position of the attachment, and in the nature and number of aliphatic or aromatic acids attached to the sugar residues. For anthocyanin structure and chemical properties see Chapter 15.

RP–HPLC Elution Characteristics

The anthocyanins exist in solution as various structural forms in equilibrium depending on the pH and temperature (see Chapter 15). Thus in order to obtain reproducible results it is essential to control the pH of the mobile phase and to work with thermostatically controlled columns. For a good HPLC resolution, anthocyanin equilibria have to be displaced toward their flavylium forms; thus peak tailing is minimized and peak sharpness improved. Furthermore, photometric detection is also facilitated, since flavylium cations are coloured structures that can be selectively detected in the visible region at about 520 nm, avoiding the interference of other phenolics and flavonoids that may be present in the same extracts. Typically the pH of elution should be below 2. Anthocyanin peaks become wider as the percentage of the apolar solvent increases in the gradient and the acidity of the mobile phase decreases. The key parameter is the p*K* of the proton transfer equilibrium; the smaller the p*K* value of the anthocyanin the lower the pH must be to provide a complete displacement of equilibria towards the flavylium form. In this sense, anthocyanin 3-glucosides, with typical p*K* values ranging from 2 to 3, are usually less affected by the increase of the pH in the mobile phase than anthocyanin 3,5-diglucosides with smaller p*K* values (1.5–2). A common strategy to reduce peak widening and improve peak resolution is acidification of the less polar solvent, so that the percentage of acid modifier is kept constant throughout the elution gradient. Furthermore, acidification of the extracts before injection (*e.g.* to pH 1.5) may help to improve resolution of the chromatograms.

After the works of Williams *et al.*[19] and Wulf and Nagel,[20] published in 1978, reversed-phase HPLC (RP-HPLC) has become the technique of choice for the separation of anthocyanins. Columns most usually employed are C_{18} alkyl bonded phases. Different elution conditions (solvents, gradient and flow rate) have been described in the literature, optimized as a function of the column and sample composition The anthocyanins are highly polar compounds and for their elution in reversed phases acid mixtures with a high proportion of water are required. Gradient elution is usually required to provide a good separation, especially if complex mixtures containing anthocyanins with similar polarities have to be resolved. Commonly binary elution systems are used consisting of an aqueous acidified polar solvent and a less polar organic solvent, most often acetonitrile (ACN) or methanol, possibly acidified. For analytical separations ACN is preferred to methanol, whose mixtures with water have relatively high viscosity which gives rise to higher pressures; furthermore, peak splitting is sometimes observed when using methanol gradients. Typical gradients run from 0–10% ACN to 35% ACN, which is usually enough for the complete elution of anthocyanins if adequate gradient stepping and flow rate are employed. Mixtures containing highly polar anthocyanins (*e.g.* with various sugar substituents) require greater polarities in the mobile phase and for a good separation it is advisable to enlarge the starting phase of the gradient at low ACN percentages (*e.g.* 0–5%).

Formic and acetic acids are the acid modifiers most usually employed for anthocyanin separation, although trifluoroacetic acid (TFA) is also currently gaining place due to it reduces problems of corrosion and improves reproducibility of retention times.[21] Nevertheless, better resolutions are usually obtained using formic acid. Other acid modifiers used are phosphoric acid, phosphate buffers or ammonium acetate, for which problems of precipitation may exist. In our experience, the use of 4.5% formic acid is usually enough to provide a good separation of anthocyanins in most cases. However, when dealing with anthocyanin showing low pK values (such as 3,5-diglycosides) or complex mixtures containing compounds with similar polarity characteristics, the use of a lower elution pH, as provided by 10% formic acid, may be necessary for obtaining narrower peaks and improving resolution. In these cases, a better washing after a set of injections is advisable in order to minimize problems of corrosion. In spite of low pHs of elution not being recommended for RP phases, in our experience no increased phase damage has been observed nor loss of column performance with a pH of elution about 1.8, but, on the contrary, extended column life is even observed, providing that usual care is taken regarding column washing and storage. A comparison of various recently developed reversed-phase columns for the separation performance of twenty wine anthocyanins including monoglucosides, diglucosides and acylated derivatives (acetyl and *p*-coumaroyl) has been made by Berente *et al.*[5] C_{18}, C_{12} and RP phenyl-bonded stationary phases were checked, better separations being obtained by the use of a C_{12} 4 μm column in combination with acetonitrile-phosphate buffer at pH 1.6 and 50 °C, as mobile phase.

In order to optimize an HPLC separation method or to check the suitability of given HPLC conditions, some plant extracts may be particularly useful. Thus, *Vaccinium* species (*e.g. V. myrtillus*) could be used as a reference for the separation of 3-glycosides, as it contains galactose, glucose and arabinose derivatives of delphinidin, cyanidin, petunidin, peonidin and malvidin.[22–25] Also, young red wines and red grape skins can be used as reference samples; they typically contain 3-glucosides and their acetyl and *p*-coumaroyl derivatives of the same five anthocyanidins, as well as the equivalent 3,5-diglucosides in the case of hybrid or American grape varieties.[5,26,27]

HPLC behaviour of the anthocyanins is closely related to their polarity and their elution pattern yields useful information on the nature of the sugar moieties, acylation and B-ring substituents. As it is well known, the order of elution of compounds in reversed phases is closely related to their polarity, the more polar compounds eluting earlier than the less polar ones. In the case of the six most widespread anthocyanidins, the polarity depends on the substitution pattern of their B-ring. The compounds become more polar as the number of hydroxyl groups in this ring increases and more apolar as the number of methoxyl groups increases. Thus, the following order of elution is observed:

delphinidin → cyanidin → petunidin → pelargonidin → peonidin → malvidin

This same elution order is maintained for anthocyanins bearing identical sugar substituents. Sugar substitution increases the polarity and, therefore, anthocyanins elute at shorter retention times than their parent anthocyanidins. The influence of the sugar moieties on the polarity varies depending on the type of sugars and their position of substitution. When considering the more usual 3-glycoside derivatives, the elution observed for anthocyanins derived from the same anthocyanidin is as follows:

Monosaccharides:

galactoside → glucoside → arabinoside → xyloside → rhamnoside

Disaccharides:

sophoroside → sambubioside → rutinoside

Although disaccharides may be in general considered more polar than monosaccharides, this is not totally true in all cases. For instance, rutinosides are less polar than glucosides, but more polar than pentosides; also, in the most usual gradient systems, the polarity of sophorosides and sambubiosides is greater than that of glucosides but close to that of galactosides, the usual order of elution observed being:

sophoroside → galactoside → sambubioside → glucoside

The anthocyanins may also bear additional sugar moieties, usually attached to their 5, 7 or 3′ position. In general, the introduction of a second sugar moiety in the anthocyanin molecule increases the polarity, and thus, for instance, 3,5-diglucosides elute before the corresponding 3-glucosides. It has also been indicated that the increase of polarity is greater in the case of 3,5-derivatives than in that of 3,7-derivatives bearing identical sugars.[28] On the other hand, disaccharide moieties in position 3 increase the polarity less than the presence of the same two monosaccharides in 3,5-positions (personal observation).

When dealing with anthocyanin mixtures containing diversely glycosylated derivatives derived from different aglycons, it is not as easy to predict the order of elution, since the substitution pattern of ring B in some cases may affect the polarity more than the presence of additional sugar molecules. Therefore, we cannot expect to find an elution order according to the sugar substitution pattern. Furthermore, the separation of compounds with close polarity characteristics is sometimes dependent on the HPLC conditions used, and changes in column and/or mobile phase composition may result in modifications in the retention characteristics of some anthocyanins. In this context useful data on anthocyanin resolution including α values in different eluting mixtures for 40 anthocyanins and anthocyanidins expressed in relation to Cy 3-glucoside were published by Goiffon *et al.*[25] in 1999.

Acylation of the sugar moieties of the anthocyanins involves a loss of polarity that is reflected in an increase of the retention time. The extent of this polarity decrease depends on the type of acyl substituent. For the most usual aliphatic acids residues, the elution order would be as follows:

malic → acetic → malonic → succinic

In the case of acylation with phenolic acids the decrease of polarity is greater for hydroxycinnamic acids than for most benzoic acids. In each of these groups, the polarity depends on the number of hydroxyl/methoxyl groups in the aromatic ring. Thus, in a hypothetical mixture of phenolic acids, the following elution order should be observed:

gallic → protocatechuic → *p*-hydroxybenzoic → vanillic → caffeic → syringic

→ *p*-coumaric → ferulic → synapic

The linkage between the acyl residue and the sugar is sensitive to mineral acids, especially in the case of aliphatic acid moieties, which may be thus partially cleaved during extraction. This aspect and the fact that many acylated anthocyanins show similar spectra to non-acylated anthocyanins may have caused its presence in some products to be confused or overlooked and, therefore, their occurrence in plant sources understimated.[29] It is necessary to take into account that the existence of acyl substitution always represents a decrease in the polarity that is reflected in the HPLC retention time; on the other hand, acylating residues can be selectively eliminated by an alkaline hydrolysis (see Chapter 15) giving place to the parent glycosylated anthocyanin. Both aspects can be used to confirm the acyl nature of the compounds.

The usual substitution position for acyl moieties is at 6″ of the sugar, although unusual 3″-acyl substituents have also been reported in the genus *Allium*.[30] These same authors indicate that the acyl linkage at the 6″-position of the sugar is more labile to acid cleavage than the corresponding linkage at the 3″-position.

When compounds with different substitution patterns occur in a product, a useful observation for identification purposes is that the same anthocyanidin distribution pattern usually exists among anthocyanins derived from different aglycones. Thus, for instance, in the case of *Vitis vinifera* grapes, a similar distribution of delphinidin, cyanidin, petunidin, peonidin and malvidin is found among 3-glucosides, 3-acetylglucosides and 3-*p*-coumaroylglucosides.

Spectral Characteristics

Further information to that derived from the HPLC elution pattern can be obtained from the UV-visible spectra of the anthocyanins. Combined information of these two characteristics as provided by the HPLC–DAD technique may allow the identification of anthocyanins in a mixture without requiring previous isolation.

Anthocyanin spectra are commonly obtained in acidic media, so that the spectrum corresponding to their flavylium cations is recorded. For convenience, reference spectra of anthocyanins are usually collected in methanol/HCl and spectral data obtained in such media are available in various publications.[17,18,28] However, the spectral characteristics of the anthocyanins are dependent on both pH and the nature of the solvent. On the one hand, the pH value influences the distribution of anthocyanin forms in equilibrium, which affects the ratio of absorbances in the UV and the visible regions. On the other hand, the use of different solvents may cause shifts in the absorption bands, and usually variable hypsochromic shifts in Band I are observed for anthocyanin spectra recorded on-line after HPLC separation in relation to those obtained in methanol/HCl. For this reason, direct comparison of spectra collected on-line with those in the literature may be inappropriate and, for a precise identification, spectra should be compared with ones obtained using the same solvent composition. An advisable strategy would be to count on a library of spectra obtained in similar elution conditions. In order to achieve this, a good and cheap starting point could be the injection of reference samples containing known anthocyanin patterns, *e.g.* extracts from *Vaccinium myrtillus* or grape skin of red *Vitis vinifera* or American varieties, as previously indicated for the HPLC optimization.

Maximum wavelength of absorption (λ_{max}) in the visible region for the six commonest anthocyanidins is related to the substitution pattern of their B rings; as the number of substituents in that ring increases the colour of the flavylium cation shifts towards larger wavelengths. Thus, λ_{max} for trisubstituted anthocyanidins (*i.e.* delphinidin, petunidin and malvidin) is 8–12 nm higher than for disubstituted ones (cyanidin and peonidin), and about 22–26 nm than for monosubstituted (pelargonidin). Typical UV-visible spectra for the flavylium forms of mono-, di- and tri-hydroxylated anthocyanidins recorded on-line using an acetonitrile/4.5% formic acid gradient are shown in the Figure 2.

Methylation of the hydroxyl groups has little effect on λ_{max} and only induces a slight hypsochromic shift of 1–4 nm. This shift is usually not observed when spectra are obtained on-line after RP-HPLC separation, since methylated

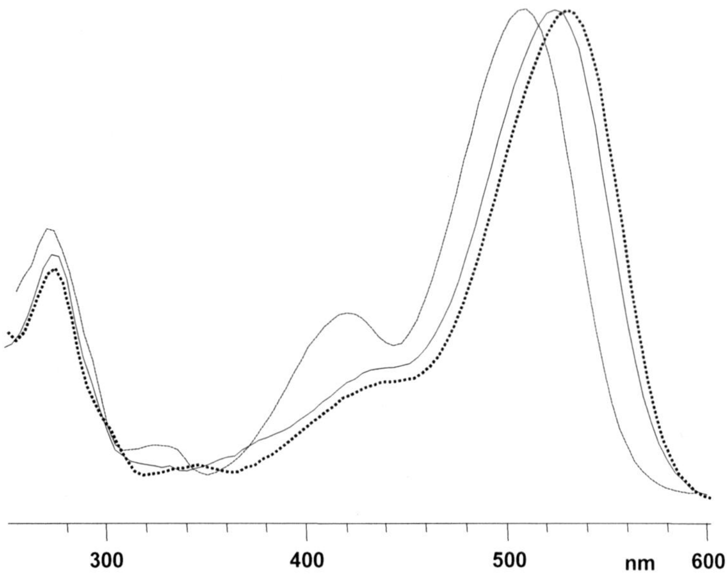

Figure 2 *UV-visible spectra of pelargonidin (λ_{max} 512 nm), cyanidin (λ_{max} 525 nm), and delphinidin (λ_{max} 530 nm), as recorded on-line after RP-HPLC separation using an acetonitrile–4.5% formic acid gradient*

derivatives elute at higher proportions of the less polar solvent, which compensates the hypsochromic effect and may even induce a slight bathochromic displacement in λ_{max}. In our experience, no significant differences in the λ_{max} are observed on changing 4.5% formic acid to 10% formic acid. Regarding less usual anthocyanidins, methylation of the A-ring hydroxyls (*e.g.* rosinidin and hirsutinidin) results in hypsochromic shifts of about 10 nm in band I.[31] 3-Deoxyanthocyanidins show band I at much lower wavelengths (up to more than 40 nm less) than 3-hydroxyanthocyanidins (Figure 3). A relevant feature of these latter compounds

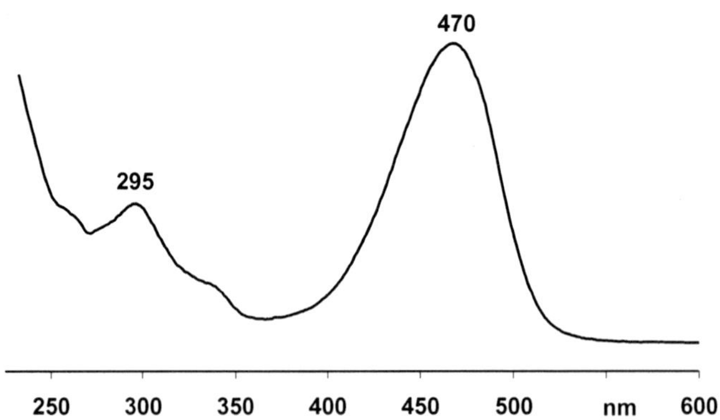

Figure 3 *UV-visible spectrum of carajurin, a 3-deoxyanthocyanidin identified in the leaves of* Arrabidea chica *(Devia et al., 2002)*

is that they are stable aglycones and, therefore, are found in their natural sources as non-glycosylated forms.[32]

Spectra of the anthocyanins (*i.e.* glycosylated anthocyanidins) exhibit Band I at about 6–10 nm lower than their parent anthocyanidins when obtained in the same solvent. For anthocyanin spectra obtained on-line this difference may be greater due to the solvent effect induced by the changes in the mobile phase occurring during a gradient elution. For anthocyanin derivatives of a given aglycone, λ_{max} in the visible region undergoes a bathochromic shift as the proportion of the apolar solvent (acetonitrile, methanol) in the mobile phase becomes higher. The extent of this solvent effect depends on the type of compound; the shift is as high as 9–11 nm in the case of the usual 3-monoglucosides and about 4–5 nm in the case of 3,5-diglucosides when the percentage of ACN increases from 0% to 35%.[26,27] The composition of the mobile phase at the time of elution of the anthocyanin should be thus taken into account for the correct identification of the anthocyanidin.

Glycosylation at 3 and 5 positions of the anthocyanidin has an important influence on the spectral characteristics. Anthocyanins showing only glycosidic substitution at position 3 show a shoulder in their spectrum at about 440 nm, which is not observed in those substituted at both positions 3 and 5 (Figure 4). For this reason, 3-glycosylated anthocyanins exhibit a ratio A_{440}/A_{max} that is almost twice as much as that of the ones with glycosidic substitution at position 5 or both positions 3 and 5.[33] In general, the nature of the sugar substitution has no effect on the anthocyanin spectrum. Usual A_{440}/A_{max} ratios for 3-glycosides of the most usual anthocyanins range from 0.2 to 0.35 depending on the solvent composition. Typical ratios for 3,5-diglycosides are around 0.1–0.15. An exception are the pelargonidin derivatives whose spectra show increased absorption at 420–440 nm (Figure 2) and, therefore, higher A_{440}/A_{max} ratios: greater than 0.4 for the 3-glycosides and up to 0.2 for 3,5-diglycosides. Ratios may be influenced by the solvent composition and, therefore, comparison for identification purposes should be made for separations run in similar mobile phase conditions.

Less common anthocyanins glycosylated in the B-ring have been also detected. The transfer of the glycosyl moiety from 3-position to 3′ gives the anthocyanins different spectroscopic and chromatographic properties: 3′-glycosides show a

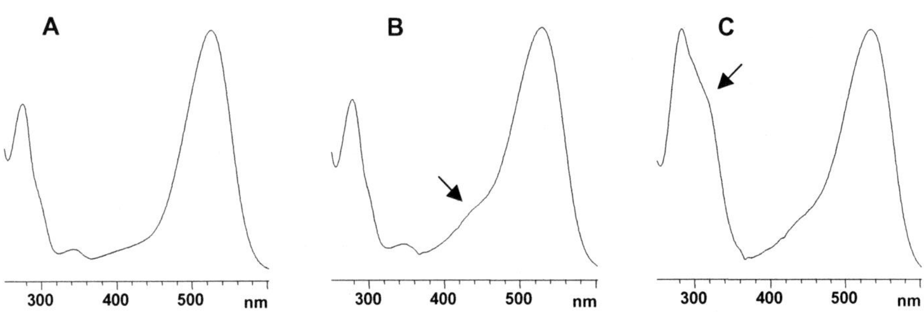

Figure 4 *UV-visible spectra of malvidin 3,5-diglucoside* (A), *malvidin 3-glucoside* (B) *and malvidin 3-cumaroylglucoside* (C), *as recorded on-line after RP-HPLC separation using an acetonitrile–4.5% formic acid gradient*

hypsochromic shift in Band I (about 13 nm), increased A_{440}/A_{max} ratio (values higher than 0.66) and shorter retention time in RP-HPLC than their analogous 3-glycosides.[34]

No information about the identity of the sugar moieties can be obtained from the UV-visible spectra of the anthocyanins.

Spectra of the acylated anthocyanins are influenced by the nature of the acyl residue. Compounds acylated with aromatic acids show increased absorption in the UV region in relation to the corresponding non-acylated derivatives. For this reason, an increase in the ratio of absorbance between the maxima in the UV and visible regions (A_{280}/A_{max}) is observed for these type of anthocyanins. Information about the number of acyl groups may also be obtained from this ratio. Typical A_{280}/A_{max} values obtained in the ACN/4.5% formic acid mobile phase gradients usually employed in our laboratory range between 0.60 and 0.75 for non-acylated anthocyanins, from 0.80 to 0.95 for monoacylated ones, and are higher than 1.0 for polyacylated derivatives. We have also observed that the type of acid modifier used may have an influence in the ratio. Lower values are usually obtained with 10% formic acid than with 4.5% formic acid, probably due to the lower pH provided for the first one, which provides a greater displacement of anthocyanin equilibria towards the flavylium form that is responsible for the absorption about 520 nm.

In the case of acylation with hydroxycinnamic acids, a shoulder or a peak appears in the spectrum of the anthocyanins at 305–325 nm characteristic of the type of acyl residue: 305–310 nm in the case of *p*-coumaric acid (Figure 4), and 320–325 nm for caffeic, ferulic and sinapic acids. This shoulder is not observed in the case of benzoic acids, whose maximum absorption is at 260–290 nm. It has been suggested that acylation with cinnamic acids also produces a bathochromic shift in the visible λ_{max} of the pigment.[33] In our experience the shift observed for spectra obtained on-line is just explained by the greater percentage of the less polar solvent that usually exists at the time of elution of the acylated anthocyanins.

In general, it is assumed that anthocyanins acylated with aliphatic acids display the same spectra as non-acylated ones and, therefore, information for their identification should be taken from their retention characteristics in reversed-phase HPLC and their sensitivity to alkaline cleavage. Nevertheless, this assumption could not be totally true, and sometimes increased UV absorption at 280 nm might also exist in the case of acylation with aliphatic acids.[21]

An interesting observation is that when no sufficiently low pH in the mobile phase exists and/or extracts are not sufficiently acid, besides the peak broadening, chalcone forms of the anthocyanins may also appear in the chromatograms at lower retention times than their flavylium forms. These forms can be detected by their characteristic spectra showing maximum absorption in the visible region at about 360 nm.

3 Flavan-3-ols (Catechins, Proanthocyanidins, Condensed Tannins)

Flavan-3-ols are flavonoids which have the heterocycle (C-ring) completely saturated. They can occur as both monomers and polyflavan-3-ol structures (*i.e.*

condensed tannins, *syn.* proanthocyanidins) exhibiting different degrees of condensation, from dimers to highly polymerized structures, C4–C8 and C4–C6 being the most usual linkages between elementary units. Compounds also exist with doubly linked units, bearing an additional ether linkage between the C2 of the upper unit and the oxygen at C7 or C5 of the lower one, in addition to the usual C4–C8 or C4–C6 bond ('A-type' linkage). Basic flavan-3-ol units differ according to their hydroxylation pattern and the stereochemistry of the asymmetric carbons of the heterocycle. The most common compounds are those displaying 3′,4′-dihydroxy (catechins) and 3′,4′,5′-trihydroxy substitutions (gallocatechins) on the B-ring. Condensed proanthocyanidins (PA) yield anthocyanidin pigments after heating in acidic media; compounds that possess upper dihydroxylated units release cyanidin, and are called procyanidins; those bearing upper tryhydroxylated units release delphinidin, *i.e.* prodelphinidins. Propelargonidins with 4′-hydroxy B rings are relatively rare and have not been reported in any major food source. PA also exist that have units with different hydroxylation degree (mixed or heterogeneous PA). Carbons C2 and C3 (as well as C4 in PA) of the flavanol heterocycle are asymmetric and occur in different configuration. With some very rare exceptions, the configuration of C2 is *R*. The stereochemistry of the C2–C3 linkage may be either *trans* (2*R*,3*S*) or *cis* (2*R*,3*R*); they correspond respectively to (gallo)catechin and epi(gallo)catechin units. In the proanthocyanidins the C3–C4 linkage is always *trans*. Flavan-3-ols mostly occur as non-glycosylated structures; heterosidic foms also exist but they are much less frequent than for other flavonoids.[35] For further description of flavanol structure see Chapters 12 and 13.

The use of HPLC–DAD for the analysis of flavanols presents more inconveniences than in the case of anthocyanins. On the one hand, only monomers and oligomers up to tetramer can be usually separated and detected as defined peaks by chromatographic techniques. Polymeric forms, which may constitute the bulk of PAs in many plant materials, are not well resolved as they elute as a continuous set of molecules, giving place to a drift in the baseline and the formation of characteristic 'humps' in the HPLC chromatograms. Such polymers have to be then characterized by chemical degradation in the presence of nucleophiles, usually phloroglucinol[36,37] or benzylmercaptan,[38–40] and more recently by tandem HPLC–MS techniques (see also Chapter 13). On the other hand, the spectral characteristics of the flavanols do not allow their easy detection and identification in the chromatograms. Different reasons exist for this. Flavan-3-ols show maximum absorption at non-specific wavelengths (270–290 nm), at which many phenolics also absorb, thus not allowing their selective detection. Furthermore, they have lower extinctions in this band of the UV spectrum than other usual accompanying phenolics, which poses problems of identification and interferes the adequate observation of their peaks in the chromatograms. For this reason, a clean-up process (*e.g.* by solvent partition or SepPak cartridges) is usually required prior to HPLC analysis in order to avoid interferences. Even if well-defined peaks are obtained, another challenge is the accurate identification of PA oligomers in the chromatograms, as very little information can be extracted from their spectra and no standards are commercially available. Thus, for their identification prior isolation is usually required, which is not always easy to achieve taking into account the quantities in which they may be present and the difficulties for their separation and further characterization. The quantification of individual flavanols by HPLC is not easy for the same reasons. The non-specificity of the UV detection and the usually low concentration of the flavanols in plant and food extracts make high purification necessary to avoid interferences in their quantitative analysis. Furthermore, the lack of reference proanthocyanidins obliges the expression of the results with regard to a reference substance, normally catechin or epicatechin. This causes concomitant errors of quantification, due to the different extinctions shown by the distinct flavanols.

Reversed-phase HPLC was first applied to flavan-3-ol separation by of Lea[41,42] and since then it has become the technique of choice for their individual analysis. Although RP columns are by far the most commonly used, normal-phase columns have been also employed;[13–15,43,44] separation of individual flavanols in these latter is not as good as for RP columns, but they allow separations to be obtained in accordance with the degree of polymerization, sometimes making chromatograms easier to interpret (see Chapter 12).

As is usual for other flavonoids, the separation of flavan-3-ols in RP columns is usually accomplished by establishing a gradient between a diluted acid and a less polar solvent. The most commonly used acids are acetic or formic, although phosphoric, citric or trifluoroacetic acids have been also employed. The solvent is normally acetonitrile or methanol, although gradients exclusively of acetic acid have been also used.[45–47] Flavan-3-ol monomers and oligomers are less retained in reversed-phase columns than most flavonoids, even if they are not glycosylated. For their elution, usually both long gradients and high percentages of the aqueous solvent are required, especially when prodelphinidin mixtures, and even catechin-based procyanidins, have to be resolved. In a typical HPLC run for the separation of proanthocyanidin oligomers, the percentage of the polar solvent (*i.e.* acetonitrile or methanol) usually does not exceed 20%. The factors that determine their elution from RP columns are complex and not well understood. Structural features with an influence in the retention of flavan-3-ols are the pattern of substitution in the B-ring, the stereochemistry of the hydroxyl group at C3, the relative position of units in the oligomers and the degree of condensation. The following tendencies are observed:

1. Trihydroxylated flavanols (*i.e.* gallocatechins and prodelphinidins) are more polar and elute earlier than the corresponding dihydroxylated ones (*i.e.* catechins and procyanidins).
2. Flavanols with 2*R*,3*S* stereochemistry (*i.e.* (gallo)catechin units) elute earlier than the equivalent 2*R*,3*R* flavanols (*i.e.* epi(gallo)catechin units).
3. In the case of oligomeric flavanols, the lower sub-unit is that with more influence on the retention characteristics and, thus, the compounds which present (gallo)catechin as terminal sub-unit elute earlier than the equivalent ones having epi(gallo)catechin units. This is probably due to the fact that the strongest bond to the RP is produced across C4 of the lower unit, which is more exposed in the case of the 2*R*,3*R* flavanols by the axial position of the 3-hydroxyl group.

4. Compounds with C4–C6 or A-type interflavan linkages eluted later than their C4–C8 isomers.
5. The presence of galloyl acyl moieties retards the elution, although the elution order of the galloyled derivatives is not easy to predict due to the strong influence of their spatial conformations on their retention characteristics.

In a procyanidin series having epicatechin (EC) as extension unit(s) the elution is produced later as the degree of polymerisation increases. Thus, for instance:

dimer EC-4,8-EC (B2) → trimer EC-4,8-EC-4,8-EC (C1)

→ tetramer EC-4,8-EC-4,8-EC-4,8-EC

or

dimer EC-4,8-C (B1) → trimer EC-4,8-EC-4,8-C

→ tetramer EC-4,8-EC-4,8-EC-4,8-C

This rule does not behave similarly when considering equivalent catechin (C) series; thus, it is usual for the trimer C-4,8-C-4,8-C (C2) to elute earlier than or overlapping with the dimer C-4,8-C (B3). Although no direct evidence has been obtained for oligomers containing (epi)gallocatechin extension units, similar behaviour might be expected.

In Table 1 the sequence of elution in a separation on C_{18} reversed-phase columns that could be expected for some common flavanols, in accordance with the previous features and the observations made in our laboratories, is shown. Some difficulties are usually found for the separations of some compounds, which may not be well resolved in some HPLC systems. Typical overlapping is observed among gallocatechin and related prodelphinidin oligomers, as well as for the following pairs of procyanidins: trimer C2 and dimer B3, dimer B1 and catechin and/or trimer EEC, dimer B7 and trimer C1, and dimers B1-3-*O*-gallate and B2-3′-*O*-gallate. Sometimes changes in the order of elution of some of these compounds are also observed depending on the method used; for instance, it is not unusual for the procyanidin B1 to elute after catechin.

Grape seeds constitute a good material for checking procyanidin separation, as they contain a large range of catechin- and epicatechin-based flavanols, both galloylated and non-galloylated, free of interfering flavonoids and, therefore, not requiring major purification of the extracts.[48] Another useful material for the identification of monomers is green tea, whose majority flavanol peaks correspond by respective order of elution to epigallocatechin, epicatechin, epigallocatechin-3,*O*-gallate, and epicatechin-3,*O*-gallate.

Most of the HPLC methods described in the literature are usually designed for the separation of procyanidins, which are not, in general, suitable for prodelphinidins that would elute mostly in the first part of the chromatograms as unresolved

Table 1 *Expected sequence of elution of some common flavan-3-ols observed in C_{18} reversed-phase columns. The higher the number the longer the retention time*

1. Gallocatechin-gallocatechin*
2. Gallocatechin
3. Gallocatechin-gallocatechin-catechin
4. Gallocatechin-catechin
5. Gallocatechin-(4,6)-catechin
6. Catechin-catechin-catechin (C2)
7. Catechin-catechin (B3)
8. Epicatechin-catechin (B1)
9. Catechin
10. Epicatechin-epicatechin-catechin (EEC)
11. Epicatechin-epicatechin-epicatechin-catechin
12. Epigallocatechin
13. catechin-(4,6)-catechin (B6)
14. Catechin-epicatechin (B4)
15. Epicatechin-epicatechin (B2)
16. Epicatechin-3,*O*-gallate-epicatechin (B2-3-*O*-gallate)
17. Epicatechin
18. Epigallocatechin-3,*O*-gallate
19. Epicatechin-3,*O*-gallate-catechin (B1-3-*O*-gallate)
20. Epicatechin-epicatechin-3,*O*-gallate (B2-3′-*O*-gallate)
21. Epicatechin-(4,6)-catechin (B7)
22. Epicatechin-epicatechin-epicatechin (C1)
23. Epicatechin-epicatechin-epicatechin-epicatechin
24. Epicatechin-3,*O*-gallate
25. Epicatechin-(4,8; 2,7)-epicatechin (A2)
26. Epicatechin-(4,6)-epicatechin (B5)
27. Epicatechin-(4,6)-epicatechin-(4,6)-epicatechin

* Unless it is indicated a 4,8-interflavan linkage must be assumed.

peaks. Separation of prodelphinidins requires running to start at high percentages (*e.g.* 100%) of the aqueous solvent and to provide later a very slow increase in the proportions of the less polar solvent. An HPLC gradient has been set up in our laboratory to allow the separation of prodelphinidins, catechin-based procyanidins and epicatechin derivatives in the same run.[49] In this method, prior to the injection of the sample, the column is thoroughly stabilized with water, and then a step gradient is established from 100% 4.5% formic acid to 100% of a mixture of 4.5% formic acid/methanol (90:10), providing a slow and modulated increase in the percentage of methanol. The separation is performed on a Spherisorb ODS2 column (150 × 46 mm ID), particle size 3 μm, using a flow rate of 0.5 ml min^{-1}. Figure 5 shows the separation of mixture of common flavanol monomers, dimers and trimers obtained using this method.

Flavan-3-ols are usually detected in the UV region at 270–280 nm. If well-defined peaks are obtained, the UV-visible spectra may help in their characterization. Gallocatechins and prodelphinidins show λ_{max} at 270 nm, and catechins and procyanidins at 280 nm. Mixed proanthocyanidins show intermediate λ_{max} values

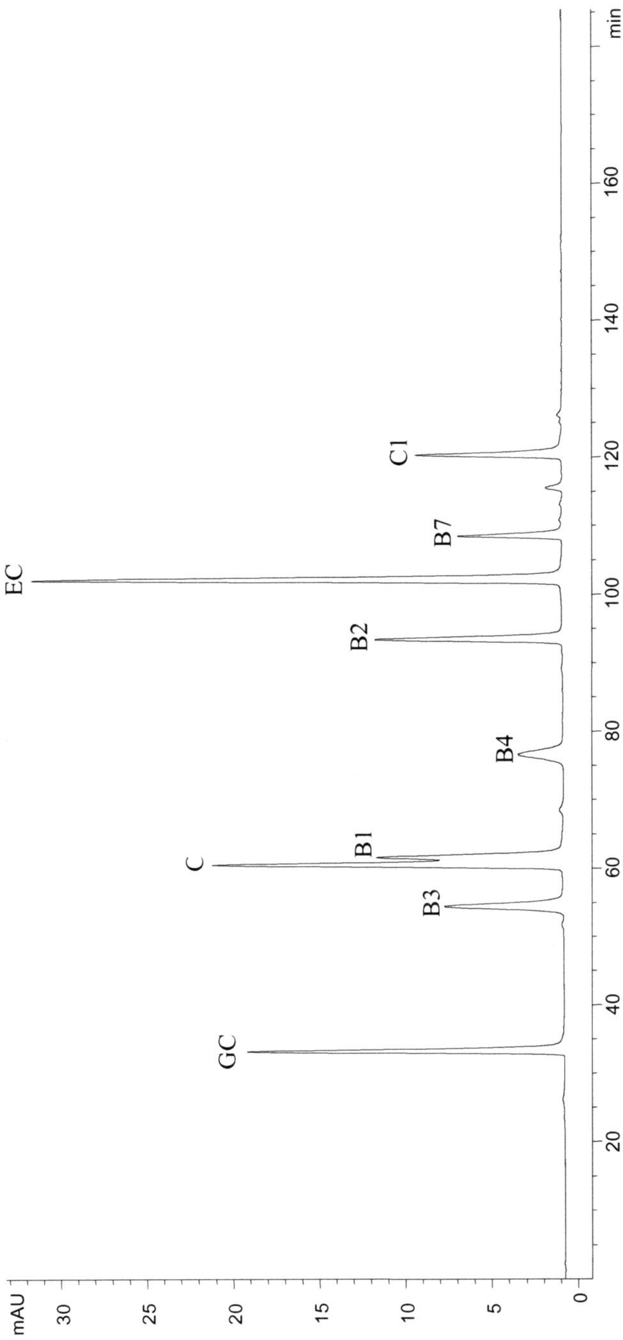

Figure 5 *HPLC chromatogram recorded at 640 nm after derivatization with DMACA of a mixture of flavanol standards. Separation performed on a Spherisorb ODS2 column (150 × 46 mm ID) 3 µm, using a step gradient between 4.5% formic acid and 4.5% formic acid/methanol (90:10), as described in reference.*[49] *See table 1 for flavanol abbreviations*

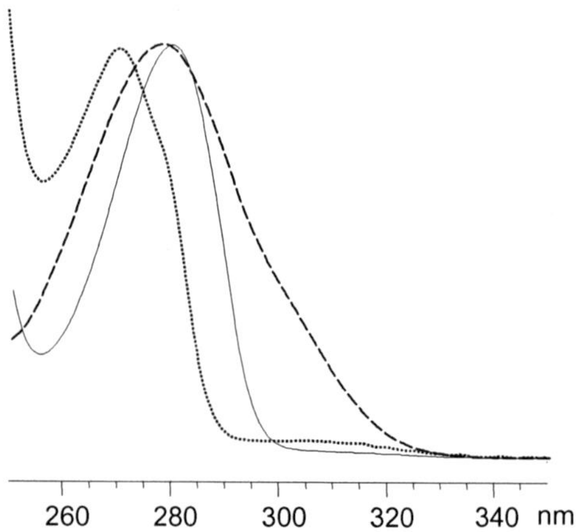

Figure 6 *UV-visible spectra of gallocatechin (·····), catechin (——) and epicatechin-3,O-gallate (- - - -) recorded on-line after RP-HPLC separation using an acetonitrile–4.5% formic acid gradient*

and the presence of galloyl acylating moieties produces a broadening of the spectra at about 310 nm (Figure 6).

An alternative proposed to overcome the problems of interference with other phenolics, as well as to facilitate the identification of flavan-3-ols, is the detection after post-column derivatization with *p*-dimethylaminocinnamaldehyde (DMACA). This method was first described by Treutter[50] in 1989 and further optimized for the selective detection of PA in plant extracts[51,52] and qualitative and quantitative analysis of flavan-3-ols in food and beverages.[49,53,54] The technique is based on the ability of the aldehydes to act as electrophiles and condense with activated aromatic rings in strongly acidic media. After reaction with DMACA flavanols form coloured adducts, showing maximum absorption between 632 and 640 nm, thus preventing the interference of other coloured compounds which might be present in the extract, such as anthocyanins, and overcoming separation problems. In this technique after injection of the sample a separation is made on a reversed-phase column and a first detection is performed in a diode array detector. Thereafter the eluent is mixed with the reagent in a T-connection and after reaction a second detection at 640 nm is performed. The fact of the first detection being accomplished in a diode array may allow the UV spectra of the compounds to be obtained, thus contributing to their identification. Although acetonitrile might provide better chromatographic separation of flavanols, in this system the use of methanol in the mobile phase is necessary, since the reaction with DMACA is inhibited by the first solvent, but favoured in alcoholic media. The reagent shows a high specificity for flavanols; that is why in the detection carried out after the reaction only peaks corresponding to these compounds are observed in the chromatograms. In addition to its selectivity, the derivatization method increases the sensitivity, when compared with detection at 280 nm, allowing peaks of substances present in low levels to be observed. A typical chromatogram recorded at 640 nm after DMACA reaction for a mixture of flavanol standards is in Figure 5.

Another interesting characteristic of the derivatization method is that, if double detection is carried out in the UV region (280 nm) and after the chemical reaction with DMACA (640 nm), information about the identity of the compounds can be obtained from the ratio of their peak areas at 640 nm and at 280 nm, since a relationship exists between the ratio of absorbances A_{640}/A_{280} and the flavan-3-ol structure.[55] The following features are encountered:

1. Gallocatechins show a higher ratios than equivalent catechins.
2. 2,3-*cis*-Compounds (*i.e.* epi(gallo)catechin) have higher ratios than 2,3-*trans* ones (*i.e.* (gallo)catechin).
3. Within a correlative series of oligomers, the ratio decreases with the increase in the degree of polymerization.
4. Proanthocyanidins with a 4,6-linkage show enhanced ratios compared with the analogous 4,8 compounds and these higher than A-type derivatives.
5. The ratio is more influenced by the type of chain extension units, thus offering complementary information to that obtained from the elution characteristics from reversed-phase columns, more influenced by the terminal subunit.

Table 2 shows the A_{640}/A_{280} ratios obtained for some flavan-3-ols after a reaction time of 2.5 min after HPLC separation using the formic acid/methanol gradient previously referred to. It is necessary to indicate that in order to obtain adequate useful ratios a good chromatographic separation of flavanols both among themselves and from other substances must exist. If not, neither adequate peak integration nor good peak spectra are going to be obtained. Both aspects are difficult to fulfil when dealing with complex flavanol mixtures, especially in the cases of prodelphinidins, which are more difficult to resolve, show lower UV extinctions, and are usually present at smaller concentrations than procyanidins.

4 Flavones and flavonols

These flavonoids are characterised by the presence of a carbonyl group at position 4 and a double bond between the C-2 and C-3 in their heterocyclic ring. The basic difference between flavones and flavonols is the presence of oxygenation at C-3 in the case of flavonols. These are much more common in nature than flavones. They both occur generally as glycosides that are mainly located in the vacuoles of epidermal tissues in the plant aerial parts. However, they can also be present as aglycones, mainly as lipophilic methyl ether derivatives. In this case they are present in plant exudates (resins, waxes, *etc.*). Flavone glycosides can be *O*-glycosides or *C*-glycosides. The former are much more common that the latter. The *O*-glycosides can be mono-, di- or tri-glycosides with mono-, di- or tri-saccharides. In addition these glycosides can be acylated with one or several aliphatic (acetic, malonic, malic, *etc.*) or aromatic (*p*-coumaric, caffeic, ferulic,

Table 2 *A_{640}/A_{280} ratios obtained for some flavan-3-ols using a reaction time with the DMACA reagent of 2.5 min after HPLC separation (adapted from de Pascual-Teresa et al.*[49]*)*

Flavan-3-ol	*Ratio A_{640}/A_{280}*
Gallocatechin (GC)	13.75
Epigallocatechin (EGC)	17.18
Catechin (C)	3.31
Epicatechin (EC)	4.40
Epicatechin-3,*O*-gallate (ECG)	1.05
GC-(4,8)-GC	4.62
GC-(4,8)-C	2.23
GC-(4,6)-C	2.52
GC-(4,8)-GC-(4,8)-C	2.17
EC-(4,8)-C (B1)	1.93
EC-(4,8)-EC (B2)	2.37
C-(4,8)-C (B3)	1.38
C-(4,8)-EC (B4)	1.69
EC-(4,6)-EC (B5)	2.76
EC-(4,6)-C (B7)	2.21
EC-(4,8)-EC-(4,8)-EC (C1)	1.36
EC-(4,6)-EC-(4,6)-EC	1.54
EC-(4,8)-EC-(4,8)-C	1.54
EC-(4,8)-EC-(4,8)-EC-(4,8)-EC	1.19
EC-(4,8)-EC-(4,8)-EC-(4,8)-C	1.19

sinapic, benzoic, gallic, *etc.*) acid molecules, rendering quite complex molecules. In some cases, these glycosides can be sulfated (as in the case of halophytic species). The natural occurrence of quite complex flavonoid molecules has been demonstrated to be rather frequent when analysis using HPLC–MS methods has become available to many phytochemical laboratories. In a recent study we have demonstrated the presence of kaempferol pentaglucosides diacylated with sinapic acid in *Brassica* species (unpublished results).

Chromatographic Behaviour

The HPLC-reversed-phase chromatographic behaviour of flavonol and flavone glycosides is very similar to that reported above for anthocyanins. This is closely related to their polarity although size can also affect the interaction with the stationary phase and therefore retention time. Their elution pattern provides useful information on the nature of the sugar moieties, presence of acylation and B-ring substituents. As it is well known, the elution order of compounds in reversed-phase HPLC is closely related to their polarity, the more polar compounds eluting earlier than the less polar ones. In the case of the most widespread flavonols (3-hydroxyflavones), the polarity depends on the substitution pattern of their B-ring. The compounds become more polar as the number of hydroxyl groups increases and more apolar as the number of methoxyl groups increases. Thus, the following order of elution is observed:

Myricetin (3,5,7,3′,4′,5′-hexahydroxyflavone)

→ quercetin (3,5,7,3′,4′-pentahydroxyflavone)

→ kaempferol (3,5,7,4′-tetrahydroxyflavone)

→ isorhamnetin (3,5,7,4′-hydroxy-3′-methoxyflavone)

→ galangin (3,5,7-trihydroxyflavone)

Something similar is observed in the case of flavones:

Tricetin (5,7,3′,4′,5′-pentahydroxyflavone)

→ luteolin (5,7,3′,4′-tetrahydroxyflavone)

→ apigenin (5,7,4′-trihydroxyflavone)

→ chrysoeriol (5,7,4′-trihydroxy-3′-methoxyflavone)

→ chrysin (5,7-dihydroxyflavone)

→ acacetin (apigenin4′-methylether)

Introduction of an additional hydroxyl on the A-ring also decreases the retention time. Thus 6-hydroxyluteolin elutes with shorter retention times than luteolin, scutellarein (5,6,7,4′-tetrahydroxyflavone) than apigenin, and baicalein (5,6,7-trihydroxyflavone) than chrysin. The same behaviour is observed with flavonols as gossypetin (3,5,7,8,3′,4′-hexahydroxyflavone) and quercetagetin (3,5,6,7,3′,4′-hexahydroxyflavone) which elute with shorter retention times than quercetin, and herbacetin (8-hydroxykaempferol) that elutes earlier than kaempferol.

The introduction of methyl ethers on the hydroxyls of a flavone molecule increases lipophilicity and retention time. The following elution order is observed in the case of quercetin methyl ethers:

Quercetin → quercetin 3-methyl ether → quercetin 3′-methyl ether

→ quercetin 3,3′-dimethyl ether → quercetin 7,3′-dimethyl ether

→ quercetin 3,7-dimethyl ether

This elution order shows that depending on the position where the methyl ether is introduced, the effect on the retention time is different and this could be used in the identification of flavonoid methyl ethers in complex extracts.[56]

This general rule is not always followed when the methyl ether is introduced on the hydroxyl in the 5 position, as in this case the internal hydrogen bond between the hydroxyl and the carbonyl at the 4 position is no longer possible and the retention time decreases. For instance, the fully methylated flavone sinensetin (5,6,7,3′,4′-pentamethoxyflavone) elutes earlier than the corresponding flavone with a free hydroxyl at the 5 position (5-hydroxy-6,7,3′,4′-tetramethoxyflavone).[57]

This same elution order is maintained for flavones bearing identical sugar substituents. Sugar substitution increases the polarity and, therefore, the glycosides elute with shorter retention times than their parent flavone aglycones. The influence of the sugar moieties on the polarity varies depending on the type of sugars and their position of substitution. When considering the more usual 3-glycoside derivatives, the elution observed for different flavonol 3-monoglycosides of the same aglycone follow the same order as described for anthocyanins previously in this chapter.

Since the introduction of HPLC–MS, the occurrence of flavonoid trisaccharides has been demonstrated to be more usual than was thought before. The main problem in this case is that sophorosides [glucosyl(1-2)glucosides] and sophorotriosides [glucosyl(1-2)glucosyl(1-2)glucosides], for instance, co-elute under most chromatographic conditions, and their presence can only be demonstrated using methods such as HPLC–ESI-MS (Tomás-Barberán, unpublished results).

Flavones and 3-hydroxyflavones may also bear additional sugar moieties, usually attached to their 5, 7 or 3′ positions. In general, the introduction of a second sugar moiety in the molecule increases the polarity and thus, for instance, 3,7-diglucosides elute before the corresponding 3-glucosides.

Flavone glucuronides are also present in plant tissues, and are also products of metabolism in mammalian organisms, and therefore can be found very frequently in biological fluids. The HPLC separation of flavone glucosides and the corresponding glucuronides is not always an easy task as they can co-elute under the usual chromatographic conditions. Examples of the co-occurrence of kaempferol and quercetin 3-glucosides and 3-glucuronides are provided by strawberries and grapes. These compounds can, however, be nicely separated using solvents without acid. Under these conditions the glucuronides elute with much shorter retention times than the glucosides. The lack of acid, however, leads to broad peaks with tailing.

Flavone *C*-glycosides generally elute with shorter retention time than the corresponding *O*-glycosides. Thus vitexin (6-*C*-glucosylapigenin) elutes with shorter retention time than apigenin 7-*O*-glucoside. Flavone di-*C*-glycosides are also present in many plant species and these elute with shorter retention times than the corresponding mono-*C*-glycosides. Flavone *C*-glycosides occur in nature as 6-*C*-glycosides, 8-*C*-glycosides and 6,8-di-*C*-glycosides. 8-*C*-glycosylflavones elute with shorter retention times than the corresponding 6-*C*-glycosylflavones. Thus, apigenin-8-*C*-glucoside elutes earlier than apigenin-6-*C*-glucoside (Ferreres, personal communication). In the case of asymmetric di-*C*-glycosides, 6-*C*-glucosyl-8-*C*-pentosyl flavones elute with shorter retention times than the isomeric 6-*C*-pentosyl-8-*C*-glucosyl flavones.

Acylation with aliphatic or aromatic acids on the sugar residue always affects the retention time of the parent flavonoid, generally increasing it. The same comments reported above for the effect of acylation with different acids on the chromatographic behaviour of anthocyanins can be applied in the case of flavone glycosides. The introduction of a sulfate residue on the flavonoid molecule always decreases its retention time.

UV-visible Spectroscopy

Flavone and flavonol spectra show two absorption peaks in the region 250–400 nm, 310–370 nm (Band I) and 250–295 nm (Band II). The UV-visible spectra recorded with a diode array detector coupled to the HPLC equipment are not as accurate as those recorded for methanol solutions of the isolated compounds, but they provide very useful information for flavonoid identification. As important as the Band I and Band II maxima is the shape of the spectra in flavonoid identification. An experienced phytochemist can as a first glance identify the type of flavonoid just with a view of its spectrum. There are some hydroxyl substituents that affect considerably the spectrum shape and wavelength of the maxima, while other substituents are less relevant.

The majority of the naturally occurring flavones have a phloroglucinol substitution on the A-ring (5,7-dihydroxyflavone). The hydroxyl substitution of ring A is reflected in the Band II absorption band of the flavonoid spectrum, while the B-ring substitution is related to the Band I. An increase in the oxygenation of the corresponding ring produces a bathochromic shift in the corresponding band maximum and sometimes an increase in its absorbance. Introduction of methyl ethers on the free hydroxyls has little effect on the UV spectrum with the exception of hydroxyls at the 3, 5, 6 and 8 positions in which methylation has marked effects on the DAD-recorded spectrum. A typical example of this effect can be observed in the spectrum of fully methylated flavones, as in the case of tangeretin (5,6,7,8,4′-pentamethoxyflavone) (Figure 7). This spectrum shows a Band II maximum at 270 nm, although it has a tetraoxygenated A-ring.

Flavonoids with an unsubstituted B-ring (chrysin, techtochrysin, galangin, baicalein), show spectra with a relevant Band II at around 267–268 nm in the case of the three first, and a shoulder corresponding to Band I at around 314–315 in the case of chrysin and techtochrysin. In galangin there is a free hydroxyl at 3-position, and this shoulder is converted to a maximum at 358 nm, although the absorbance of Band I is lower than that of Band II. The introduction of an additional hydroxyl on A-ring of chrysin, as is the case of baicalein, leads to a bathochromic shift of Band II to give a maximum at 282 nm, although the Band I maximum is not affected. Flavonoids with a single oxygenation on ring B at the 4′ position show a single Band II response in the spectrum, while 3′,4′-dioxygenated flavonoids have two responses on the Band II (two maxima or one maximum and one shoulder)[17] (see examples in Table 3).

Introduction of a methyl ether (or a glycoside) on the hydroxyls at positions 7, 3′, or 4′ have little effect on the wavelength maxima and the spectrum shape. Thus, the spectra of apigenin, acacetin and genkwanin are very similar. Something similar occurs with the spectra of luteolin, chrysoeriol, diosmetin (5,7,3′-

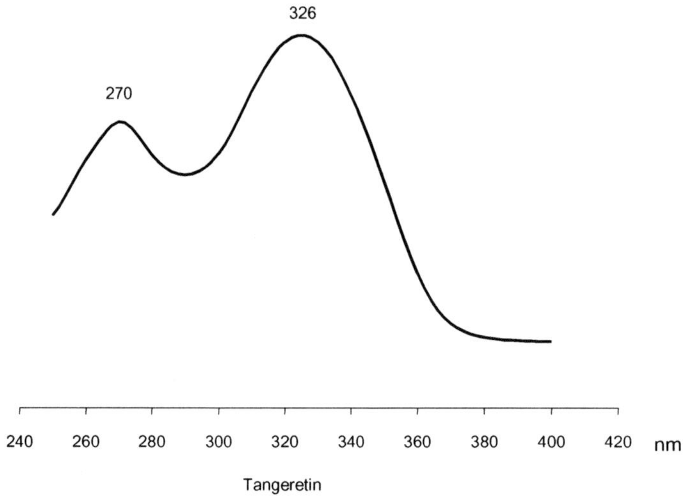

Figure 7 *UV spectrum of tangeretin (5,6,7,8,4′-pentamethoxyflavone) recorded with a DAD detector*

trihydroxy-4′-methoxyflavone) and di- or tri-methyl ether derivatives of luteolin in which the hydroxyl at 5-position remains free. As observed in the case of baicalein and chrysin, the introduction of an additional hydroxyl in position 6 of apigenin and luteolin leads to spectra showing a bathochromic shift of Band II to give maxima above 280 nm (see the spectra of scutellarein and 6-hydroxyluteolin) (Table 3). When the additional hydroxyl is introduced at position 8, Band II of the spectrum is also shifted towards higher wavelengths, but in this case the spectral shape is different, especially in the case of isoscutellarein, where a new maximum between Band I and Band II appears at around 305 nm which is the band with higher absorbance in the spectrum. This is also the case of 8-hydroxycirsimaritin and 8-hydroxysalvigenin, which show identical spectra with maxima at 304 (Band II) and 333 (Band I) nm.[58]

This maximum can be associated to the presence of the *para*-dihydroxy grouping corresponding to the 5,8-dihydroxy substitution. The introduction of methyl ether on the 5-, 6- or 8-hydroxyl decreases markedly the Band II maximum wavelength [see the following pairs of compounds: ladanein and salvigenin, thymonin and 8-methoxycirsilineol, 6-hydroxyluteolin and nepetin, thymusin and cirsimaritin, *etc.* (Table 3)].

Another dramatic effect on the UV spectrum of the introduction of methyl ethers (or glycosides) is the case of blocking the hydroxyl at the 3 position. Thus conversion of flavonols into 3-methoxyflavones causes a decrease in 15–20 nm of the wavelength maximum of Band I. This remarkable difference is generally used as a diagnostic tool in the identification of free hydroxyls at the 3 position of the flavonoid molecule.

The introduction of sugar residues on free hydroxyls to give flavonoid glycosides also affects the UV spectra of flavonoids. The same behaviour

Table 3 *UV spectra of flavone aglycones recorded with a DAD coupled to an HPLC equipment*

Common name	*Structure*	*UV spectrum in mobile phase (nm)*
Luteolin	5,7,3′,4′-tetrahydroxyflavone	255, 267sh, 349
Chrysoeriol	5,7,4′-trihydroxy-3′0-methoxyflavone	253, 267sh, 347
Tricetin	5,7,3′,4′,5′-pentahydroxyflavone	248, 267sh, 302sh, 351
6-Hydroxyluteolin	5,6,7,3′,4′-pentahydroxyflavone	255, 283, 342
Hypolaetin	5,7,8,3′,4′-pentahydroxyflavone	255, 278, 266sh, 339
6-Hydroxychrysoeriol	5,6,7,4′-tetrahydroxy-3′-methoxyflavone	254, 284, 339
Nepetin	5,7,3′,4′-tetrahydroxy-6-methoxyflavone	256, 274, 345
Cirsiliol	5,3′,4′-trihydroxy-6,7-dimethoxyflavone	254sh, 273, 346
Cirsilineol	5,4′-dihydroxy-6,7,3′-trimethoxyflavone	254sh, 274, 343
5-Demethylsinensetin	5-hydroxy-6,7,3′,4′-tetramethoxyflavone	255, 275, 341
Sideritoflavone	5,3′,4′-trihydroxy-6,7,8-trimethoxyflavone	259, 278, 349
Thymonin	5,6,4′-trihydroxy-7,8,3′-trimethoxyflavone	256, 290, 345
8-Methoxycirsilineol	5,4′-dihydroxy-6,7,8,4′-tetramethoxyflavone	257, 280, 346
5-Demethylnobiletin	5-hydroxy-6,7,8,3′,4′-pentamethoxyflavone	258, 281, 342
Apigenin	5,7,4′-trihydroxyflavone	267, 338
Acacetin	5,7-dihydroxy-4′-methoxyflavone	266, 333
Genkwanin	5,4′-Dihydroxy-7-methoxyflavone	267, 338
Scutellarein	5,6,7,4′-tetrahydroxyflavone	283, 335
Isoscutellarein	5,7,8,4′-tetrahydroxyflavone	280, 306, 330sh, 366sh
Hispidulin	5,7,4′-trihydroxy-6-methoxyflavone	272, 336
Cirsimaritin	5,4′-dihydroxy-6,7-dimethoxyflavone	275, 334
Salvigenin	5-hydroxy-6,7,4′-trimethoxyflavone	276, 333
Ladanein	5,6-dihydroxy-7,7′-dimethoxyflavone	284, 334
Thymusin	5,6,4′-trihydroxy-7-methoxyflavone	294, 335
Pebrellin	5,6-dihydroxy-7,8-dimethoxyflavone	296, 334
8-OH-cirsimaritin	5,8,4′-trihydroxy-6,7-dimethoxyflavone	304, 333
Xanthomicrol	5,4′-dihydroxy-6,7,8-trimethoxyflavone	289, 290sh, 333
8-OH-salvigenin	5,8-dihydroxy-6,7,4′-trihydroxyflavone	304, 333
5-Demethyltangeretin	5-hydroxy-6,7,8,4′-tetramethoxyflavone	288, 291sh, 332
Tangeretin	5,6,7,8,4′-pentamethoxyflavone	270, 326
Chrysin	5,7-dihydroxyflavone	268, 315
Tectochrysin	5-hydroxy-7-methoxyflavone	267, 314
Baicalein	5,6,7-trihydroxyflavone	284, 313
Galangin	3,5,7-trihydroxyflavone	266, 300sh, 358

mentioned for the introduction of methyl ethers can be followed in the case of glycosides.

The introduction of a *C*-glycosyl residue at the 6 or 8 positions of a flavone also have a clear effect on the spectrum. It produces a bathochromic shift on Band II (270–271 nm in mono-*C*-glycosides and 273–274 in the case of di-*C*-glycosides). This shift is less intense than that produced by the introduction of an oxygenation in the same positions, but it is clearly observed in the spectrum and can be used as a diagnostic feature of this type of compound. In Figure 8 the UV spectrum or isoorientin (6-*C*-glucosyl-luteolin) is shown.

The type of sugar has no effect on the UV spectrum, although acylation with aromatic acids does have a direct effect on the UV spectrum of the flavonoid. The resultant spectrum is an addition of that of the flavonoid to that of the aromatic acid. The most common naturally occurring aromatic acid esters are those derived of the hydroxycinnamic acid (*p*-coumaric, caffeic, ferulic and sinapic acids) although benzoic acid esters have also been reported (benzoic, *p*-hydroxybenzoic, gallic). Acylation with hydroxycinnamic acids makes Band I of the spectrum shift to lower wavelengths with an increase in intensity depending on the number of

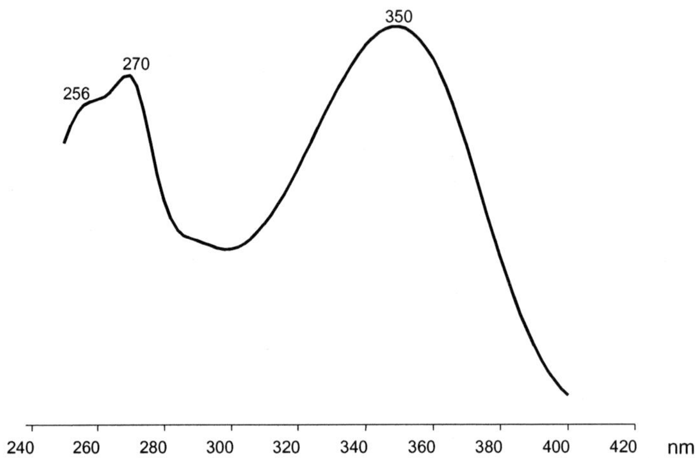

Figure 8 *UV-visible spectrum of isoorientin (6-C-glucosyl luteolin) recorded with a DAD detector*

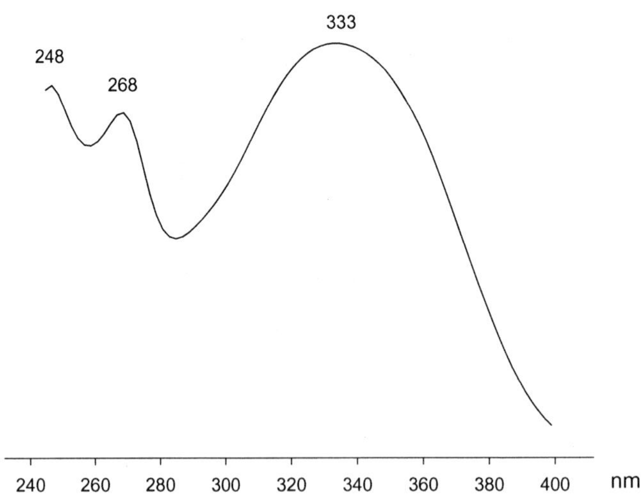

Figure 9 *UV-visible spectrum of kaempferol 3-sinapoyl sophorotrioside 7-glucoside recorded with a DAD detector*

acyl residues present in the molecule. The maximum of Band I and the shape of the resultant flavonoid depends on the type of acid and the number of acyl residues linked to the flavonoid molecule. This effect is illustrated in Figures 9 and 10, with the spectra of a monosinapoyl derivative of kaempferol sophorotrioside 7-glucoside and a disinapoyl derivative of the same flavonoid. The increased intensity of the Band I maximum in the case of the diacylated derivative can be used to identify this type of compound.

In Figure 11, the UV spectrum of a caffeoyl derivative of kaempferol 3-sophoroside 7-glucoside is shown to show the differences in the spectrum that

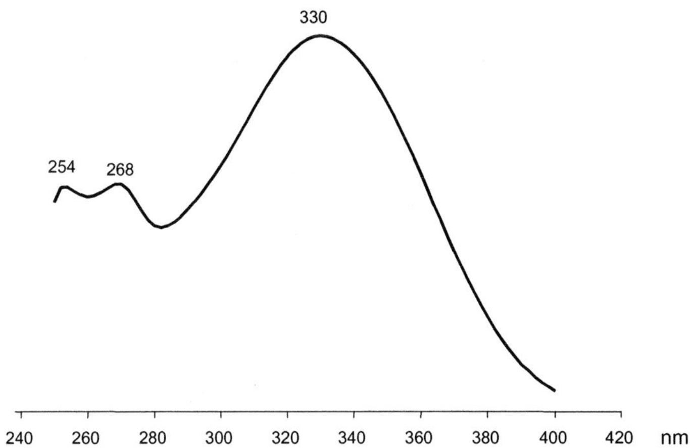

Figure 10 *UV-visible spectrum of kaempferol 3-disinapoyl sophorotrioside 7-glucoside recorded with a DAD detector*

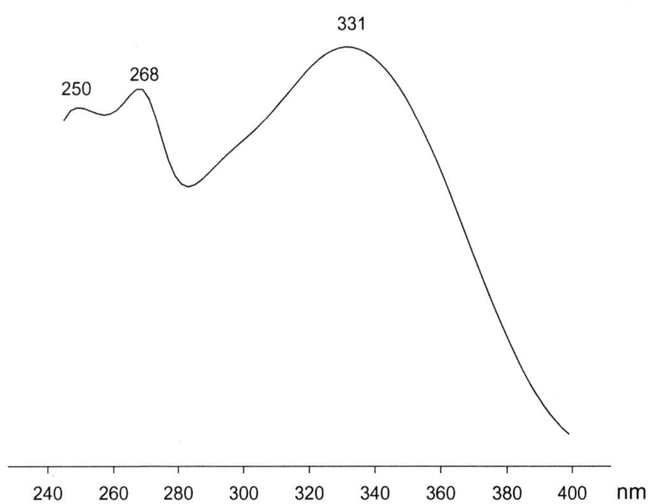

Figure 11 *UV-visible spectrum of kaempferol 3-caffeoyl sophoroside 7-glucoside recorded with a DAD detector*

could be associated with differences in the type of acid that is acylating the flavonoid. The shoulder observed at around 300 nm can be associated with the caffeoyl residue as this shoulder is not observed in sinapoyl or *p*-coumaroyl derivatives. This shoulder is, however, also observed in the case of feruloyl derivatives.

5 Flavanones, Dihydroflavonols and Isoflavonoids

Flavanones

These compounds are widely distributed within the *Citrus* genus in the Rutaceae family, and consequently used for differentiation of species or varieties of this genus. They also abound in the genus *Prunus*, in the Rosaceae and more rarely have been found in some species of *Mentha*[59] and *Cyclopia*.[60] However, in comparison with other flavonoids these compounds are of restricted occurrence. These compounds usually occur as glycosides and neohesperidosides (2-*O*-α-L-rhamnosyl-D-glucosides) attached at position 7, and generally present the 2*S* configuration. Flavanones have a saturated C-ring. The numbering system for the flavanone nucleus is the same as that for flavonols. Flavanones can easily be converted into isomeric chalcones in alkaline media (or *vice versa* in acidic media) provided that here is a hydroxyl substituent at position 2′ (or 6′) of the chalcone.

The flavanones are less soluble than the chalcones and are easily precipitated at low pH. These precipitates remain insoluble in water, methanol, ethanol or acetone, but could be re-dissolved in strong solvents such as dimethyl sulfoxide, dimethyl sulfoxide mixed with methanol, or dimethylformamide. This solubility aspect is of analytical significance.[61]

HPLC can be used for separation, quantitation and identification of these compounds. Reversed-phase HPLC is commonly used, so that glycosides are eluted before aglycones and aglycones with more OH groups before those with less. Most separation systems include C_{18} columns and generally the dimensions are 250 × 4.6 mm ID (5 μm particle size). Mobile phases (eluting solvents) could vary, the most commonly used being those that include mixtures of acidified water (*e.g.* formic, acetic or phosphoric acid) with methanol or acetonitrile. Nogata *et al.*,[62] in 1994, used a gradient increasing methanol, while Mouly *et al.*[63] used another one increasing acetonitrile. Nevertheless, the elution order is the same for both systems. Thus, those flavanones sustituted with a rutinoside (6-*O*-L-rhamnosyl-β-D-glucoside) elute before the neohesperidosides (2-*O*-L- rhamnosyl-β-D-glucoside). Sustitutions in the B-ring also render differences in retention time, eluting first those that present two hydroxyl groups in the B-ring (eriodictyol derivatives), followed by those that present an unsubstituted B-ring (naringenin derivatives), those that present a methoxy and a hydroxy group (hesperetin derivatives), and finally those that only present a methoxy group (isosakuranetin derivatives) (Figure 12).

Flavanones elute before the counterpart flavones due to the effect of unsaturation between positions 2 and 3. One factor operating to make flavones less polar

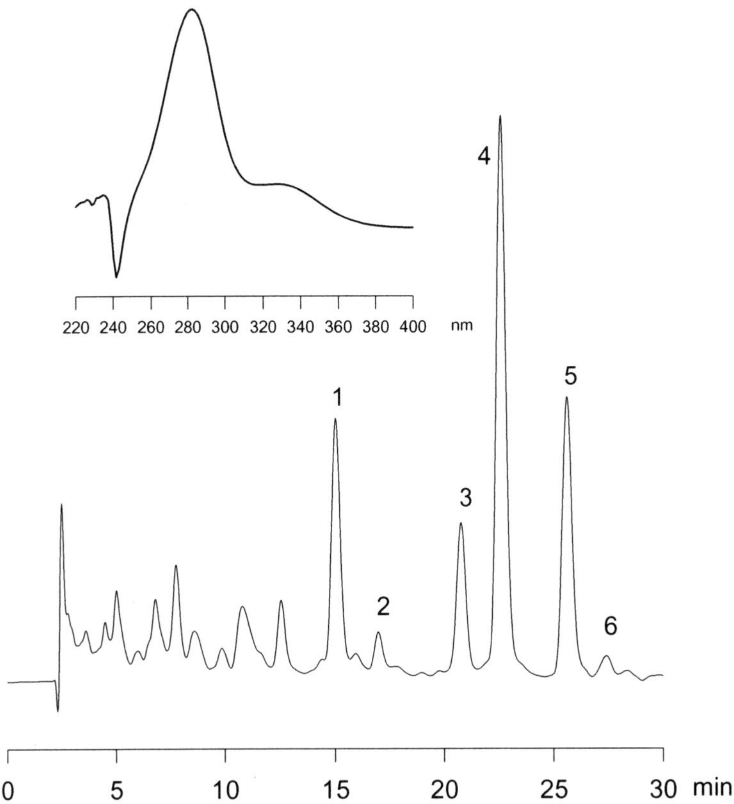

Figure 12 *HPLC chromatogram recorded at 280 nm of a mixture of citrus juices. Separation performed on a Spherisorb ODS2 column (250 × 46 mm ID) 5 μm, using a gradient (A = 5% formic acid, B = MeOH) starting 20%B, 35%B at 25 min and 90%B at 40 min. Peaks: 1, eriocytrin; 2, neoeriocytrin; 3, narirutin; 4, naringin; 5, hesperidin and 6, neohesperidin. Representative UV spectrum of flavanones (e.g. naringin) recorded with a DAD detector*

than flavanones is a large electron density on the oxygen atom of the 4-keto group of flavones resulting from resonance structures where the keto oxygen assumes a negative charge.[64] Flavones because of the their total planarity may be more difficult to solvate than the partially planar flavanone

Unfortunately, the UV spectra of these compounds would not give much information, as all absorb at similar wavelength, exhibiting a very strong maximum at 285 nm (Band II) and a small peak or mere inflection (shoulder) at 320 (band I). This distinguishes them clearly from the corresponding flavones, which have a UV spectra with a strong Band I. However, it is less easy to conclude from UV spectra whether the compound is a flavanone or a dihydro-flavonol and how the B-ring is substituted (Figure 12). Nevertheless, Mabry *et al.*[17] indicated that removal of the 5-hydroxyl group causes a 10–15 nm hypsochromic shift of Band II.

Dihydroflavonols

These compounds occur in over 50 plant families, and are characteristic of many genera of Compositae and Leguminosae.[65]

The HPLC elution order follows the same general pattern as for the other flavonoids.

The UV spectra of dihydroflavonols have a major absorption peak in the range 270–290 nm (Band II) and are almost identical with those obtained for the equivalent flavanones. As with flavanones, removal of the 5-OH group causes a hypsochromic shift of Band II, while increasing oxygenation in the B-ring has no noticeable effect on the UV spectra.[17]

Isoflavonoids

These compounds are divided into several groups: isoflavones, with over 234 known aglycones forming the largest group; the much rarer isoflavanones; pterocarpans (the second largest class of isoflavonoids); isoflavans and coumestans (representing the highest possible level of oxydation of the isoflavonoid skeleton). As isoflavones represent the most important group found in vegetables, we shall here discuss only this group of compounds (for further details of other isoflavonoids' distribution, spectral characteristics, *etc.*, see Williams and Harborne[66]).

Isoflavones show a very limited distribution in the plant kingdom, being confined to the Leguminosae, chiefly in soy and soy products,[67–69] but a few isoflavones are found in other legumes such as red clover.[70,71]

Different HPLC methods for the analysis of these compounds have been discussed elsewhere,[6] presenting a reversed-phase C_{18} column as stationary phase and mixtures of methanol, acetonitrile, trifluoroacetic acid and acidified water mixtures or buffers as mobile phase.

As in flavones, the strongest hydrogen bond in a isoflavone is the carbonyl group at C4 which bears a partial negative charge. If an OH group is present at position 5 a strong internal hydrogen bond in formed between this group and the carbonyl group. As a result the retention time of the 5-OH flavonoids are higher than those of their counterparts not possessing a free 5-OH group [*e.g.* genistein (7,4′-OH) elutes later than daidzein (5,7,4′-OH), and biochanin A (5,7-OH, 4′-OCH_3) after formononetin (7-OH, 4′-OCH_3)].[69–70] This generalization only accounts for aglycones, as glycosides are able to form several hydrogen bonds that would interfere with this rule.[72] Also, as for the rest of the flavonoids, glycosylation speeds up the elution and methoxylation slows it down (*e.g.* formononetin elutes later than its counterpart daidzein[69]). Another interesting study has been carried out where the authors have developed a method for the separation of lupin isoflavones (genistein, 2′-hydroxygenistein, their 7-*O*-glucosides and the respective 6-, 8-, 3′-mono- and 6,3′-di-prenylated derivatives) by HPLC.[73] HPLC was carried out with a C_{18} column, using a mixture of aqueous acetic acid and methanolic acetic acid. The order of elution of the isoflavonoid derivatives was inversely proportional to their degree of polarity, with the 2′-hydroxygenistein glucoside eluting first and the 6,3′-diprenylgenistein eluting last.

More recently Griffith and Collison,[74] have developed an improved HPLC method for the analysis of isoflavones from soy and soy derivatives. Minor forms of malonyl (with shorter elution times) and acetyl isoflavones (with longer elution times) were also detected.

HPLC coupled with diode array detection is valuable for the identification of isoflavones, since their spectra differ in the absorption properties from most of the other flavonoids. This is because they exhibit a C2–C3 double bond, and the location of the B-ring at C3 prevents conjugation of the phenyl group with the pyrone carbonyl group. This results in a significantly diminished contribution of the B-ring to the spectrum, relatively unaffected by increased hydroxylation. The spectra consists of a prominent band in the 250–270 nm range (Band II) with a peak of very low intensity, or shoulder, in the 300–330 nm range (Band I). Nevertheless, Band II is shifted bathochromically by increased oxygenation in the A-ring.[17,67,68]

6 Chalcones and Aurones

Chalcones

Native chalcone glycosides tend to transform to flavanone glycosides during extraction. Due to this fact chalcones *per se* are of restricted occurrence in foods (Solanaceae such as tomato).[61,75] This has been attributed to the acid hydrolysis, used prior to HPLC in many routine analysis that converts the chalcone into the corresponding flavanone.[61] Nevertheless, Krause and Galensa[75] determined naringenin-chalcone in almost all the tomato skin samples analysed, but not the isomeric flavanone naringenin, concluding that the extraction procedure used in previous investigations caused spontaneous cyclization of the chalcone. Other authors have developed HPLC methods for separation of anthocyanin chalcones by HPLC[76] or have obtained six chalcone forms of anthocyanins in the skin of *Vitis vinifera*.[77] In these last cases a pH over 3 and/or high temperature were used.

It should be noted that chalcones have a numbering scheme which differs from that used for other flavonoids (Figure 13).

HPLC systems include C_{18} columns and gradients including methanol and acidic water. Elution order is the same as that described above for the corresponding anthocyanins. Also the oxygenation on the B-ring influences the elution order of these compounds, together with the degree of methoxylation and glycosylation.[78]

As a general rule the spectrum of the chalcone consists of a very intensive Band I absorption in the 345–390 nm range, and a lower intensity absorption for Band II in the range 240–270 nm.[79] Chalcones lacking B-ring oxygenation may present a shorter wavelength for Band I. As stated above for other flavonoids, increased oxygenation of the A- or B-ring results in a bathochromic shift in Band I (Figure 14) . In previous work[80] it was demostrated that there is a good correlation between B-ring hydroxylation and Band I maxima. Moreover, this

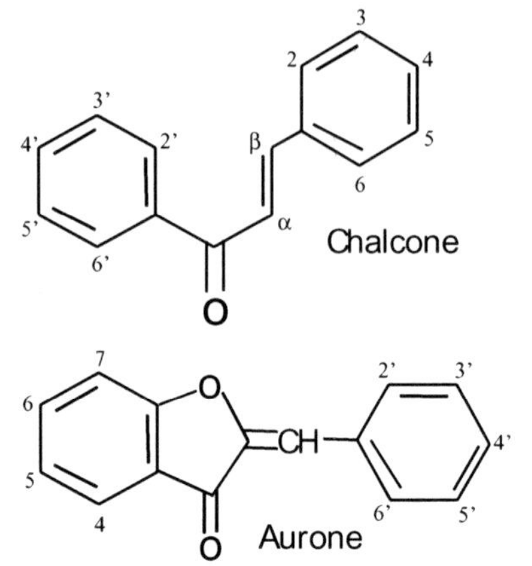

Figure 13 *Basic structure and numbering system of chalcones and aurones*

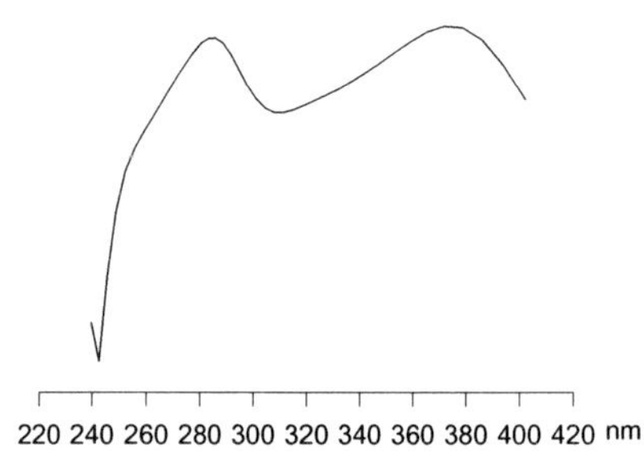

Figure 14 *UV spectrum of hesperidin chalcone (2′,3-dihydroxy, 4-methoxy, 4′-rutinoside chalcone) recorded with a DAD detector*

author revealed that chalcones with maxima below 300 nm might have a very high proportion of *O*-alkyl groups on the A-ring and most often an unsubstituted B-ring.

Aurones

Aurones are highly coloured flavonoids based upon the 2-benzylidenecoumaronone structure. The conventional numbering system has positions 4–7 on the A-ring, where position 4 is equivalent to position 5 in normal flavonoids (Figure 13). Chemotaxonomic surveys have disclosed the presence of known aurones and aurone glycosides in several groups of plants (*e.g.* Ciperaceae, Leguminosae, Compositae, *etc.*).[81]

As for other flavonoids the HPLC elution order is dependent on hydroxylation, methoxylation and glycosylation.

UV spectral properties of aurones have been discussed by many authors since 1956.[82] In general it has been proposed that the spectra are characterized by a very intensive Band I absorption in the 390–430 nm range, and a Band II absorption of much lower intensity in the range 240–270 nm.[18] Methylation or glycosylation of hydroxyl groups has little effect on the UV spectrum. Thus, spectra of the glycosides are very similar to those of their aglycones.[17] Nevertheless, hydroxylation produces a wide range of absorbance values in Band I.[18]

7 Dihydrochalcones

The greatest variety of structures have been found in the Leguminosae, Compositae and Annonaceae.[65] In the remaining families dihydrochalcones occur only sporadically; *e.g.* in the Rosaceae phloretin derivatives are characteristic of apples (*Malus*) and derived products (juice, jam, cider, *etc.*). They are present preferentially in the seeds, but could also be detected in the skin and pulp. These compounds could also be found in other members of the Rosaceae, such as *Pyrus* and *Prunus.*

The numbering system for the dihydrochalcone nucleus follows that of the chalcones (Figure 13).

As for other flavonoids the HPLC elution order is dependent on hydroxylation, methoxylation and glycosylation. For example phloretin xylosyl glucoside (4,4′,6′-trihydroxy-2′-xylosylglucoside) elutes before phlorizdin (4,4′,6′-trihydroxy-2′-glucoxydihydrochalcone)[83] and this elutes before asebotin, which has a 4′-methoxy substitution, while the aglycone, phloretin (4,2′,4′,6′-tetrahydroxydihydrochalcone), elutes with the highest retention time.[72]

These compounds can be easily identified and quantified, by reversed-phase HPLC on a C_{18} column using an acidified water (pH 2.8)–methanol solvent gradient, and well separated from the other neutral phenolic compounds (procyanidins and flavonol glycosides) characteristic of apple juice.[84]

The characteristic spectra of these compounds exhibit a strong maxima at 280 nm. (Figure 15)

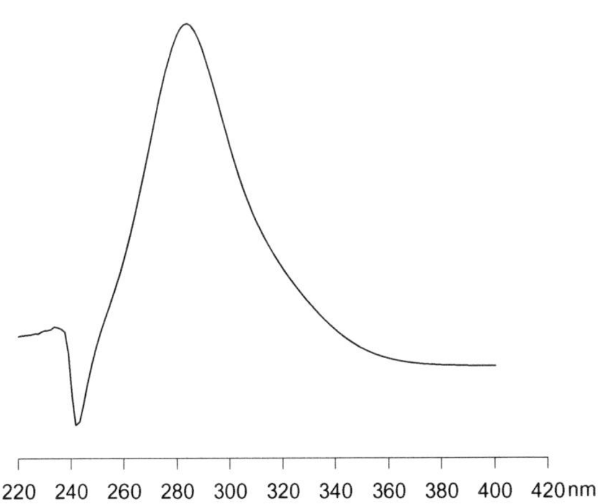

Figure 15 *Representative UV spectrum of dihydrochalcones (e.g. phlorizdin) recorded with a DAD detector*

References

1. J. B. Harborne and T. J. Mabry (eds), *The Flavonoids: Advances in Research*, Chapman and Hall, New York, 1982.
2. J. B. Harborne (ed.), *The flavonoids: Advances in Research since 1980*, Chapman and Hall, New York, 1988.
3. J. B. Harborne (ed.), *The flavonoids: Advances in Research since 1986*, Chapman and Hall, New York, 1994.
4. C. Rice-Evans, N. J. Miller and G. Paganga, *Free Radic. Biol. Med.*, 1996, **20**, 933.
5. B. Berente, M. Reichenbächer and K. Danzer, *Fresenius' J. Anal. Chem.*, 2001, **371**, 68.
6. H. M. Merken and G. R. Beecher, *J. Agric. Food Chem.*, 2000, **48**, 577.
7. D. J. Daigle and E. J. Conkerton, *J. Liq. Chromatogr.*, 1988, **11**, 309.
8. Y. Nogata, H. Ohta, K. I. Yoza, M. Berhow and S. Hasegawa, *J. Chromatogr. A*, 1994, **667**, 59.
9. K. Robards and M. Antolovitch, *Analyst*, 1997, **112**, 11R.
10. X. G. He, *J. Chromatogr. A*, 2000, **880**, 203.
11. K. Hostettmann and M. Hostettmann, in *The Flavonoids: Advances in Research*, eds J. B. Harborne and T. J. Mabry, Chapman and Hall, New York, 1982, p. 1.
12. R. Galensa and K. Herrmann, *J. Chromatogr.*, 1980, **189**, 217.
13. J. F. Hammerstone, S. L. Lazarus, A. E. Mitchell, R. Rucker and H. H. Smith, *J. Agric. Food Chem.*, 1999, **47**, 490.
14. J. Rigaud, M. T. Escribano-Bailon, C. Prieur, J. M. Souquet and V. Cheynier, *J. Chromatogr. A*, 1993, **654**, 255.
15. A. L. Waterhouse, S. Ignelzi and J. R. Shirley, *Am. J. Enol. Vitic.*, 2000, **51**, 383.
16. J. B. Harborne, *Comparative Biochemistry of Flavonoids*, Academic Press, London, 1967.
17. T. J. Mabry, K. R. Markham and M. B. Thomas, *The Systematic Identification of Flavonoids*, Springer, New York, 1970.
18. K. R. Markham, *Techniques of Flavonoid Identification*, Academic Press, London, 1982.
19. M. Williams, G. Hrazdina, M. M. Wilkinson, J. G. Sweeny and G. A. Iacobucci, *J. Chromatogr.*, 1978, **155**, 389.
20. L. W. Wulf and C. W. Nagel, *Am. J. Enol. Vitic.*, 1978, **29**, 42.
21. F. Lopes-Da-Silva, S. de Pascual-Teresa, J. C. Rivas-Gonzalo and C. Santos-Buelga. *Eur. Food Res. Technol.*, 2002, **214**, 248.
22. O. M. Andersen, *Phytochemistry*, 1987, **26**, 1220.
23. J. L. Cabrita and O. M. Andersen, *Phytochemistry*, 1999, **52**, 1693.
24. P. Dugo, L. Mondello, G. Errante, G. Zapia and G. Dugo, *J. Agric. Food Chem.*, 2001, **49**, 3987.
25. J. P. Goiffon, P. P. Mouly and E. M. Gaydou, *Anal. Chim. Acta*, 1999, **382**, 39.
26. E. Hebrero, C. Santos-Buelga and J. C. Rivas-Gonzalo, *Am. J. Enol. Vitic.*, 1988, **39**, 227.

27. E. Hebrero, C. Garcia-Rodriguez, C. Santos-Buelga and J. C. Rivas-Gonzalo, *Am. J. Enol. Vitic.,* 1989, **40,** 283.
28. J. B. Harborne, *Biochem. J.*, 1958, **70**, 22.
29. F. C. Stintzing, A. S. Stintzing, R. Carle and R. E. Wrolstad, *J. Agric. Food Chem.*, 2002, **50**, 396.
30. O. M. Andersen and T. Fossen, *Phytochemistry*, 1995, **40**, 1809.
31. D. Strack and V. Way, in *The Flavonoids: Advances in Research since 1986*, ed. J. B. Harborne, Chapman and Hall, New York, 1994, p. 1.
32. B. Devia, G. Llabres, J. Wouters, L. Dupont, M. T. Escribano-Bailon, S. de Pascual-Teresa, L. Angenot and M. Tits, *Phytochem. Anal.*, 2002, **13**, 114.
33. M. Giusti, L. E. Rodriguez-Saona and R. E. Wrolstad, *J. Agric. Food Chem.*, 1998, **47**, 4631.
34. T. Fossen, and O. M. Andersen, *Phytochemistry*, 1999, **50**, 1185.
35. E. Haslam, *Practical Polyphenolics. From Structure to Molecular Recognition and Physiological Action*, Cambridge University Press, Cambridge, 1998.
36. L. Foo and L. Porter, *J. Sci. Food Agric.,* 1981, **32,** 711.
37. J. A. Kennedy and G. P. Jones, *J. Agric. Food Chem.*, 2001, **49**, 1740.
38. Z. Czochanska, L. Y. Foo, R. H. Newman, L. J. Porter, W. A. Thomas and W. T. Jones, *J. Chem. Soc. Chem. Commun.*, 1979, 375.
39. S. Matthews, I. Mila, A. Scalbert, B. Pollet, C. Lapierre, C. Herve du Penhoat, C. Rolando and D. M. X. Donnelly, *J. Agric. Food Chem.*, 1997, **45**, 1195.
40. C. Prieur, J. Rigaud, V. Cheynier and M. Moutounet, *Phytochemistry*, 1994, **36,** 781
41. A. G. H. Lea, P. Bridle, C. F. Timberlake and V. L. Singleton, *Am. J. Enol. Vitic.*, 1979, **30,** 289.
42. A. G. H. Lea, *J. Chromatogr.*, 1980, **194**, 62.
43. G. E. Adamson, S. L. Lazarus, A. E. Mitchell, R.L. Prior, G. Cao, P. H. Jacobs, B. G. Kremers, J. F. Hammerstone, R. Rucker, K. A. Ritter and H. H. Smith, *J. Agric. Food Chem.*, 1999, **47**, 4184
44. E. L. Wilson, *J. Sci. Food Agric.*, 1981, **32**, 257.
45. M. Boukharta, M. Girardin and M. Metche, *J. Chromatogr.*, 1988, **455**, 406.
46. E. Revilla, M. Bourzeix and E. Alonso, *Chromatographia*, 1991, **31**, 465.
47. B. S. Sun, T. Pinto, M. C. Leandro, J. M. Ricardo-da-Silva and M. I. Spranger, *Am. J. Enol. Vitic.*, 1999, **50**, 179.
48. M. T. Escribano-Bailon, Y. Gutierrez-Fernandez, J. C. Rivas-Gonzalo and C. Santos-Buelga, *J. Agric. Food Chem.*, 1992, **40**, 1794.
49. S. de Pascual-Teresa, D. Treutter, J. C. Rivas-Gonzalo and C. Santos-Buelga, *J. Agric. Food Chem.*, 1998, **46**, 4209.
50. D. Treutter, *J. Chromatogr.*, 1989, **467**, 185.
51. D. Treutter, W. Feucht and C. Santos-Buelga, *Acta Horticulturae*, 1994, **381**, 789.
52. D. Treutter, C. Santos-Buelga, M. Gutmann and H. Kolodziej, *J. Chromatogr. A*, 1994, **667**, 290.
53. S. de Pascual-Teresa, C. Santos-Buelga and J. C. Rivas-Gonzalo, *J. Agric. Food Chem.*, 2000, **48**, 5331

54. S. de Pascual-Teresa, J. C. Rivas-Gonzalo and C. Santos-Buelga, *Int. J. Food Sci. Technol.*, 2000, **35**, 33.
55. C. Santos-Buelga and D. Treutter, *Polyphénols actualités,* 1995, **12**, 17.
56. F. A. Tomás-Barberán, F. Tomás, L. Hernández and F. Ferreres, *J. Chromatogr.*, 1985, **347**, 443.
57. F. Ferreres, F. Tomás-Lorente and F. A. Tomás-Barberán, in *Studies in Natural Products Chemistry*, Volume 5, *Structure elucidation (Part B)*, ed. Att-ur-Rhaman, Elsevier, Amsterdam, 1989, p. 621.
58. R. J. Grayer, N. C. Veitch, G. C. Kite, A. M. Price and T. Kokubun, *Phytochemistry*, 2001, **56**, 559.
59. D. J. Guedon and B. P. Pasquier, *J. Agric. Food Chem.*, 1994, **42**, 679.
60. D. Ferreira, B. I. Kamara, E. V. Brandt and E. Joubert, *J. Agric. Food Chem.*, 1998, **46**, 3406.
61. F. A. Tomás-Barberán and M. N. Clifford, *J. Sci. Food Agric.*, 2000, **80**, 1073.
62. Y. Nogata, H. Ohta, K. I. Yoza, M. Berhow and S. Hasegawa, *J. Chromatogr.*, 1994, **667**, 59.
63. P. Mouly, E. M. Gaydou and A. Auffray, *J. Chromatogr.*, 1998, **800**, 171.
64. L. W. Wulf and C. W. Nagel, *J. Chromatogr.*, 1976, **116**, 271.
65. R. J. Grayer, in *Methods in Plant Biochemistry*, Volume 1, eds P. M. Dey and J. B. Harborne, Academic Press, London, 1989, p. 283.
66. C. A. Williams and J. B. Harborne, in *Methods in Plant Biochemistry*, Volume 1, eds P. M. Dey and J. B. Harborne, Academic Press, London, 1989, p. 357.
67. H. J. Wang and P. A. Murphy, *J. Agric. Food Chem.*, 1994, **42**, 1666.
68. H. J. Wang and P. A. Murphy, *J. Agric. Food Chem.*, 1994, **42**, 1674.
69. A. A. Franke, L. J. Custer, C. M. Cerna and K. K., Narala, *J. Agric. Food Chem.*, 1994, **42**, 1905.
70. X. G. He, L. Z. Lin and L. Z. Lian, *J. Chromatogr.*, 1996, **755**, 127.
71. S. A. Bingham, C. Atkinson, J. Liggins, L. Bluck and A. Coward, *Br. J. Nutr.*, 1998, **79**, 393.
72. K. van de Casteele, H. Geiger and C. F. Van Sumere, *J. Chromatogr.*, 1982, **240**, 81.
73. H. Gagnon, S. Tahara, E. Bleichert and R. K. Ibrahim, *J. Chromatogr.*, 1992, **606**, 255.
74. A. P. Griffith and M. W. Collison, *J. Chromatogr.*, 2001, **913**, 397.
75. M. Krause and R. Galensa, *Z. Lebensm. Unters. Forsch.*, 1992, **94**, 29.
76. N. W. Preston and C. F. Timberlake, *J. Chromatogr.*, 1981, **214**, 222.
77. V. Cheynier and J. Rigaud, *Am. J. Enol. Vitic.*, 1986, **37**, 248.
78. K. van de Casteele, H. Geiger and C. F. van Sumere, *J. Chromatogr.*, 1982, **240**, 81.
79. K. R. Markham and T. J. Mabry, in *The Flavonoids*, ed. J. B. Harborne, T. J. Mabry and H. Mabry, Chapman and Hall, New York, 1975, p. 45.
80. B. A. Bohm, in *Methods in Plant Biochemistry*, Volume 1, ed. P. M. Dey and J. B. Harborne, Academic Press, London, 1989, p. 237.
81. B. A. Bohm, in *The Flavonoids: Advances in Research Since 1986,* ed. J. B. Harborne, Chapman and Hall, New York, 1994.

82. T. A. Geisman and J. B. Harborne, *J. Am. Chem. Soc.*, 1956, **78**, 832.
83. F. A. Tomás-Barberán, C. García-Viguera, J. L. Nieto, F. Ferreres and F. Tomás-Lorente, *Food Chem.*, 1993, **46**, 33.
84. B. Suarez-Vallés, J. Santamaria-Victorero, J. J. Mangas-Alonso and D. Blanco-Gomis, *J. Agric. Food Chem.*, 1994, **42**, 2732.

CHAPTER 6

Application of LC–NMR in the Structure Elucidation of Polyphenols

JEAN-LUC WOLFENDER, KARINE NDJOKO AND KURT HOSTETTMANN

1 Introduction

The coupling of HPLC with nuclear magnetic resonance spectroscopy (LC–NMR) is one of the most powerful methods for the separation and structural elucidation of unknown compounds in mixtures.[1] The recent progress in pulse field gradients and solvent suppression, the improvement in probe technology and the construction of high-field magnets have given a new impulse to this technique which has emerged since the mid-1990s as a very efficient method for the on-line identification of organic molecules. LC–NMR thus represents a potentially interesting complementary technique to LC–UV–MS in phytochemical analysis for detailed on-line structural analysis of natural products. Recent applications have demonstrated the usefulness of this technique.[2–7]

For the analysis of polyphenols, and especially of flavonoids, xanthones or related compounds, LC–NMR affords useful information on the nature of the polyphenol skeletons and on their substitution patterns. As will be discussed below, the information provided by LC–NMR consists mainly of simple ^{1}H-NMR spectra or ^{1}H–^{1}H correlation experiments. Access to ^{13}C-NMR information is possible but is restricted only to a very limited number of cases where the concentration of the LC–peak of interest is very important and ^{13}C-NMR data can be deduced indirectly from inverse detection experiments.[8] Indeed, due to the low natural abundance of the ^{13}C isotope (1.1%) the sensitivity for direct measurement in the LC–NMR mode is not sufficient. Approaches involving transfer of spin polarisation from immobilised nitroxide radicals, known as 'dynamic nuclear polarisation' have been mentioned allowing the LC–^{13}C-NMR detection of a mixture of halogenated hydrocarbons, but at present this technique

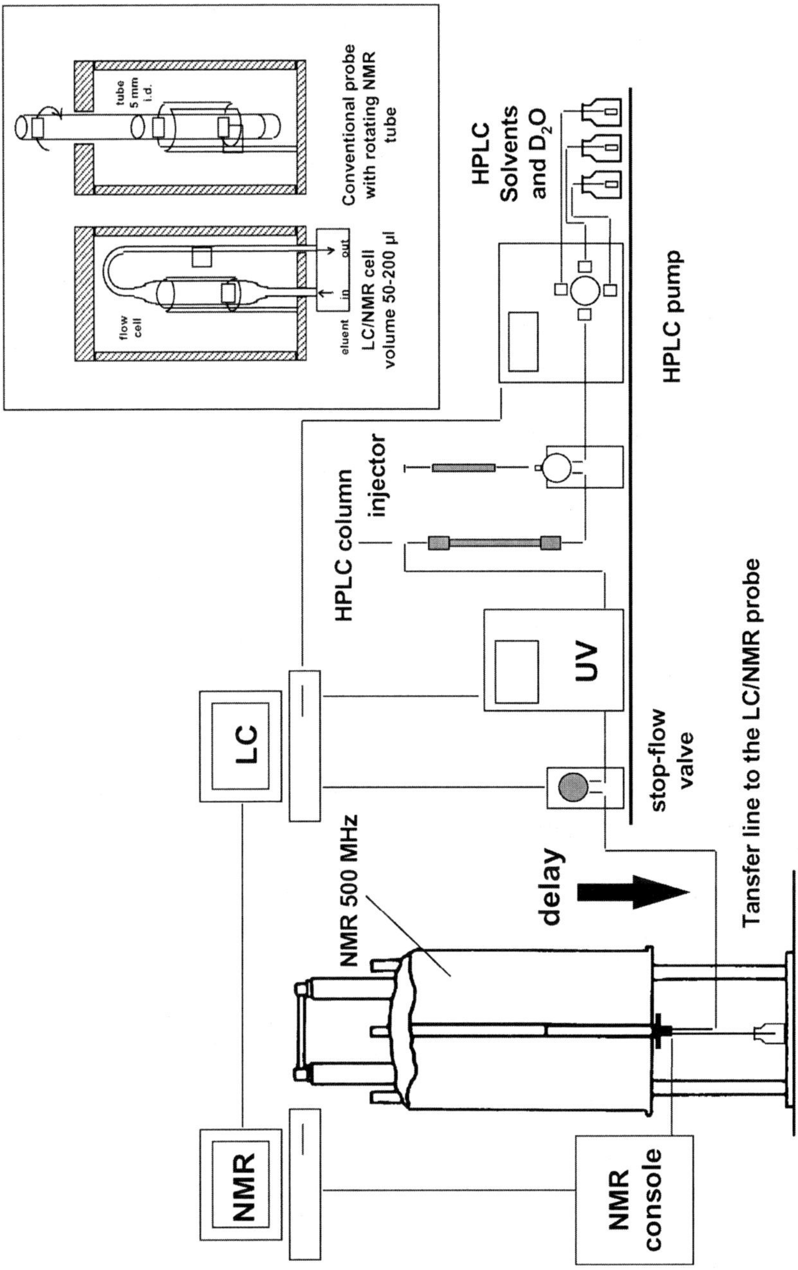

Figure 1 *Typical setup used for on-flow and stop-flow LC–NMR experiments. The control of the stop-flow valve is achieved by the computer of the HPLC–UV system which triggers both the valve of the HPLC pump and the NMR acquisition computer. In the insert, a comparison between a conventional NMR probe and a LC–NMR flow cell is given*

has not been applied to our knowledge to LC–^{13}C-NMR analysis of natural products.[9]

Because of this lack of ^{13}C-NMR information, as will be shown in the different examples treated, on-line identification of polyphenols can only very rarely rely on LC–^{1}H-NMR data alone but usually is based on the interpretation of a combined set of spectroscopic data recorded on-line by complementary LC–UV–DAD and LC–MS–MS analyses. Several recent studies have emphasised the need of multiple hyphenation for the on-line identification of organic molecules.[10]

In polyphenol chemistry, the combination of UV, MS and NMR spectroscopic data of pure constituents has often permitted their unambiguous structure determination. Other techniques such as IR or X-ray crystallography have been used less often and mainly when the other spectroscopic methods failed to give a complete structure assignment. Thus by analogy, the combination of the LC-hyphenated techniques LC–UV, LC–MS and LC–NMR should ideally enable the complete structure characterisation of any plant metabolites directly in an extract, provided that its corresponding LC peak is clearly resolved. In practice, however, such a statement is not fully true and many factors may hinder on-line detection and structure determination of a plant metabolite. Often only partial structure information will be obtained, but these on-line data will already provide very precious information for targeting the isolation of new compounds or for the dereplication of known constituents.[11]

Several examples demonstrating the type of information that can practically be obtained by LC–NMR will be discussed in this paper. In the following section, more emphasis is put on the basic principle of LC–NMR, its different modes of operation and its possibilities and limitations.

Figure 1 shows a typical experimental set-up and Figure 2 shows the types of information that can be obtained.

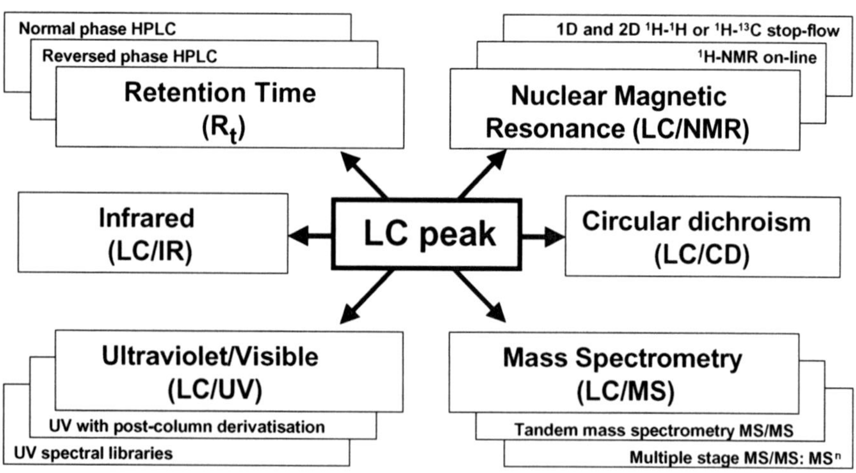

Figure 2 *Type of information that can be obtained on a given LC peak with the different LC hyphenated techniques available*

2 Principle of Operation of LC–NMR

Solvent Suppression

The main problem of LC–NMR is the difficulty in observing analyte resonances in the presence of the much larger resonances of the mobile phase.[12] This problem has even worsened in the case of typical LC reversed-phase operating conditions, where more than one protonated solvent was used and where the resonances changed frequencies during the analysis in gradient mode. Furthermore, the continuous flow of sample in the detector coil complicated solvent suppression. These problems have now been overcome thanks to the development of fast, reliable and powerful solvent suppression techniques such as WET (Water suppression Enhanced through T_1 effects).[13] These techniques consist of a combination of pulsed field gradients, shaped rf pulses, shifted laminar pulses and selective ^{13}C decoupling and are much faster than classical presaturation techniques previously used in this field.[14] Thus, in typical reversed HPLC conditions, non-deuterated solvents such as MeOH or MeCN can be used. Water is usually replaced by D_2O, which is a relatively 'cheap' deuterated solvent, and gives better quality spectra. The main drawback of this procedure is that analyte signals localised under the solvent resonances will also be suppressed.

LC–NMR is not a sensitive LC detection technique and often requires the accumulation of many transients for achieving a satisfactory detection of the compounds of interest. Consequently LC–NMR can be operated in two different modes, the on-flow and stop-flow modes.

Operation in the On-flow Mode

In the on-flow mode, the LC–NMR spectra are acquired continuously during the separation and are stored as a set of scans in discrete increments. The on-flow data are processed as a 2D NMR experiment. One dimension of the contour plot represents the NMR ppm scale and the other the time scale (Figure 3). According to the analysis, a compromise between the number of scans per increment and the LC resolution has to be made. The main drawback of on-flow LC–NMR is its inherent relatively low sensitivity. In our experience, a rough estimation of the detection limit in the on-flow mode with a 60 μl cell on a 500 MHz instrument was realised for the secoiridoid glycoside swertiamarin (MW 374). An injection of 20 μg of this compound on-column was necessary in order to obtain a signal/noise (S/N) ratio of 3 for the olefinic proton H-3 (16 scans/increment 1 ml/min).[2] These limits can be reduced, however, if proper digital treatment of signals is achieved with state-of-the-art techniques available for modern NMR instruments. Thus, practically on-flow LC–NMR measurements will be mainly restricted to the direct measurement of the main constituents of a crude extract and this often under overloaded LC-conditions. Typically 1–5 mg of crude plant extract will have to be injected on-column. The detection limits can be improved by performing the analysis at low-flow[15,16] or by running time slice experiments over a whole chromatogram,[17] where the flow is stopped at defined time intervals. Both

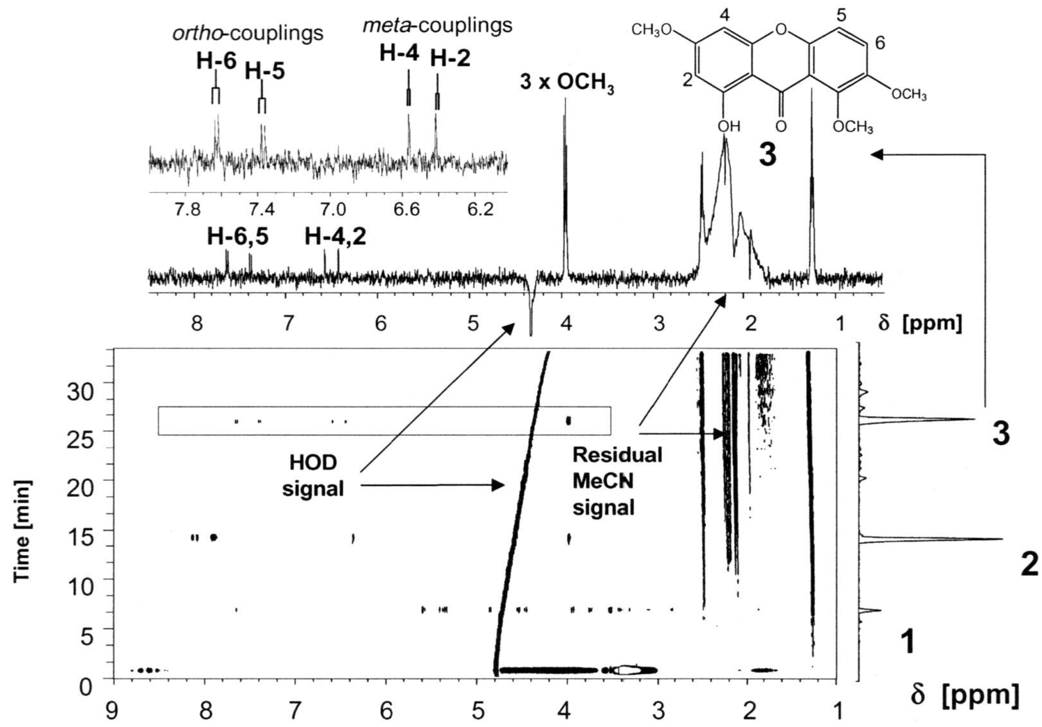

Figure 3 *Bidimensional LC–^{1}H-NMR chromatogram of the crude CH_2Cl_2 extract of* Swertia calycina *with the spectrum obtained for the decussatin* **3**. *Methoxyl groups and aromatic proton signals of* **2** *and* **3** *are clearly visible together with all the resonances of the monoterpene glycoside* **1**. *The signal of HOD is negative (see spectrum of* **3**) *and was continually skifted during the LC gradient. LC–UV chromatogram at 254 nm (vertical display). HPLC conditions: Inj. 1 mg; Col. Nova-Pak C_{18} (150 × 3.9 mm i.d., 4 μm); MeCN-D_2O gradient (5:95 to 65:35 in 50 min); 0.05% TFA; 1 ml min^{-1}. LC–NMR conditions: 24 scans/increment, flow cell (60 μl, 3 mm i.d.), 500 MHz*

modes of operation enable a higher number of transients per increments to be recorded and thus a significantly better S/N ratio is obtained. Another way to improve sensitivity is to perform on-line trace enrichment. Such a procedure has been proposed, for example, for the analysis of the hop bitter principles, where one-line solid-phase extraction prior to LC–NMR analysis was achieved.[18] In spite of its low sensitivity LC–NMR has not only been used for the study of small molecules. It has been demonstrated that saponins with molecular weights ranging between 890 and 930 Da could be successfully analysed by this means.[19]

A simple example of the type of data that can be obtained in the on-flow mode is shown for the on-line identification of the constituents of the CH_2Cl_2 extract of *Swertia calycina* (Gentianaceae). This extract had a relatively simple composition and permitted a loading of 1 mg on-column for the on-flow run. The separation was achieved at 1 ml min^{-1} with a MeCN–D_2O gradient (5:95–65:35 in 50 min). The LC–NMR detection was achieved by applying 64 transients per increment.[20] As shown in Figure 3, on the LC–NMR contour plot, ^{1}H-NMR spectra were successfully recorded for all the three major constituents (**1**, **2** and **3**). The important trace starting from 4.8 ppm (at 0 min) and ending to 4 ppm (at 30 min) was due to the change of the chemical shift of the residual negative water (HOD) signal during the LC gradient. The traces between 1 and 2.6 ppm were due to residual MeCN signals and solvent impurities.

In this extract, the only polyphenolic constituent was compound **1**, a xanthone according to its UV spectrum. The on-flow spectra recorded for **1** (Figure 3) corresponded to a slicing of this bidimensional plot at the time corresponding to the elution of this compound. The resonances recorded for xanthone **1** – three methoxy groups at 3.92, 3.93 and 3.95 ppm, one pair of *meta* (1H, δ 6.55, *d*, $J = 2.4$, H-4; 1H, δ 6.41, *d*, $J = 2.4$, H-2) and one pair of *ortho* (1H, δ 7.36, *d*, $J = 9.2$, H-5; 1H, δ 7.62, *d*, $J = 9.2$, H-6) coupled aromatic protons – showed that **1** was decussatin as already suggested by the LC–UV–MS data. The two other constituents were identified as 2-methoxy-1,4-napththoquinone (**2**), the first naphthoquinone to be reported in a Gentianaceae species, and sweroside (**3**) a widespread bitter principle.[20]

Operation in the Stop-flow Mode

Another way to improve the detection limit in LC–NMR is to work in the stop-flow mode. Operation in stop-flow requires that the retention times of the analytes of interest are known or that a sensitive method of detection such as LC–UV or LC–MS is used prior to LC–NMR to trigger the detection.[21,22] In practice, one of these detectors is connected on-line before the NMR instrument and the signal of the analyte of interest passing through this detector is used to trigger a valve which stops the LC flow exactly when the peak reaches the NMR cell after a calibrated delay. The stop-flow mode gives the possibility to acquire an important number of transients on a given LC-peak, and satisfactory LC–^{1}H-NMR spectra of compounds present in the low μg range can be obtained. In this mode, various 2D correlation experiments such as COSY, NOESY, HSQC and HMBC are also possible provided that the concentration of the metabolite is high enough (generally more than 100 μg of sample is required).

Sensitivity

The sensitivity of the stop-flow mode is much better than that of the on-flow mode. By combining the use of new techniques of NMR data acquisition, such as digital filtering and oversampling, with microbore HPLC, it appears that detection limits for structural characterisation are in the region of 5 ng on 500 MHz or 600 MHz instruments in the stop-flow mode. Absolute $^{1}H-^{13}C$ detection limits of a 100 μl commercially available LC–NMR probe was estimated for 3′-deoxy-3′-azidothymidine (AZT) (MW 252) on a 500 MHz magnet in the stop-flow mode. The S/N ratio of the H-1 proton of the ribose moiety of this molecule was used to define the detection limits. A S/N ratio of 3.0 for this signal was obtained for 1 μg of AZT after 64 scans (5 min) and for 85 ng after an overnight experiment. With 21 μg of AZT, a $^{1}H-^{1}H$ 2D TOCSY was acquired in 2.6 h and all the cross peaks of $^{1}H-^{13}C$ HMQC could be measured overnight.[23] Recently, it was demonstrated that it was possible with a new capillary microcoil probe to detect amounts of α-pinene as low as 37 ng in the on-flow mode and that detection of an impurity in the sub-nmol range is possible in the stop-flow mode.[24]

Limitations

If LC–NMR is now practically applicable for the direct observation of metabolite resonances in LC reversed-phase systems, this does not mean that all the information obtained by conventional measurement in standard deuterated NMR solvents will be obtained. One problem linked to the use of solvent suppression is that the signals of the analytes of interest, which reside under the solvent peak, will be suppressed together with the solvent signal. This can be a major drawback when dealing with unknown constituents. In order to detect all analyte signals, one alternative is to carry out the solvent suppression in two independent solvent systems such as MeCN–D_2O and MeOH–D_2O.[25] Another problem is that the chemical shifts recorded in a reversed-phase solvent will slightly differ from those reported in standard deuterated NMR solvents. This can be a drawback if precise comparisons with literature data have to be performed.[26]

In spite of these practical problems, the progress realised in LC–NMR during recent years is impressive and this technique provides invaluable on-line information in crude plant extract analysis.

Hardware Set-up

A modern LC–NMR system consists of a high resolution NMR instrument (400–800 MHz) with a dedicated LC–NMR flow probe connected to an HPLC system equipped with a valve or a loop collector for stop-flow experiments (see Figure 1).[6] The HPLC control unit is connected with the data acquisition system of the

NMR for the synchronisation of the different operations. A sensitive detector such as UV is usually coupled up-stream to trigger the stop-flow measurements. The HPLC system is located at a distance of *ca.* 3 m from the magnet. A long capillary of polyetherketone (PEEK) with the smallest possible internal diameter is used for connection to the LC–NMR probe. The flow cell (40–200 μl) consists of a non-rotating glass tube surrounded by the RF coil and connected at both ends with HPLC tubing.[14] The homogeneity is good in the small cell volumes of the flow-probes, notably because modern computer methods for optimising the field homogeneity reduce the requirement for spinning.[27]

3 Generalities on the ^{1}H-NMR of Polyphenols

The ultimate spectroscopic procedure in the structure elucidation of polyphenols is NMR and many studies have permitted the elaboration of rules of interpretation of both ^{1}H- and ^{13}C-NMR spectra of polyphenols.[28,29] In the LC–NMR mode however, the information obtained is mainly limited to ^{1}H-NMR.

^{1}H-NMR data provide information on the frequencies at which protons resonate. In the case of polyphenols, this range is limited to δ 1–10 ppm since in typical LC–NMR conditions all phenolic protons that might resonate at lower field (7–18 ppm) are exchanged due to the presence of D_2O in the LC solvent system. However, a great deal can be told about a proton simply from the δ value at which it resonates.[30] Much interesting information on the proton's surroundings can be deduced from the chemical shifts recorded, such as the presence of methyl, methoxy or methylene groups as well as various types of aromatic protons. Calculation of δ values from empirical values have, however, to be treated with caution since, for example, for flavonoids, they have to take into account steric effects, which may, for example, prevent coplanarity of the chromone and phenyl rings. Besides chemical shifts, ^{1}H-NMR coupling patterns, especially those of the aromatic protons of the polyphenolic nucleus, *meta* ($J = 1$–3 Hz) and *ortho* ($J = 7$–9 Hz) can easily be observed in LC–NMR. *Para* coupling ($J < 1$ Hz) is usually not resolved, but causes broadening of the signal only. This type of information is complementary to mass spectrometric information where A- or B-ring substitution can be deduced from the fragmentation pattern, but where A- or B-ring hydroxylation patterns cannot be assessed precisely. For example, for a compound like apigenin where only one hydroxy is located on the B-ring, ^{1}H-NMR coupling will immediately solve the B-ring substitution determination since each of the three possibilities of localisation of the OH group will give a unique series of splitting pattern. On isolated olefinic bonds, relationships between *cis* or *trans* coupled protons are readily distinguished by coupling constants.

In the case of glycosides, the anomeric proton (H-1) alone can provide much useful information regarding the saccharide moiety. For example, the H-1–H-2 coupling constant can at once indicate which signal relates to which sugar in a polyglycoside polyphenol, and frequently is also indicative of the α or β linkage of the glycosidic bond. For example β-linked glucopyranosides which exhibit H-1–H-3 coupling constants of 7–8 Hz are readily distinguishable from the α-linked glucopyranoside with 3–4 Hz couplings. The same is true for many other glycosides.[29] Chemical shift data related to the glycosidic H-1 signal can also be of diagnostic value. For example, the H-1 signal of a sugar attached to another sugar can usually be distinguished from that of a sugar directly attached to the polyphenol aglycone in that it resonates upfield relative to the latter. Exceptions to this are known in the case where the primary glycoside is a *C*-glycoside (higher field H-1 signals compared with *O*-glycosides) or when the terminal sugar is apiofuranoside (relatively low field H-1 signals).

In the case of flavonoids, rules based on the chemical shift of the H-1 signal allow a good prediction of their attachment position on the aglycone moiety.[29]

Complementary bidimensional experiments such as $^{1}H-^{1}H$ correlation experiments (COSY) may clarify the assignment of the coupled aromatic protons or the sugar protons in the case of glycosides. Such experiments can relatively easily be performed in the stop-flow LC–NMR mode.

Through-space interactions can be observed by study of the nuclear Overhauser effect (NOE). This type of experiment is more demanding in sample amount but can be performed either as selective 1D-LC–NMR experiments where the proton of interest is selectively irradiated while solvent suppression is maintained or as 2D correlation experiments such as NOESY [8] or ROESY.[31]

4 Strategies for the On-line Identification of Polyphenols

LC–NMR plays a central role in the on-line identification of constituents of crude plant extracts. This technique alone, however, will not provide sufficient spectroscopic information for a complete identification of natural products, and other hyphenated methods such as LC–UV–DAD and LC–MS–MS or even LC–IR[32] and LC–CD[4] are needed for providing complementary information (see Figure 2). The importance of the integration of all these spectroscopic information will be underlined in the examples presented.

Ideally, the integration of all these hyphenated techniques in a single set-up, with centralised acquisition of the spectroscopic data, should permit the complete spectroscopic characterisation of the different metabolites in a single analysis. The practical creation of such a 'total analysis device' has been recently demonstrated in the case of on-line HPLC–UV(DAD)-FT-IR-NMR-MS analyses.[33] The problems associated with trying to combine these different spectroscopic techniques in a single set-up, however, hamper an efficient use of each of them separately since the most demanding technique, generally LC–NMR, compromises the others. Indeed practical problems encountered in simpler couplings such as NMR–MS are known.[21,34,35] The main drawback in using MS directly coupled to NMR is that all the masses recorded will be shifted if the natural products of interest possess protons exchangeable with deuterium and molecular assignment of unknown compounds will thus be seriously compromised. Deuteration may bring valuable additional information,[36] but D–H back exchange experiments have nevertheless to be performed for the recording of the genuine molecular weight. Another problem is that according to the type of ionisation needed, the buffer used in MS may interfere with LC–NMR detection by adding more ^1H signals that will have to be suppressed. The amount of sample needed for LC–NMR will also overload the LC–MS interface as well the LC–UV–DAD detector and the quality of detection will deteriorate. This type of problem might be solved in the near future by the integration of comprehensive dynamic splitters that may be used for reducing problems of overloading and performing efficient D–H back exchange.

Because of all these different problems, in our view, the LC–NMR analysis of the extract should be separated from the LC–UV–MS run. Furthermore, as will be discussed, LC–NMR experiments are time consuming and this type of measurement has to be performed on the LC peak of interest identified by pre-screening with LC–UV–MS.

Usually the separation is performed on a RP C_{18} column and broad acetonitrile or methanol gradients are applied, TFA is used for improving the resolution of polyphenols. UV spectra are recorded for all constituents with the aid of a photodiode array detector. Polyphenols can be efficiently localised by this means since they exhibit characteristic chromophores (see Chapter 4).

According to the type of polyphenol screened, various LC–MS interfaces can be used and the selection of correct ionisation conditions is crucial in order to obtain valuable molecular weight information. When fragment information is required, tandem MS–MS or multiple stage MS^n experiments are performed on-line.

Based on LC–UV and LC–MS information, a first dereplication step is performed. UV spectra are compared with a home-made library of spectra or with literature data. The polyphenols are also screened by checking whether their molecular weights match with those of known natural products. This type of search is performed with a library of natural products.[37] It usually generates a large number of hits. However, a cross search performed by adding botanical information (genus or family) or information about the type of constituents narrows down the possibilities. Furthermore, if a high-resolution mass spectrometer is used on-line, the molecular weight can be measured with a high accuracy (<5 ppm) and the molecular formula can be assessed directly, reducing even more the number of possibilities.[16] Widespread polyphenols are efficiently dereplicated by this means.

If additional information is required, LC–NMR is performed in a second step. The choice of the mode of operation of LC–NMR will vary according to the type of information requested and the concentration of the analyte. Several situations might be encountered:

- As a complement to the LC–UV–MS screening, several peaks of interest might need complementary information for dereplication purposes. In this case, stop-flow LC–NMR experiments can be performed on the LC-peak of interest and the corresponding LC–^1H-NMR spectra will be recorded. For known polyphenols, these data are often sufficient for their on-line identification. Such a dereplication study will be exemplified for the analysis of the xanthones and flavonoids from *Gentiana ottonis* (Gentianaceae).

- For the on-line identification of new polyphenols, complementary information to 1D LC–^{1}H-NMR is necessary. In this case, if the analyte is sufficiently concentrated in the LC–NMR flow cell (generally more than 100 μg of sample is required), various 2D stop-flow correlation experiments such as COSY, NOESY, HSQC and HMBC can be recorded. This was the case for the characterisation of a flavanone from *Monotes englerii* (Dipterocarpaceae).
- Finally if a full metabolite profiling of a given extract has to be performed, LC–NMR can be run in the on-flow mode. Generally in order to achieve the most sensitive detection of all constituents the analysis is performed in the low-flow mode if a complex extract has to be screened. As the sample amount injected on-column has to be consequently increased compared with typical analytical LC conditions, this gives the possibility to check the biological activity of the different peaks after HPLC microfractionation. This type of approach is exemplified for the analysis of isoflavones from isoflavanones from *Erythrina vogelii* (Leguminosae).

5 Stop-flow LC–^{1}H-NMR of the Flavonoids and Xanthones from *Gentiana ottonis*

In the course of our on-going search for new xanthones in members of the Gentianaceae family, the polyphenol constituents of *Gentiana ottonis*, a Gentianaceae species from Chile, were analysed. This study was motivated because xanthones are interesting for their potential antidepressive activity as specific inhibitors of monoamineoxidase, MAO.[38]

In this case, the root methanolic extract of *G. ottonis* was submitted to LC–UV–DAD and thermospray LC–MS (LC–TSP-MS) prior to LC–NMR analysis. The LC–UV analysis of this extract presented peaks with UV spectra characteristic of secoiridoids (**11**), flavones (**8** and **9**) and xanthones (**4–7** and **10**) (Figure 4). The LC–TSP-MS analysis of this extract allowed the attribution of molecular weights of all these compounds. Compounds **8–10** presented fragments characteristic for *C*-glycosides (losses of 90 and 120 Da).[39] According to their UV spectra, **9** and **8** (MW 448 and 446) were respectively tri- and tetra-oxygenated flavone *C*-glycosides.

In order to obtain complementary information for the characterisation of the different polyphenols found in this extract, LC–NMR was carried out with the same HPLC conditions used for LC–UV–MS. In this case, however, water was replaced by D_2O and the amount injected was increased to 0.4 mg which did not cause a noticeable degradation of the loss of LC resolution. In order to obtain satisfactory LC–^{1}H-NMR spectra, this extract was analysed in the stop-flow mode (Figure 5). LC–NMR spectra were recorded for all the main peaks and the number of transients for the recording of spectra with a good S/N varied between 128 and 2048 according to the concentration of the peak considered.

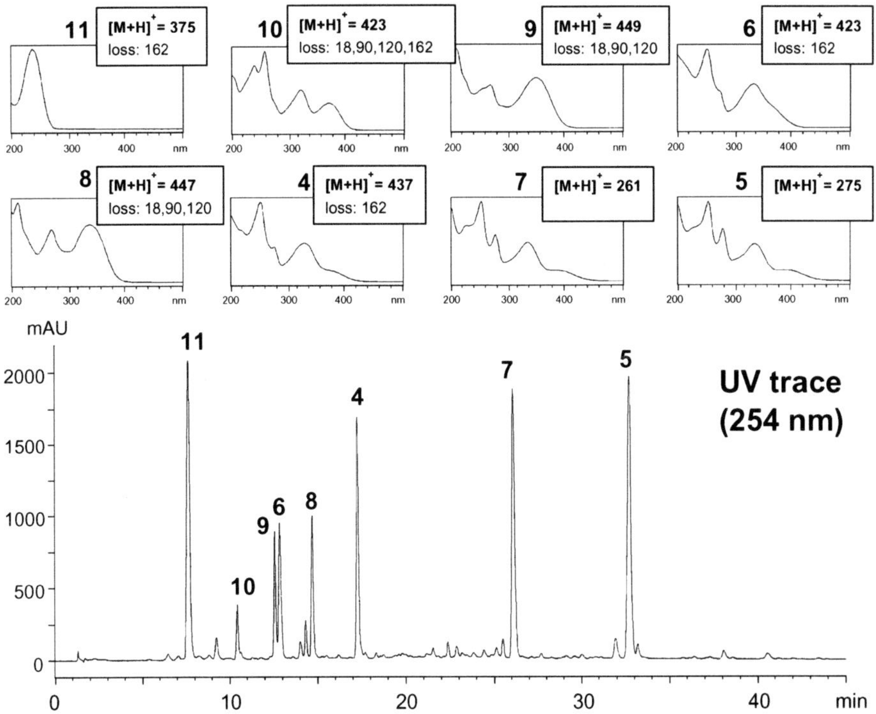

Figure 4 *LC–UV chromatogram of the root MeOH extract of* Gentiana ottonis *(Gentianaceae) with UV–DAD spectra and protonated molecules obtained for the main constituents by LC–TSP-MS. HPLC conditions: Inj. 0.4 mg; Col. Nova-Pak* C_{18} *(150 × 3.9 mm i.d., 4 μm); MeCN–*H_2O *gradient (5:95 to 65:35 in 50 min); 0.05% TFA; 1 ml* min^{-1}.

In Figure 5, the stop-flow LC–NMR spectra of two xanthones **4** and **5** are displayed. According to the LC–UV–MS data, it could be deduced that one of these xanthones (**4**) was an *O*-glycoside and the other its respective aglycone (**5**).

Xanthone *O*-glycoside **4** (MW 436) presented the characteristic proton anomeric signal of a glucosyl moiety (^{1}H, δ 5.20 *d*, *J* = 7.6, H-1′) as well as other signals characteristic for an *O*-glucosyl moiety (δ 3.3–5.1. H2′–H6′). On the aglycone moiety, one methoxy group (δ 3.96) was present. A pair of *meta* coupled aromatic protons (^{1}H, δ 6.70, *d*, *J* = 2.0, H-4 and ^{1}H, δ 6.47, *d*, *J* = 2.0, H-2) indicated a 1,3-disubstituted A-ring. The B-ring exhibited a pair of *ortho* coupled protons (^{1}H, δ 7.34, *d*, *J* = 8.8, H-6 and ^{1}H, 7.17, *d*, *J* = 8.8, H-7). This information, together with UV data, confirmed a 1,3,5,8-substitution pattern for **3**. The LC–^{1}H-NMR of **5**, the aglycone of **4**, also exhibited a pair of *meta* and *ortho*-coupled protons, as well as a methoxy group. However, with respect to **4**, one of the *ortho* protons was shifted upfield, indicating glycosylation on the B-ring at C-8. This finally allowed the identification of **5** and **4** respectively as 1,5,8-trihydroxy-3-methoxyxanthone (bellidifolin) and 1,5-dihydroxy-3-methoxy-8-*O*-glucosylxanthone (8-*O*-glucosylbellidifolin). Similar deductions allowed the

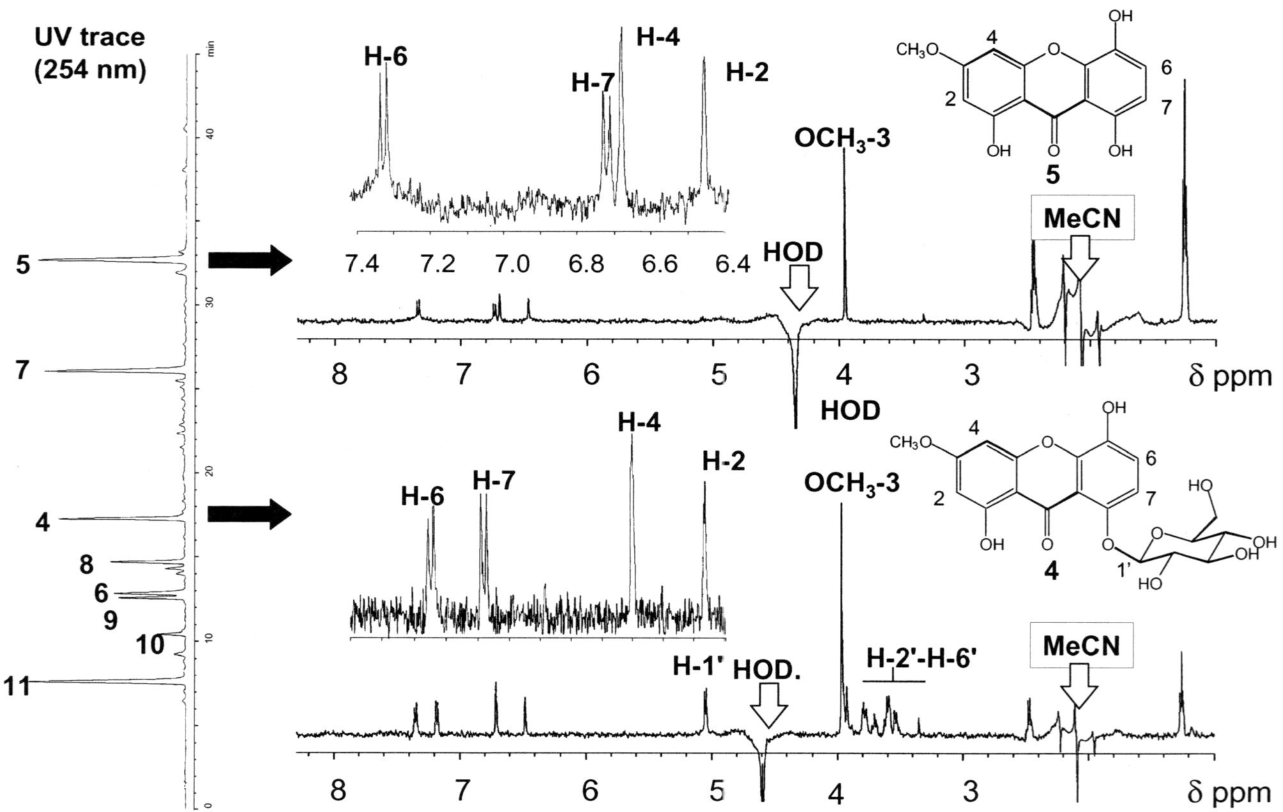

Figure 5 *LC–UV chromatogram of the root MeOH extract of* Gentiana ottonis *(Gentianaceae) (vertical display) with the stop-flow LC–*^{1}*H-NMR spectra of the xanthones aglycone and glycoside of* **4** *and* **5**. *On these LC–*^{1}*H-NMR spectra, the coupling constants and the integration of the aromatic signals were easily measured. HPLC conditions: same as in Figure 4 except that* H_2O *was replaced by* D_2O *LC–NMR conditions: 1024 transients, flow cell (60 µl, 3 mm i.d.), 500 MHz*

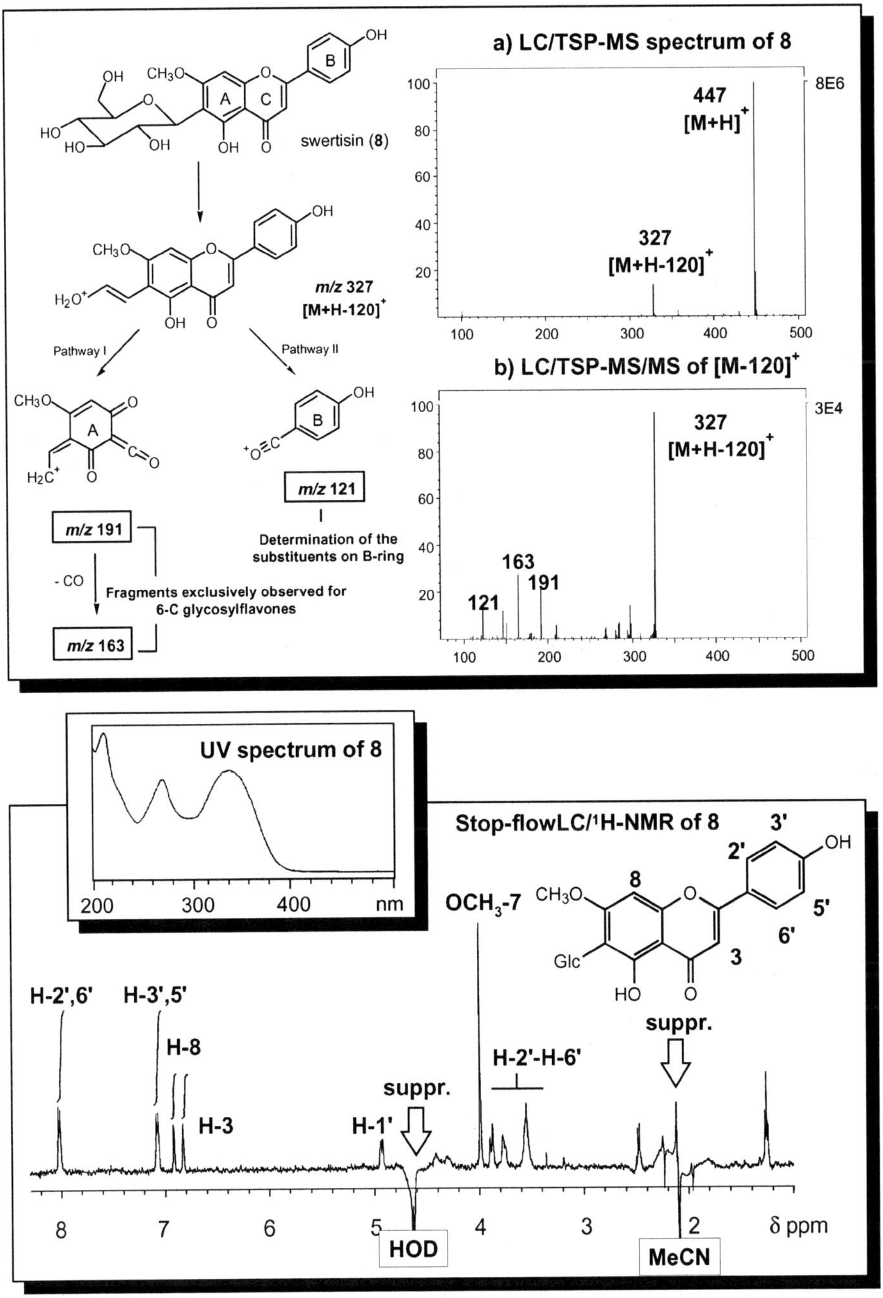

Figure 6 *Stop-flow LC–^{1}H-NMR spectrum of swertisin* (**12**)*, from the root MeOH extract of* Gentiana ottonis*, together with its LC–TSP-MS (a) and LC–TSP-MS–MS (b) spectra. The LC–MS–MS analysis was performed using the fragment [M + H – 120]+ (spectrum a) as parent ion. Characteristic daughter ions at m/z 121, 163 and 191 were observed, indicating the substitution on the A- and B-ring of the C-glycosylflavone* (**12**)

identification of **7** and **6** as 1,3,5,8-tetrahydroxyxanthone (desmethylbellidifolin) and 1,3,5-trihydroxy-8-*O*-glucosylxanthone (8-*O*-glucosyldesmethybellidifolin).

The stop-flow LC–^{1}H-NMR of the flavone *C*-glycoside **8** (MW 446) (Figure 6) gave signals for six aromatic protons (δ 6.8–8.1) and one methoxyl group (δ 4.0), together with those of the C-glycoside moiety (δ 3.5–5.0). A pair of symmetric *ortho* coupled protons (2H, δ 7.06, *d*, $J = 8.3$, H-3′,5′ and 2H, δ 8.00, *d*, $J = 8.3$, H-2′,6′) was characteristic for a B-ring substituted in C-4′. The singlet at δ 6.8 was attributable to H-3. The other singlet at δ 6.9 was due to a proton either at position C-6 or C-8 on the A-ring. LC–UV–MS and stop-flow LC–NMR data were not sufficient in this case to fully ascertain the structure of **8**.

In order to ascertain the position of *C*-glycosylation, a LC–MS–MS experiment was performed by choosing [M + H – 120]+ as parent ion which gave fragments at *m/z* 191 and 163, characteristic for 6-*C*-glycosylated flavones. Indeed it is known that TSP-MS–MS spectra of the [M + H – 120]+ fragments of isomeric *C*-glycosylflavones show different specific daughter ions.[39,40] Furthermore, the fragment observed at *m/z* 121 was indicative of a monohydroxylated B-ring, confirming the position of the methoxy group on the A-ring. These UV, MS, stop-flow LC–^{1}H-NMR and MS–MS data allowed the identification of **8** as the known 5,4′-dihydroxy-7-methoxy-6-*C*-glucosylflavone (swertisin). Similarly, the flavone *C*-glycoside **9** (MW 448) was identified as the common 5,7,3′,4′-tetrahydroxy-6-*C*-glucosylflavone (isoorientin). Based on its UV and MS spectra, compound **10** was readily identified as the widespread 1,3,6,7-tetrahydroxyxanthone-2-*C*-glucoside mangiferin.[20]

6 Stop-flow LC–^{1}H-NMR with 2D Correlation Experiments for the Structure Elucidation of a Flavanone of *Monotes englerii*

The combination of LC–UV, LC–MS and LC–^{1}H-NMR data permits in many cases the structure elucidation of various known compounds. These data, however, are not always sufficient for a full on-line identification of new constituents. If a complete structural identification has to be carried out, information on the ^{13}C signals is necessary and different 2D correlation experiments have to be performed. These experiments can be performed by LC–NMR in the stop-flow mode, provided that the sample concentration in the LC–NMR cell is high enough (typically hundreds of μg are needed). In this mode, ^{1}H–^{1}H correlations such as COSY or NOESY can be measured. Information on ^{13}C can be extracted from ^{1}H–^{13}C correlation experiments such as HSQC or HMBC.

The possibilities of LC–NMR in the stop-flow mode are exemplified by the structure determination of a flavanone from an enriched fraction of the CH_2Cl_2 extract of *Monotes engleri* (Dipterocarpaceae).[8] This plant showed antifungal activity against the yeast *Candida albicans* in a bioautographic TLC assay.[41] The activity was linked to a compound that gave a strong UV absorbing spot on TLC at 254 nm. On the LC–UV analysis at 254 nm, this corresponded to the main compound **12** of the extract. The UV spectrum of this compound was character-

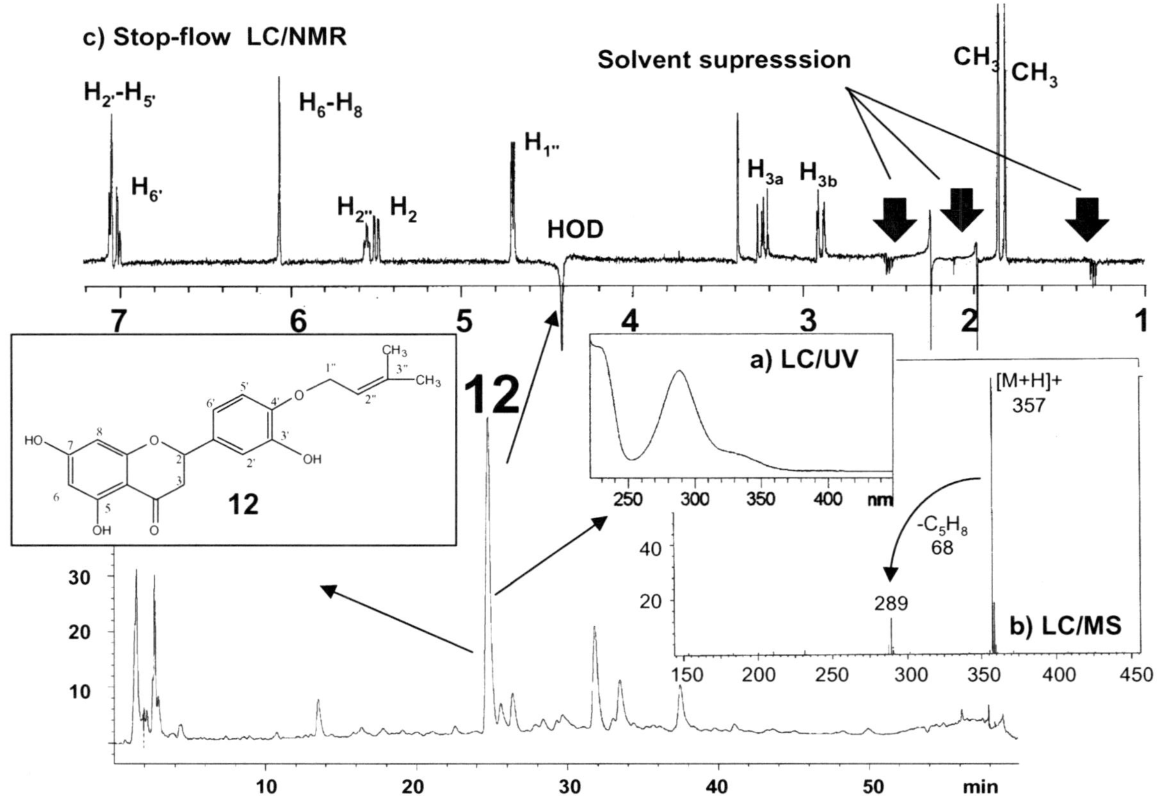

Figure 7 *LC–UV chromatogram (254 nm) of the crude CH_2Cl_2 extract of* Monotes engleri *(Dipterocarpaceae). Insert: (a) UV spectrum of the flavanone* **12***; (b) LC–TSP-MS spectrum of* **12***; (c) stop-flow LC–^{1}H-NMR of* **12**. *HPLC conditions: Inj. 1 mg; Col, Nova-Pak C_{18} (150 × 3.9 mm i.d., 4 μm); MeCN–D_2O gradient (20:80 to 70:30 in 50 min); 1 ml min^{-1}. LC–NMR conditions: 24 scans/increment, flow cell (60 μl, 3 mm i.d.), 500 MHz*

istic of a flavanone and its LC–TSP-MS spectrum showed a protonated molecule [M + H]+ at *m/z* 357 which exhibited a loss of 68 Da ($[M + H - C_4H_8]^+$), due to an *O*-prenyl substituent. Based on these data and according to the molecular weight of a flavanone skeleton, **12** was most probably a flavanone with one *O*-prenyl unit and three hydroxy substituents. This hypothesis was confirmed by the stop-flow LC–^{1}H-NMR spectra measured on this peak (Figure 7).

The presence of a flavanone skeleton with a prenyl unit could indeed be established: two methylene protons at δ 2.9 and δ 3.2 ppm suggested the presence of a γ-dihydropyrone ring. The deshielded doublet for the methylene of the prenyl unit at δ 4.7 ppm confirmed the *O*-connection of the prenyl unit, as deduced from the LC–MS data. These LC–^{1}H-NMR data were, however, not sufficient for a complete identification and the position of the prenyl unit could not be deduced. Stop-flow LC–2D-NMR correlation experiments were thus performed on an enriched fraction containing 1 mg of **12**: GDQCOSY (gradient double quantum filtered cosy), GHSQC (gradient heteronuclear single quantum coherence), GHMBC (gradient heteronuclear multiple quantum coherence), 1D TOCSY and WET-NOESY (2D nuclear Overhauser effect) NMR experiments were recorded in a total acquisition time of 9.6 hours. The LC–GDQCOSY confirmed the coupling of the different protons especially between H-2 and H-3 in the C-ring and H-1″ and H-2″ on the prenyl chain. The LC–GHSQC showed all C–H connectivities (Figure 8). In the LC–GHMBC, the long-range couplings of the γ-dihydropyrone ring were visible and the position of the prenyl unit on the B-cycle could be determined. The combination of the data obtained with these two inverse ^{1}H–^{13}C correlation experiments allowed assignment of all the ^{13}C resonances of **12** on-line, directly in the LC eluent. In order to determine the exact position of the

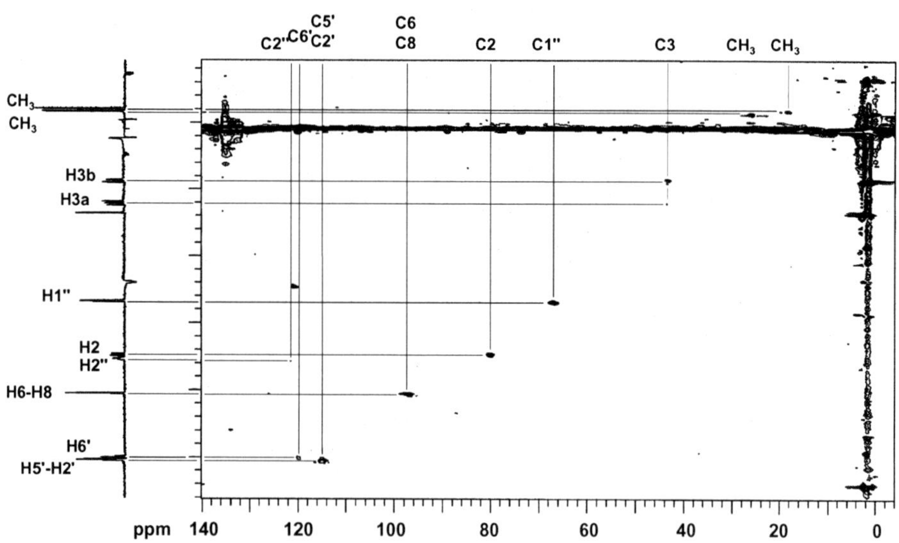

Figure 8 *Stop-flow LC–GHSQC spectra of the flavanone* **12**. *1 mg of the enriched fraction injected, solvent MeCN–D_2O. LC–GHSQC: transients 4, increments 128, time 30 min*

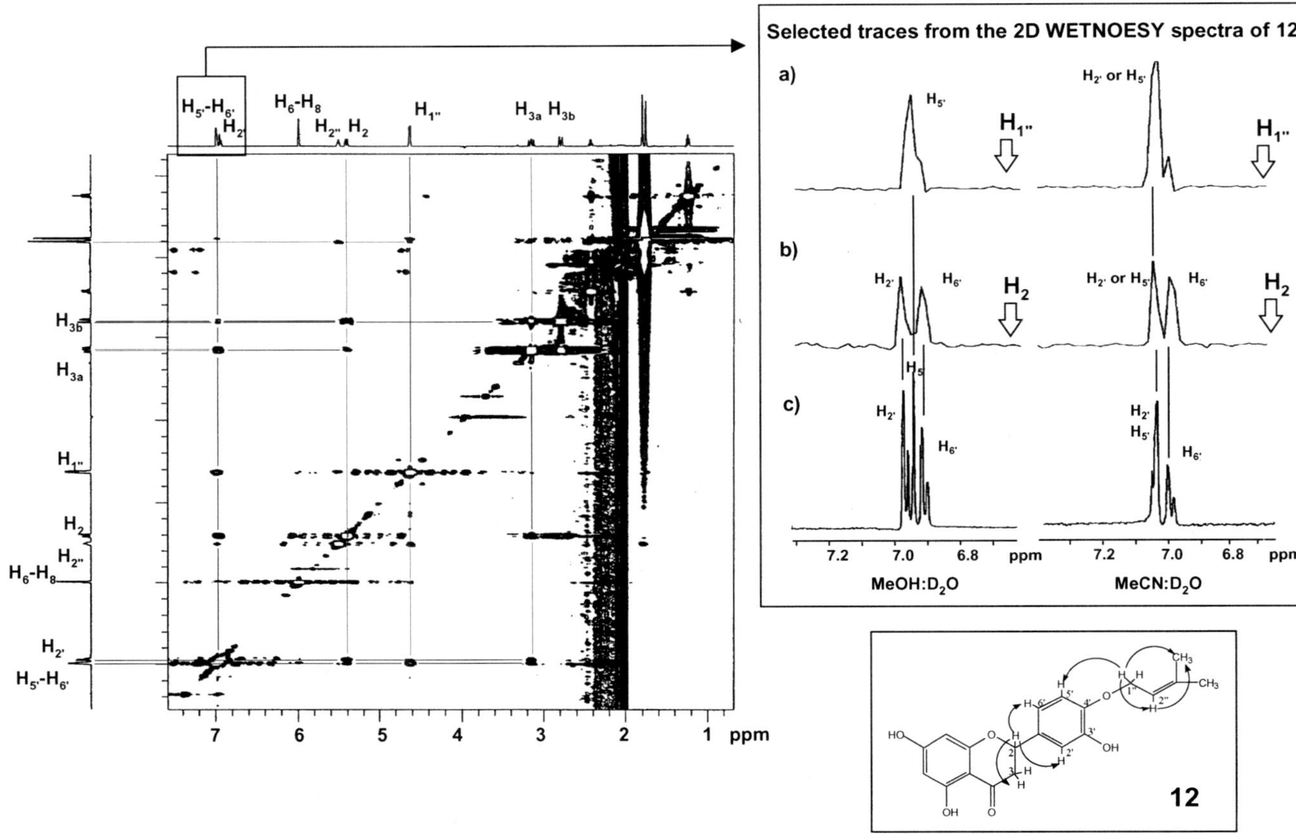

Figure 9 *Stop-flow LC–WET-NOESY spectrum of the flavanone* **12**. *1 mg of the enriched fraction injected, solvent MeOH–D$_2$O. Insert: Selected traces showing the main NOE correlations in both MeCN–D$_2$O and MeOH–D$_2$O solvent systems. LC–WET-NOESY: transients 8, increments 256. time: 2.5 hr. quantity injected: 1 mg*

prenyl unit on the flavanone skeleton of **12**, a 2D LC–WET-NOESY experiment was undertaken. The sensitivity was good enough for the observation of clear NOE effects. However, the weak resolution of the H-pattern of the B-cycle in the MeCN:D_2O solvent system did not permit the differentiation of H–C(2′) and H–C(5′), preventing the full structural assignment of **12** (Figure 9). A second stop-flow LC–NMR analysis of the active fraction was then performed in MeOH:D_2O. In this latter solvent system, the three aromatic protons of the B-cycle were observed at distinct chemical shifts. Two doublets ($J = 7.8$ Hz) at δ 6.90 and δ 6.95 were attributable to the *ortho*-coupled protons H–C(6′) and H–C(5′), and one broad singlet at δ 6.97 was due to H–C(2′). A second WET-NOESY was performed in this latter solvent system (Figure 9). First, an irradiation of H–C(2) gave enhancement of the two H–C(2′) and H–C(6′) aromatic protons at δ 6.97 and δ 6.90. A correlation with the H–C(2) and the two protons on C(3) was also shown (Figure 9). Secondly, irradiation of H–C(1″) gave enhancement of the *ortho*-coupled proton at δ 6.95 (H–C(5′)). Other NOE effects were also measured on the prenyl unit, particularly between H–C(1″) and H–C(2″), H–C(1″) and H–C(5″), as well as between H–C(2″) and H–C(4″) (Figure 9). According to these observations, the attachment of the *O*-prenyl substituent at position C(4′) was confirmed and **12** could thus be identified on-line as 2,3-dihydro-5,7-dihydroxy-2-[3-hydroxy-4(3-methyl-2-butenyl)oxyphenyl]-4*H*-1-benzopyran-4-one. It was found to be a new natural product and its complete isolation was finally performed for the measurement of its physical data, as well as the determination of its absolute configuration. Pure compound **12** showed a negative optical rotation ($[\alpha_D] = -24.7$), and the absolute configuration at carbon C(2) has been determined by CD experiments.[8]

It should be noted, however, that absolute configuration determination is possible on-line by HPLC–CD. Bringmann and co-workers[4] have demonstrated from the on-line structural investigation of an isoquinoline and a naphthylisoquinoline alkaloid from the crude extract of a Dioncophyllaceae species *Habropetalum dawei* that 2D LC–ROESY and LC–TOCSY allow the constitution of their relative configuration. The absolute configuration was determined on-line by subsequent stop-flow HPLC–CD experiments on these compounds in the extract and by empirical analysis of their CD data.

7 On-flow LC–^{1}H-NMR Characterisation of the Isoflavone and Isoflavanone Constituents of *Erythrina vogelii*

In our search for new antifungal agents from plant origin, the investigation of *Erythrina vogelii* (Leguminosae), a plant used in the traditional medicine of Ivory Coast to treat various infectious ailments, was undertaken. In a preliminary TLC bioautography screening, several constituents of the dichloromethane extract of the roots showed significant antifungal activity against a plant pathogenic fungus *Cladosporium cucumerinum*. In order to rapidly identify the active principles, the crude extract was analysed by low-flow LC–^{1}H-NMR which gave a sensitive detection of all the main peaks. LC microfractionation was performed just after LC–NMR detection and all peaks collected were submitted to antifungal bioautography assays against *C. cucumerinum.* By this means, antifungal activity could be efficiently linked to three of the LC peaks. In order to obtain complementary on-line structural information for all peaks of interest, high-resolution LC–MS–MS together with LC–UV with post-column addition of UV shifts reagents was undertaken on the crude extract.

This example shows well how biological data and chemical screening can be used for the on-line identification of bioactive polyphenols. As will be discussed in this example, the structural investigation of the different peaks relies not only on on-flow LC–NMR; the structure assignment is based on various sets of on-line spectroscopic data.

Preliminary LC–UV–DAD and High Resolution LC–APCI–Q-TOF–MS–MS Analyses

LC–UV–DAD and positive ion high resolution LC–APCI-MS–MS analyses were carried out by reversed-phase HPLC with a broad MeCN–H_2O gradient on the crude dichloromethane extract of *E. vogelii.* The LC–MS–MS analyses were performed on a Q-TOF instrument. In order to obtain molecular and fragment information in a single run, scan dependent MS–MS autoswitch experiments were carried out sequentially on the three main ions recorded by LC–APCI-MS at four different energy levels.[42]

The LC–UV trace revealed the presence of a dozen major peaks. These different constituents shared rather similar types of UV spectra with two main absorption bands of decreasing intensity. For a first group of peaks (**13**, **15** and **16**), three maxima were recorded at *ca.* 210, 260 and 290(sh) nm, for the other group (**14**, **17–20**) these bands were present at *ca.* 210, 290 and 333(sh) nm. Most of the minor peaks in the chromatogram presented also one of these two types of chromophore.

From the high-resolution LC–MS and LC–MS–MS data, the molecular formula could be assigned precisely thanks to the on-line high-accuracy measurements performed on a Q-TOF instrument with post-column addition of a reference compound for lock mass. Molecular weights were obtained in all cases with a precision of less than 5 ppm. The presence of fragments due to losses of 56 and 69 Da in the LC–MS–MS spectra of all constituents, with the exception of compound **17**, revealed the possible presence of prenyl chains. From this preliminary LC–UV–MS screening it could be concluded that the CH_2Cl_2 extract of *E. vogelii* most probably consisted of a combination of various prenylated isoflavanones or isoflavones.

Low-flow LC–NMR with Microfractionation

In order to obtain more information and to tentatively identify the compounds responsible for the antifungal activity of the extract, an on-flow LC–^{1}H-NMR analysis of the extract followed by HPLC microfractionation was carried out. A C_{18} radial compression column having a large internal diameter (8 mm) was preferred to classical analytical columns because of its high loading capacity and an important amount of crude extract (10 mg) was injected on-column. A satisfactory LC resolution of the main peaks was obtained at 1 ml min^{-1} flow rate. For the LC–NMR analysis, the HPLC conditions were changed in order to achieve the same separation over 19 hours (Figure 10) with a low flow rate (0.1 ml min^{-1}). As mentioned above this low-flow mode was needed in order to improve the sensitivity. At the flow rate chosen (0.1 ml min^{-1}), the number of scans per increment was increased to 256 and the limit of detection was thus consequently lowered. This experiment permitted the acquisition of the LC–NMR spectra of more than 10 LC–peaks in the crude extract over 19 h overnight.

During this LC–NMR analysis, the fractions were collected every 10 min (1 ml) in Eppendorf tubes and submitted to bioautography with *C. cucumerinum* after evaporation to dryness. This post-chromatographic antifungal assay revealed that the fractions associated with the LC–peaks eluting at 8.4, 10.8, 11.3, 12.2 and 14.6 h in the low-flow LC–NMR experiment display distinguishable antifungal activities (Figure 10).

LC–UV–DAD with Post-column Derivatisation

A first evaluation of the LC–NMR and LC–MS data revealed that complementary spectroscopic information was lacking for a precise localisation of the hydroxy groups of the different polyphenols detected. Thus, LC–UV–DAD analyses of the extract with post-column addition of UV shift reagents were performed according to a protocol already tested for flavonoids or xanthones.[43,44] The crude extract was analysed twice, once after post-column addition of a weak base and then with acidic $AlCl_3$. For each peak, the shifted DAD UV spectra were superimposed on the original spectra and compared (Figure 11). The shifts observed were interpreted according to the rules previously established for the analysis of pure polyphenols.[45] These analyses provided useful complementary information for the on-line identification of the peaks of interest as will be discussed below.

Interpretation of the On-line Data

The ^{1}H-NMR data extracted from the low-flow LC–NMR analyses of the crude extract confirmed that all the major constituents of the crude extract of *E. vogelii* were prenylated isoflavones and isoflavanones. In all peaks, aromatic signals between 5.9 and 6.4 ppm were indicative of an A-ring oxygenated in positions 5 and 7. This was in good agreement with chemotaxonomic data which indicated a 5,7,4′ oxygenation for biosynthetic reasons. For compounds **13**, **15** and **16**, methylene protons at 4.39–4.57 ppm were characteristic of an isoflavanone nucleus while for peaks **14** and **17–20** the presence of an aromatic singlet at *ca.* 8 ppm was indicative of isoflavones. This was also confirmed by the presence of

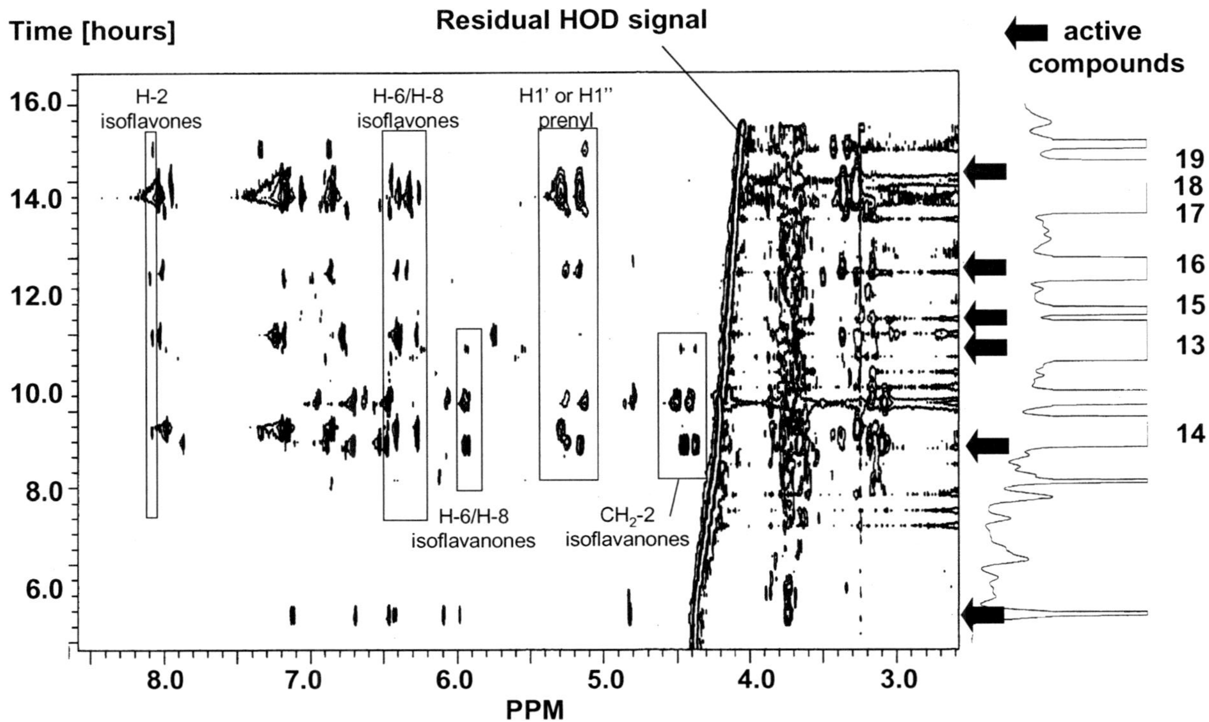

Figure 10 *On-flow LC–NMR contour plot of the crude CH_2Cl_2 extract of* Erythrina vogelii. *On this contour plot all ^{1}H-NMR resonances of the analytes are clearly visible. A sensitive detection was obtained thanks to the low flow rate used. Several characteristic region for the resonances of isoflavanones and isoflavones are highlighted. The signals at ca 4.5 ppm clearly indicate that the isoflavanones are eluting first on the chromatogram. HPLC conditions: Inj. 10 mg; Col. μ?Bondapak C_{18} (100 × 8 mm i.d., 10 μm); MeCN–D_2O gradient (5:95 to 100:0; 19 hours).; 0.05% TFA; 0.1 ml min^{-1}. LC–NMR conditions: 256 scans/increment, flow cell (60 μl, 3 mm i.d.), 500 MHz*

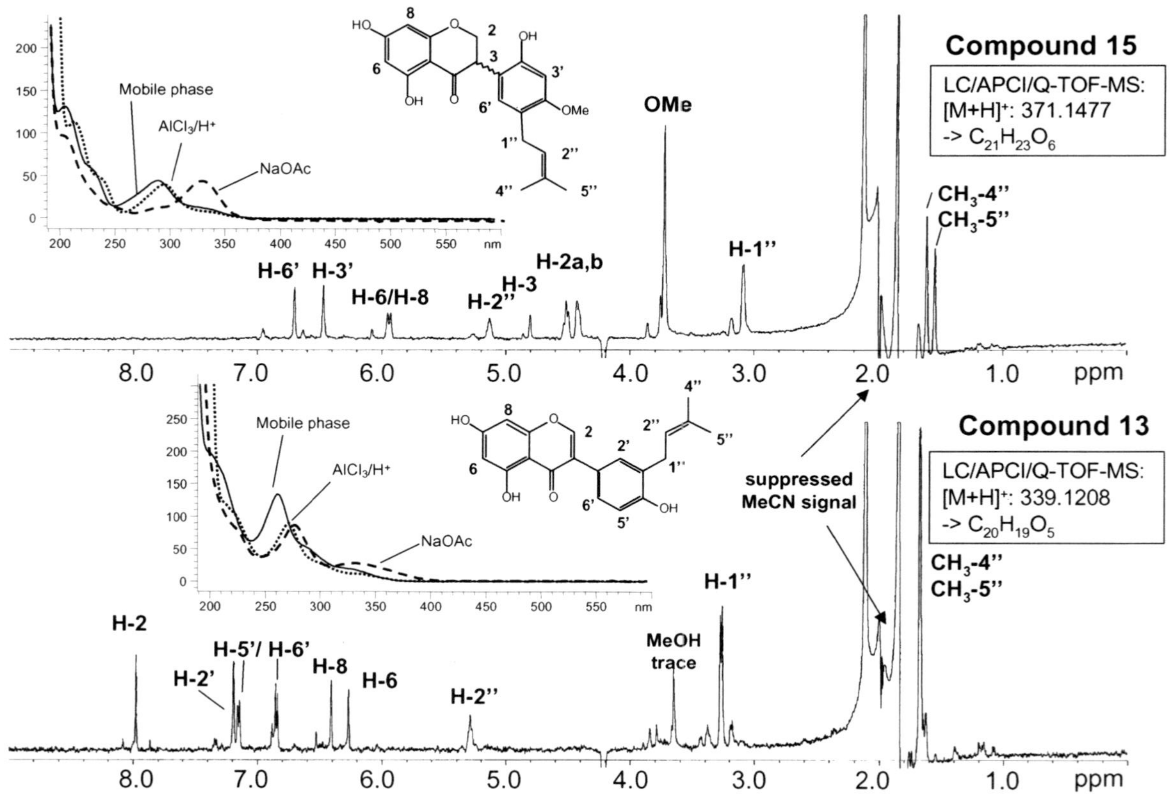

Figure 11 *On-flow LC–NMR spectra of an isoflavanone (***13***) and an isoflavone (***15***). Complementary UV–DAD and shifted UV–DAD spectra recorded on-line as well as exact molecular formula assignment based on LC–APCI-QTOF-MS data. Same LC–NMR conditions as in Figure 10*

two distinct types of chromophore for each group of constituents in the UV–DAD spectra recorded (see Figure 11). The presence of one prenylated group was detectable by signals at 3.0–3.3 ppm (H-1‴), 5.1–5.3 ppm (H-2‴) and 1.5–1.6 ppm (2 CH_3) for all compounds except **17**. In the case of **16** and **18–20**, the splitting of these latter signals revealed the presence of a second prenyl group. These diagnostic signals were already observable on the contour plot of the on-flow LC–NMR experiment and most of the characteristic resonances are highlighted in Figure 10. Compound **13**, eluting at 9.8 h in the LC–NMR analysis and at 57.4 min in the LC–UV–MS–MS run, presented the strongest zone of inhibition in the post-chromatographic antifungal bioautography assays. As discussed above and based on LC–UV and LC–NMR data, the skeleton of **13** could be attributed to an isoflavone. The high resolution LC–APCI-Q-TOF–MS spectrum of **13** exhibited a protonated molecule at *m/z* 339.1208 $[M + H]^+$, which indicated a molecular formula of $C_{20}H_{19}O_5$. The fragments at *m/z* 283.0667 $[M - 56]^+$ and *m/z* 271.0683 $[M - 68]^+$ were characteristic of losses of C_4H_7 or C_5H_9 from one prenyl moiety. From the molecular formula, it could be deduced that three hydroxy groups were also present on the isoflavone nucleus. The LC–^{1}H-NMR spectrum showed two singlets at δ 6.26 and 6.40 attributable to two aromatic protons at H-6 and H-8. Three signals corresponding to an ABX system at δ 6.85 (*d*, $J = 8.0$), 7.14 (*d*, $J = 8.0$), and 7.19 (br s), were attributed to the aromatic protons at H-5′, H-6′ and H-2′, respectively in the B-ring. Finally, a singlet at δ 7.9 was assigned to H-2, a proton characteristic of the isoflavone nucleus. The position of the three hydroxy groups and the prenyl unit was deduced from the complementary UV shifted spectra and the MS–MS experiments. The 5,7-dihydroxylation of the A-ring was deduced from the UV shifts observed upon addition of $AlCl_3$ and NaOAc.[45] The bathochromic shift of Band II with $AlCl_3$ was due to the formation of a complex between the OH in position 5 and the carbonyl in 4. On the other hand, the shift of Bands I and II upon addition of the weak base suggested the presence of OH in the most acidic positions 7 and 4′ (Figure 11). In the CID MS–MS spectra obtained at 40 eV, a fragment at *m/z* 153.0450 $[A_1]^+$ resulting from a typical RDA cleavage confirmed the presence of two hydroxy groups on the A-ring[42] (Figure 11). All these data suggested that **13** was isowighteone, a known prenylated isoflavone which was dereplicated by this means.

Another compound of the extract (**15**) (rt: 11.2 hrs LC–NMR and 27.0 min in the LC–UV–MS–MS) showed a UV spectrum and ^{1}H-NMR signals of a three-proton spin system at δ 4.39 (H-2a), 4.47 (H-2b) and 4.80 (H-3), typical of an isoflavanone nucleus. The protonated molecule at *m/z* 371.1477 $[M + H]^+$ suggested a molecular formula of $C_{21}H_{22}O_6$ for **15** and consequently, a substitution by 3 OH, 1 OMe and 1 prenyl was deduced. The presence of the OMe was confirmed by a singlet at δ 3.66 (3H) in the LC–NMR spectrum.

As in the case of **13**, UV shifted and MS–MS data confirmed the presence of a 5,7-dihydroxylation of A-ring. The two broad singlets corresponding to H-6 and H-8 were, however, slightly high-field shifted compared with **13** at δ 5.93 and 5.95. In the LC–^{1}H-NMR spectrum, two broad singlets at δ 6.47 and 6.70 were attributed to the aromatic protons of the B-ring (Figure 11). The position of the prenyl unit and that of the methoxy group could not, however, be precisely assigned based on ^{1}H-NMR data only since the resolution of the aromatic signals did not allow distinction between *meta* or *para* coupled protons. These two substituents could be either at 2′ and 5′ or at 3′ and 6′. A computer simulation indicated that the best matching for H-3′ (δ_{calc} 6.40) and H-6′ (δ_{calc} 6.64) was found for a B-ring substitution by a prenyl at 5′ and a methoxy at 4′. It was not possible to ascertain fully the structure on these indications only. However, this information suggested that **15** was likely to be a new natural product and its targeted isolation was undertaken. Its structure was confirmed to be 5,7,2′-trihydroxy-4′-methoxy-5′-(3-methylbut-2-enyl)-isoflavanone.

The antifungal compound **14** (rt: 10.8 LC–NMR and 25.0 min in the LC–UV–MS–MS) was found to be an isomer of **15** with similar UV, ^{1}H-NMR and MS–MS spectra, but it did not display any activity in the antifungal assay. The only striking difference between **14** and **15** was that for **14** the UV Band I was changed upon addition of the weak base, indicating probably hydroxylation at 2′ and thus methoxylation at 4′. As for **14**, this compound was presumably a new natural products and its complete structure elucidation was confirmed after targeted isolation.[46]

By following the same approach, eight polyphenols were thus identified partially or totally on-line in the crude extract of *E. vogelii*. The interpretation of these on-line data has permitted the dereplication of a known antifungal agent and also revealed that two additional antifungal isoflavanones were potentially new natural products, and their isolation was undertaken. The on-line spectroscopic data of the LC-peaks which did not present activity in the antifungal assay revealed the presence of several new natural products.[16]

8 Other Applications of LC–NMR for the Study of Polyphenols

At present the use of LC–NMR for the on-line identification of polyphenols is still rather scarce in the literature.[6] It can be noted, however, that LC–NMR–MS was applied successfully for the analysis of the extract of *Hypericum perforatum*. Various products were tentatively identified on-line. They included flavonoids such as the galacturonide, rutinoside, glucoside, arabinoside, rhamnoside and galactosides of quercetin, and the aglycone itself, the biflavonoid I5,II8-biapigenin and the naphthodianthrone hypericin, amongst others. The arabinoside and galacturonide of quercetin were tentatively identified by this means and were not previously described as constituents of this plant.[47]

Another application has been described for the analysis of tetraglycosylated flavonoids in the aqueous and hydroalcoholic extracts of the leaves of *Maytenus aquifolium* (Celastraceae). In this case, the separation was achieved on a C_{30} reversed-phase column using a MeOH–D_2O as solvent system. The challenge was to suppress MeOH signals at δ 3.3 and 4.7 without affecting the characterisation of the sugar signals, mainly those corresponding to the anomeric protons. The LC–^{1}H-NMR measurements were performed in the stop-flow mode at 600 MHz and allowed the characterisation of these flavonoids by detection of both aromatic and anomeric proton signals. In the aliphatic region, in particular, the anomeric signals were recorded and coupling constants ranging from 1.5 to 7.5 Hz were determined successfully. This has permitted the assignment of the nature of the sugars of the tertrasaccharide moiety.[48] The same group has reported very recently the on-line identification of the three main di- and tri-glycoside flavonols from *Sorocea bomplandii* (Celastraceae) by stop-flow LC–NMR under HPLC conditions similar to those described for *M. aquifolium*. All aromatic protons as well as the characteristic anomeric proton signals were recorded directly from the leaf methanolic extract.[49]

9 Conclusions

LC–NMR plays a central role in the on-line identification of polyphenols since it allows the recording of precious complementary on-line structure information when LC–UV–MS data are often insufficient for unambiguous peak identification. The applications of this new LC detection method are still scarce but without doubt LC–NMR will become more and more widely used in this field. Indeed, LC–NMR has proven to be very effective in obtaining 1D spectra on both flowing and non-flowing samples, as well as stop-flow 2D spectra. However, compared with UV or MS, NMR remains a rather insensitive detection method and the need for solvent suppression has restricted the observable NMR range. Improvements in sensitivity are expected in the years to come but the main effort for the efficient use of LC–NMR in phytochemical analysis as regards the actual technology has to be made on the chromatographic side where strategies involving efficient pre-concentration, high loading, stop-flow, time slicing or low-flow procedures have to be developed. The inability of LC–NMR to provide ^{13}C NMR spectra with reasonable amounts of sample is still an important limitation of this technique for natural product analysis and for full on-line identification of unknowns. However, LC–^{1}H-NMR information may be sufficient for the on-line identification of previously known constituents or for studying derivatives of a known core skeleton which are likely to be present in a given extract.

At present LC–NMR has become a well established hyphenated technique, but further optimisation is needed. New developments are expected in the foreseeable future, including the use of higher magnetic field strengths. In addition the development of NMR probes and preamplifiers cooled with cryogenic liquids will provide lower detection limits and higher sensitivities to a degree surpassing any arising from increase in magnetic fields.[50] In this respect horizontal detector coils or solenoidal coils with susceptibility corrected wire represent promising approaches.[1,51]

With respect to solvent suppression, microbore methods or capillary separation methods such as capillary LC–NMR, CE-NMR and CEC-NMR will be employed more and more often since their operation at low flow rates will enable the use of fully deuterated solvents. The potential of these capillary techniques has been demonstrated in some scarce applications on rather simple mixtures.[24,27,52]

Developments in multihyphenated systems are expected in the near future and several groups have now demonstrated the feasibility of such 'total organic analysis devices'.[10]

The chemical screening of extracts with various hyphenated techniques generates a huge amount of information. In order to rationalise this approach and use it efficiently with a high sample throughput, the next challenge will be to find a way to centralise all these data for rapid pattern recognition with reference to standard compounds. In particular LC–UV and LC–MS–MS or LC–MS^n databases have to be built up and LC–NMR analysis should be used as a complement for studying in more detail polyphenols which are not matched in these databases. Integration of high-resolution on-line LC–bioassays is crucial in this global strategy since they will permit an efficient localisation of the bioactive compounds in the crude extracts and their subsequent on-line full or partial identification

Acknowledgments

Financial support was provided by the Swiss National Science Foundation (Grant No 2000-063670.00 to Kurt Hostettmann).

References

1. K. Albert, *J. Chromatogr. A*, 1999, **856**, 199.
2. J. L. Wolfender, K. Ndjoko and K. Hostettmann, *Curr. Org. Chem.*, 1998, **2**, 575.
3. B. Vogler, I. Klaiber, G. Roos, C. U. Walter, W. Hiller, P. Sandor and W. Kraus, *J. Nat. Prod.*, 1998, **61**, 175.
4. G. Bringmann, K. Messer, M. Wohlarth, J. Kraus, K. Dumbuya and M. Rückert, *Anal. Chem.*, 1999, **71**, 2678.
5. S. C. Bobzin, S. Yang and T. P. Kasten, *J. Chromatogr. B*, 2000, **748**, 259.
6. J. L. Wolfender, K. Ndjoko and K. Hostettmann, *Phytochem. Anal.*, 2001, **12**, 2.
7. G. Bringmann, M. Wohlfarth, H. Rischer, M. Heubes, W. Saeb, S. Diem, M. Herderich and J. Schlauer, *Anal. Chem.*, 2001, **73**, 2571.
8. E. Garo, J. L. Wolfender, K. Hostettmann, W. Hiller, S. Antus and S. Mavi, *Helv. Chim. Acta*, 1998, **81**, 754.
9. S. Stevenson and H. C. Dorn, *Anal. Chem.*, 1994, **66**, 2993.
10. I. Wilson, *J. Chromatogr. A*, 2000, **892**, 315.
11. K. Hostettmann, J. L. Wolfender and C. Terreaux, *Pharm. Biol.*, 2002, **39**, 18.
12. M. Spraul, M. Hoffmann, J. C. Lindon, J. K. Nicholson and I. D. Wilson, *Anal. Proc.*, 1993, **30**, 390.
13. S. H. Smallcombe, S. L. Patt and P. A. Keiffer, *J. Magn. Reson. Ser. A*, 1995, **117**, 295.
14. K. Albert, *J. Chromatogr. A*, 1995, **703**, 123.

15. M. Sandvoss, L. H. Pham, K. Levsen, A. Preiss, C. Mugge and G. Wunsch, *Eur. J. Org. Chem.*, 2000, **7**, 1253.
16. E. F. Queiroz, J. L. Wolfender, K. Atindehou, D. Traore and K. Hostettmann, *J. Chromatogr. A*, 2002, **974**, 123.
17. J. Ferrari, C. Terreaux, J. L. Wolfender and K. Hostettmann, *Chimia*, 2000, **54**, 406.
18. J. A. Koning, A. C. Hogenboom, T. Lacker, S. Strohschein, K. Albert and U. A. T. Brinkman, *J. Chromatogr. A*, 1998, **813**, 55.
19. T. Renukappa, G. Roos, I. Klaiber, B. Vogler and W. Kraus, *J. Chromatogr.*, 1999, **847**, 109.
20. J. L. Wolfender, S. Rodriguez, K. Hostettmann and W. Hiller, *Phytochem. Anal.*, 1997, **8**, 97.
21. R. M. Holt, M. J. Newman, F. S. Pullen, D. S. Richards and A. G. Swanson, *J. Mass Spectrom.*, 1998, **32**, 64.
22. G. J. Dear, R. S. Plumb, B. C. Sweatman, J. Ayrton, J. C. Lindon, J. K. Nicholson and I. M. Ismail, *J. Chromatogr. B*, 2000, **748**, 281.
23. B. C. Sweatman, R. D. Farrant, P. N. Danderson, I. Philippe, S. R. Salman, J. K. Nichloson and J. C. Lindon, *J. Magn. Reson. Anal.*, 1995, **1**, 9.
24. M. E. Lacey, Z. J. Tan, A. G. Webb and J. V. Sweedler, *J. Chromatogr. A*, 2001, **922**, 139.
25. K. Ndjoko, J. L. Wolfender, E. Röder and K. Hostettmann, *Planta Med.*, 1999, **65**, 562.
26. A. Cavin, O. Potterat, J. L. Wolfender, K. Hostettmann and W. Dyatmiko, *J. Nat. Prod.*, 1998, **61**, 1497.
27. B. Behnke, G. Schlotterbeck, U. Tallarek, S. Strohschein, L. H. Tseng, T. Keller, K. Albert and E. Bayer, *Anal. Chem.*, 1996, **68**, 1110.
28. V. M. Chari, in *Flavonoids Biflavonoids*, L. Farkas, F. Kallay, M. Gabor and H. Wagner, eds, Elsevier, 1982, p. 279.
29. K. R. Markham and H. Geiger, in *Flavonoids*, Chapman and Hall, 1983, p. 441.
30. P. G. Waterman and S. Mole, *Analysis of Phenolic Plant Metabolites*, Blackwell Scientific Publications, Oxford, 1994.
31. G. Bringmann, C. Günther, J. Schlauer and M. Rückert, *Anal. Chem.*, 1998, **70**, 2805.
32. T. Visser, M. J. Vredenbergregt, G. J. ten Hove, A. P. J. M. de Jong and G. W. Somsen, *Anal. Chim. Acta*, 1997, **342**, 151.
33. D. Louden, A. Handley, S. Taylor, E. Lenz, S. Miller, I. Wilson and A. Sage, *Anal. Chem.*, 2000, **72**, 3922.
34. E. Clayton, S. Taylor, B. Wright and I. D. Wilson, *Chromatographia*, 1998, **47**, 264.
35. I. D. Wilson, D. E. Morgan, F. Lafont, J. P. Shockcor, J. C. Lindon, J. K. Nicholson and B. Wright, *Chromatographia*, 1999, **49**, 374.
36. G. Bringmann, M. Wohlfarth and M. Heubes, *J. Chromatogr. A*, 2000, **904**, 243.
37. C. Hall, *Dictionary of Natural Products on CD-ROM*, Chapman & Hall, 2000.

38. C. Gnerre, U. Thull, P. Gaillard, P. A. Carrupt, B. Testa, E. Fernandes, F. Silva, M. Pinto, M. M. M. Pinto, J. L. Wolfender, K. Hostettmann and G. Cruciani, *Helv. Chim. Acta*, 2001, **84**, 552.
39. P. Waridel, J. L. Wolfender, K. Ndjoko, K. R. Hobby, H. Major and K. Hostettmann, *J. Chromatogr. A*, 2001, **926**, 29.
40. G. Rath, A. Touré, J. L. Wolfender and K. Hostettmann, *Chromatographia*, 1995, **41**, 332.
41. L. Rahalison, M. Hamburger, K. Hostettmann, M. Monod and E. Frenk, *Phytochem. Anal.*, 1991, **2**, 199.
42. J. L. Wolfender, P. Waridel, K. Ndjoko, K. R. Hobby, H. Major and K. Hostettmann, *Analusis*, 2000, **28**, 895.
43. K. Hostettmann, B. Domon, D. Schaufelberger and M. Hostettmann, *J. Chromatogr.*, 1984, **283**, 137.
44. J. L. Wolfender and K. Hostettmann, *J. Chromatogr.*, 1993, **647**, 191.
45. K. R. Markham, *Techniques of Flavonoid Identification*, Academic Press, London, 1982.
46. K. K. Atindehou, E. F. Queiroz, D. Terreaux, D. Traore and K. Hostettmann, *Planta Med.*, 2002, **68**, 181.
47. S. H. Hansen, A. G. Jensen, C. Cornett, I. Bjornsdottir, S. Taylor, B. Wright and I. D. Wilson, *Anal. Chem.*, 1999, **71**, 5235.
48. W. Vilegas, J. H. Vilegas, M. Dachtler, T. Glaser and K. Albert, *Phytochem. Anal.*, 2000, **11**, 317.
49. F. D. P. Andrade, L. C. Santos, M. Datchler, K. Albert and W. Vilegas, *J. Chromatogr. A*, 2002, **953**, 287.
50. T. M. Logan, G. S. Murali, G. S. Wang and C. Jolivet, *Magn. Reson. Chem.*, 1999, **37**, 512.
51. J. C. Lindon, J. K. Nicholson and I. D. Wilson, *J. Chromatogr. B*, 2000, **748**, 233.
52. P. Gfrörer, J. Schewitz, K. Pusecker and E. Bayer, *Anal. Chem.*, 1999, **1**, 315A.

CHAPTER 7

Polyphenols: Measurement and Calculation of Their Physical Chemistry Properties

CECILE CREN-OLIVE AND CHRISTIAN ROLANDO

1 Introduction

The determination of the physical chemistry properties of polyphenols is essential to understand the mechanism by which the polyphenols exert their protective effects on human health. However, the most fundamental physical chemistry constants (pK_a and $E°$) for most polyphenols are still not firmly established or are the subject of controversy. Indeed, due to the polyphenolic nature of the studied compounds, the three ways generally used to determine pK_a and $E°$ (experimental measurements, estimation by free energy relationships like the Hammett relation and *ab initio* or semi-empirical calculation) gave poor access to these thermodynamic constants. Macroscopic pK_a values are determined by titration of polyphenol analysed by UV-vis spectroscopy or fluorescence spectroscopy. In the case of polyphenols this method gives access to the successive pK_as but does not allow determination of the microscopic dissociation patterns of individual hydroxy groups, which is required for predicting reactivity. The measurement of the redox potential of polyphenolic compounds by electrochemistry or pulse radiolysis is a real challenge since the methods must be faster than the following reactions induced by the oxidation of the phenol group in order to get the thermodynamic value. In such a situation, the Hammett relationship is, for most of the time, a good method for estimating acidity constants and standard potentials. However, most Hammett correlations involving polyphenols are badly behaved for three reasons: (i) most of the phenol groups bear an *ortho* substituent in which the Hammett relationship is not relevant; (ii) in water the interesting substituents for polyphenols (mainly hydroxy and methoxy groups) behave abnormally and cannot be correlated to gas-phase or DMSO solutions and (iii) most of the time, hydroxy and methoxy groups correspond to a slope breaking

Table 1 *Structures of catechin and of selectively protected analogues*[1]

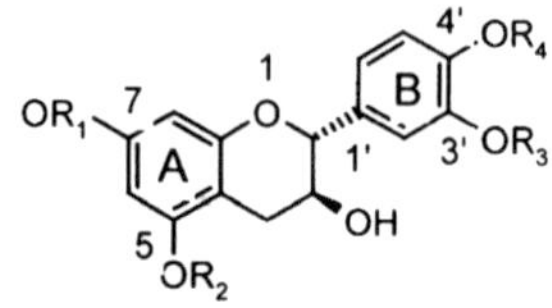

Selectively protected flavan-3-ols	R_1	R_2	R_3	R_4
I	H	H	H	H
II	Me	Me	H	H
III	H	H	Me	Me
IV	H	H	$C(Ph)_2$	
V	Me	Me	Me	H
VI	Me	Me	H	Me
VII	H	Me	$C(Ph)_2$	
VIII	Me	H	$C(Ph)_2$	
IX	H	H	Me	H
X	H	H	H	Me

zone in Hammett correlations. Besides the difficulty of estimating the variation of properties from *in vacuo* to water affects also calculations made by *ab initio* (mainly DFT) or semi-empirical-calculation. Therefore, the only solution to determine the physico-chemical properties of polyphenols and to get precise values is to use selectively protected model compounds. In this chapter, we will show particularly how the use of selectively protected flavan-3-ols[1] (Table 1) allows the determination of different physico-chemical properties of these polyphenols and finally we will compare these results with calculations.

2 Free Radical Chemistry of Flavan-3-ols

The characterization of phenoxyl radicals from flavan-3-ols, which are very reactive species implicated in the various biological properties of catechin and its derivatives, appears very delicate due to the fast evolution of these radicals at near diffusion-controlled rate and so needs the use of techniques with high temporal resolution.

Generation of Phenoxyl Radical by Pulsed Radiolysis and Flash Photolysis

Three different methods, the first one using pulsed radiolysis and the two others based on flash photolysis, are used for the generation and characterization of these phenoxy radicals by their UV-visible spectra at very short times after their formation (down to 200 ns to 2 μs depending on the technique): (i) monoelectronic oxidation of the phenolic derivative by a very powerful oxidant such as $SO_4^{\bullet -}$, $N_3^{\bullet}$ or $Br_2^{\bullet -}$, generated *in situ* by irradiation;[2–14] (ii) direct photoionization of the phenolic compound in polar solution;[8,15,16] and (iii) H-atom abstraction from the phenolic OH group in aprotic media.[15–17]

In the radiolytic experiments, phenols can be oxidized to phenoxyl radicals in irradiated aqueous solution saturated with N_2O as shown in Scheme 1. Under these conditions, all hydrated electrons are efficiently converted into hydroxyl radicals. The hydroxyl radicals react with phenols rapidly to form hydroxycyclohexadienyl adducts, most of which undergo rapid water elimination to yield phenoxyl radicals.[3] The hydrogen atoms also add to phenols but the adducts do not yield phenoxyl radicals. Furthermore, $OH^{\bullet}$ may add to positions on the side chain which will not yield phenoxyl radicals.[3]

Therefore, to minimize these side reactions, other chemical oxidants such as $SO_4^{\bullet -}$, $N_3^{\bullet}$ or $Br_2^{\bullet -}$ are preferred. Considering the redox potential of these inorganic radicals[18] ($E°$/ENH) $E°(SO_4^{-\bullet}/SO_4^{2-}) = 2430$ mV, $E°(N_3^{\bullet}/N_3^{-}) = 1370$ mV, $E°(Br_2^{\bullet -}/2Br^{-}) = 1630$ mV, the oxidation of phenols ($E°(PhO^{\bullet}, H^{+}/PhOH) = 1340$ mV) as well as phenolates ($E°(PhO^{\bullet}/PhO^{-}) = 790$ mV)[19] is possible. Scheme 2 summarizes the various possibilities of formation of phenoxyl radicals from phenols or phenolates. The monoelectronic oxidation of phenols implies an intermediary radical cation which after deprotonation gives the phenoxyl radical (Scheme 2).[7,16,20–22] As in the gas phase,[23] the more stable radical cation, which a recent study[16] has characterized by its UV-vis spectra, is the radical localized on the aromatic ring rather than on the oxygen atom. So the experimental conditions and especially the choice of the pH determine the reactivity observed during the pulsed radiolysis experiments. The choice of stronger oxidants therefore appears not necessarily judicious if the objective of the work is to study the monoelectronic oxidation of phenolates, the most probable phenomenon occurring under physiological conditions.

With this aim, a flash photolysis set-up can be used to investigate direct photoionization of the polyphenol derivatives under their basic form (Scheme 2), but also H-atom abstraction from phenolic OH by *tert*-butoxyl radicals generated by laser photolysis of *tert*-butyl peroxide in aprotic media (Scheme 2). Indeed, phenolates which present an absorption band at the irradiation wavelength of the laser currently used in these experiments ($\lambda = 308$ nm), can be oxidized by direct photo-irradiation On the contrary, the phenol does not absorb at this wavelength, leading the neutral form to be unreactive. On the other hand di-*tert*-butyl peroxide is cleanly cleaved into *tert*-butylperoxyl radical at this irradiation wavelength leading to pure H abstractraction.

$$H_2O \longrightarrow e^-_{aq} + H^{\bullet} + OH^{\bullet} + H_2 + H_2O_2$$

$$e^-_{aq} + N_2O + H_2O \longrightarrow N_2 + OH^- + OH^{\bullet}$$

Scheme 1

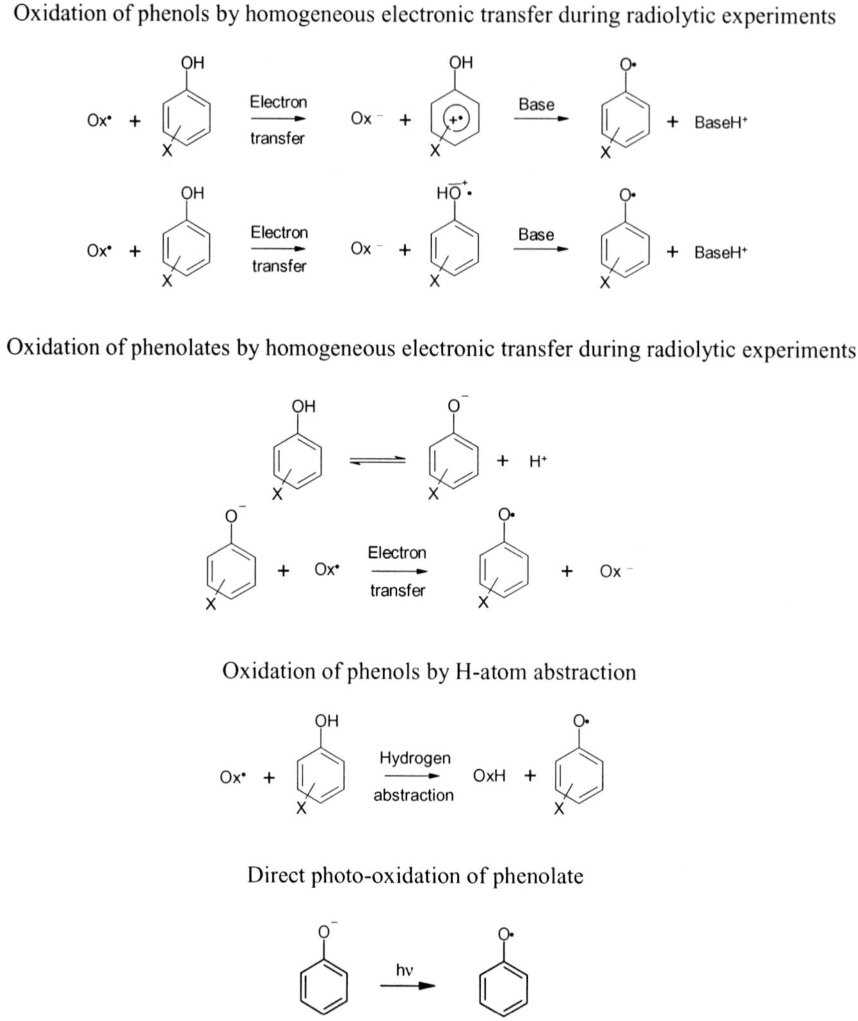

Scheme 2 *Oxidation mechanism of phenols and phenolates*

Radiolytic Experiments on Simple Phenols

The first comprehensive characterization of phenoxyl radicals[6] was realized by pulsed radiolysis: the phenoxyl radicals were obtained by oxidation of differently substituted phenolates (*ortho*-, *meta*- or *para*-hydroxy, amino or alkoxy phenols) by the radical $^{\bullet}CH_2CHO$ in strongly basic media. The authors showed that the absorption spectra (λ, ε_{max}) of the generated phenoxyl radical depend very strongly on the nature of the substituent present on the aromatic ring and on their position on the cycle. Similar results have been obtained by oxidation of 5-methoxyresorcinol and of 4-methylcatechol by the azydil radical in neutral media.[8,9]

More precisely, the authors conclude than the phenoxyl radicals having *ortho*-

hydroxyl, *ortho*-amino or *ortho*-alkoxy groups absorb principally between 300 and 350 nm while their *meta* or *para* analogues show absorptions at about 400–500 nm. If the results, in this case, of simple phenols appear reproducible and well established, the characterization of the phenoxyl radical generated from more complex polyphenols is more delicate and still subject to discussion.

Radiolytic Experiments on Flavan-3-ols

Spectra of the Phenoxyl Radicals Formed upon Oxidation of Flavan-3-ols

The case of the flavan-3-ols is, in fact, more complex since spectra and interpretations differ among the authors.[7–9,13,14,24] Usually, Simic *et al.*[8] observe that the monoelectronic oxidation of the flavan-3-ols by the radical azydil $N_3^{\bullet}$ at pH = 7 leads to a spectrum showing two absorption bands, the first one appearing around $\lambda = 320–380$ nm, and the second one at $\lambda = 550$ nm. In the case of compounds with a gallate ester, a new absorption band appears at $\lambda = 420$ nm. In contrast, the spectra obtained by Bors *et al.*[14] after oxidation of the flavan-3-ols by the hydroxyl radical $OH^{\bullet}$ in basic media (pH around 9) present four characteristics: (i) bands appearing between 280 and 320 nm, that are, in part, outside the spectral range observed by Simic *et al.*;[8] (ii) a bleaching of the 320–380 nm zone except in the case of the catechin and epicatechin; (iii) a band at $\lambda = 420$ nm, (iv) more scarcely an absorption band at $\lambda = 550$ nm, absent for example in the spectra of catechin, epicatechin gallate and epigallocatechin gallate.

Thus the results obtained in the case of the flavan-3-ols seem far from being reproducible. We must emphasize that two factors could explain at least in part the observed differences: the pH of the media and the nature of the oxidant used. Bors *et al.*[14] have, in fact, used very delicate experimental conditions. All the experiences are realized in basic medium (pH around 9). However, since the two first successive pK_as. pK_{a1}/pK_{a2}, of catechin (respectively of epicatechin, and epigallocatechin gallate) are 8.79/9.44,[25] (8.72/9.49[26], 7.75/8.00[8]), the irradiated solution at pH = 9 is composed not only of a mixture of the four monophenolates but also, in most cases, of dianions and finally of unionized flavan-3-ol. It is necessary also to take into account the high instability of the studied compound in basic environment (epimerization and rearrangement of catechin, for example). It is then difficult to know exactly which species react and which is the principal reaction occurring during the attack of $OH^{\bullet}$ on this mixture, and therefore which are the contributions of the different species to the final spectrum. Moreover, these authors choose to oxidize the flavan-3-ols by the hydroxyl radical. As noticed above, the mechanism of this kind of oxidation is not a direct electronic transfer but an addition of the $OH^{\bullet}$ on the aromatic ring leading to the formation of hydrocyclohexadienyl that undergoes a loss of water. So many secondary reactions can take place giving many secondary species which pollute the experimental medium. The differences observed on the spectra of these phenoxyl radicals lead then inevitably to different interpretations.

Different Interpretations of the UV-vis Spectra of the Phenoxyl Radicals Formed upon Oxidation of Flavan-3-ols

Simic *et al.*[8] attribute the existence of the two absorption bands at $\lambda = 380$ and 495 nm to the formation of two distinct radicals: a first short-lived radical absorbing at $\lambda = 550$ nm, and the second one more stable and characterized in neutral media by an absorption band between 350 and 400 nm depending on the studied flavan-3-ol. They then identify the two radicals formed by analogy with the radical of a simple phenol such as 5-methoxyresorcinol, methyl gallate or gallic acid, generalizing Neta *et al.*'s conclusions:[6] the short-lived radical which absorbs at 550 nm is thus attributed to the resorcinol-like phenoxyl radical, while the second more stable radical, presenting an absorption between 340 and 420 nm, is attributed to a catechol-like phenoxyl radical, *i.e.* an *ortho*-semiquinone (Scheme 3). More precisely, they propose that the radical azidyl $N_3^{\bullet}$ attacks preferentially the B ring of the catechin and of its derivatives but A-ring resorcinol-like radicals are also generated;[24] the ratio is 70/30 in favour of the B-ring attack.

Simic *et al.*[8,24] suggest that these A-ring resorcinol-like radicals present a redox potential higher than that of the B-ring *ortho*-semiquinone, which allows them to oxidize the B-ring catechol groups by inter- and intra-molecular electronic and proton transfer (Scheme 4). This irreversible process leads, therefore, to a single species of radicals after the second electronic transfer, the *ortho*-semiquinones. In addition, the ratio of the two radicals formed during the oxidation by the azidyl radical is independent of the reduction potential and charge of the oxidant since similar spectra are obtained oxidizing epigallocatechin gallate with neutral

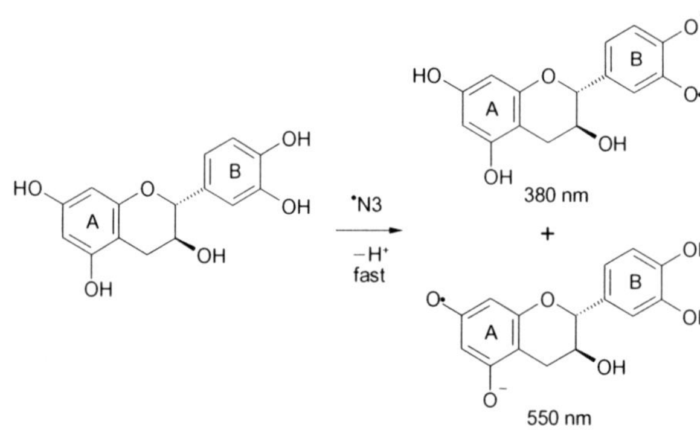

Scheme 3 *Oxidation of catechin by azidyl radicals*

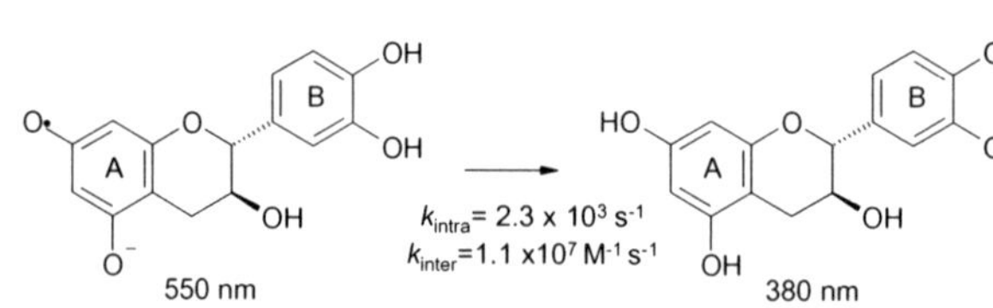

Scheme 4 *Inter- and intra-molecular electron transfer in catechin*

azidyl $N_3^\bullet$ ($E_{pH7} = 1.33$ V) and formyl ($E_{pH7} \sim 1.00$ V) radicals as well as by negatively charged thiocyanate $(SCN)_2^{\bullet-}$ ($E_{pH7} = 1.25$ V) and bromide $Br_2^{\bullet-}$ ($E_{pH7} = 1.70$ V) radicals. The interpretation of Simic *et al.*[8,24] leads to the hypothesis of the formation of two phenoxyl radicals: a catechol-like long-lived phenoxyl transient and a resorcinol-like short-lived one, which is reduced to a catechol-like phenoxyl radical.

Bors *et al.*[14] clearly disagree with this interpretation and put three different opposing arguments: first, a false interpretation of the absorption bands, second, a too high instability of the resorcinol-like phenoxyl radical (an argument that seems, nevertheless, in contradiction with the spectra obtained from simple phenols such as resorcinol and 5-methoxyresorcinol), and finally determining a kinetics constant of decay of the radicals which does not reflect the second electronic transfer. So Bors *et al.*[14] propose (i) that the absorption band at 270 nm is characteristic of an *ortho*-semiquinone structure, and (ii) that the negative absorbances at 320 nm represent the bleaching of the absorbance of the ester group of the parent molecules that would be correlated to the apparition of the absorption band at 400–420 nm representing the cetyl radical, indicating thus the participation of this group in the electronic delocalization. Finally, to explain the case of epigallocatechin, where the absorption bands move to 290 and 550 nm, they evoke a 'mesomeric' equilibrium involving an oxonium structure. They conclude therefore, in contrast to Simic *et al.*,[8] that the azidyl or hydroxyl radical oxidizes only the catechol and/or pyrogallol groups, which therefore only leads to the formation of *ortho*-semiquinones that participate in different 'mesomeric' equilibria resulting in the formation on the one hand of cetyl radical, by delocalization on the esters groups, and on the other hand of oxonium-like radicals.

So the structure of the first radicals formed upon oxidation of flavan-3-ols is under discussion. In fact, both theories presented above present some drawbacks. Indeed, in the hypotheses advanced by Simic *et al.*,[8,24] the second intermolecular transfer of electron and protons does not seem very probable: the structure of the flavanols does not allow the spatial transfer between the two sites. Concerning the interpretation of Bors *et al.*,[14] one can doubt the validity of the interpretations of the absorption bands: associating the bands appearing at 370–420 nm with the formation of a cetyl radical seems in contradiction with the apparition of similar bands with identical absorbances in compounds that do not possess any ester groups such as catechin, epicatechin, and epigallocatechin. Besides, the stability of the oxonium structure envisaged to interpret the band at 550 nm cannot explain the reactivity observed for this radical since it disappears in less than 1 ms. The only common point between the different studies, even if the advanced arguments to justify it differ (different interpretations of the absorption bands), lies in the presence, after some hundreds microseconds, of only one kind of radical: the catechol-like phenoxyl radical or *ortho*-semiquinone. However, the identification of the radicals formed upon oxidation of flavan-3-ols required more investigation and it is why we decided to study the free-radical chemistry of selectively protected flavan-3-ols using a flash photolysis experiment for the photochemical generation of radicals and their characterization through the monitoring of their UV-visible spectra.

Flash Photolysis Experiments on Selectively Protected Flavan-3-ols

It is particularly interesting to see how the study of the chemical reactivity of selectively protected flavan-3-ols allows one to understand the chemical reactivity of the native forms. Indeed, the synthesis of selectively protected catechins (Table 1) afforded more appropriate models which preserve the skeleton of these polyphenols and allow the determination of the reactivity of each ring.[15,27]

The results obtained with selectively protected flavan-3-ols during direct photo-ionization under their basic form and H-atom abstraction by *tert*-butylalkoxyl radicals show that the phenoxyl radicals of the A- or B-ring compounds can be, without any ambiguity, distinguished by their different spectral characteristics (Table 2, Scheme 5).[15,27] In fact, as described for simple phenols, the catechol-like phenoxyl radicals generated from 4-methylcatechol, and 5,7-dimethylcatechin absorb at 380 nm while the resorcinol-like phenoxyl radicals obtained from the methoxyresorcinol, 3′,4′-dimethylcatechin, 3′,4′-diphenylmethylenedioxycatechin and 5-methyl-3′,4′-diphenylmethylenedioxycatechin absorp at 495 nm. During the photo-oxidation experiments, another band at 550 nm is visible for the radicals from 5-methoxyresorcinol, 3′,4′-dimethylcatechin and 3′,4′-diphenylmethylene-dioxicatechin, and can be ascribed to a fast further deprotonation of the neutral resorcinol radical to the corresponding radical-anion in basic medium.

Catechin exhibits a more complex behaviour. H-abstraction of catechin by *tert*-butoxyl radicals leads to a spectrum characterized by two absorption bands respectively at $\lambda = 380$ and 495 nm. The 308 nm laser-induced photo-oxidation of catechin phenolate in the presence of one equivalent of base (when the two rings can be partially ionized due to the close pK_a values for rings A and B) produced a similar spectrum ($\lambda = 380$ and 495 nm), whereas in the presence of one half an equivalent of base only one absorption band at $\lambda = 380$ nm was obtained. By comparison with the obtained spectra of the compound models, it is then possible to identify precisely the radicals and to discuss the reactivity of the catechin.

Table 2 *Characteristics of phenoxyl radicals generated by photo-oxidation and H-abstraction. Delay after laser pulse respectively: 200 ns, 5 μs*

	Photo-oxidation on phenolate	*H-abstraction on phenol*
Compounds	λ_{max} *(nm)*	λ_{max} *(nm)*
Catechin **I**	380	380/495
5-Methoxyresorcinol	495/550	495
III	495/550	495
IV	495/550	495
VII	495	495
II	380	380
4-Methylcatechol	380	380
3′-Methylcatechin **IX**	380/495	380/495
4′-Methylcatechin **X**	495	495

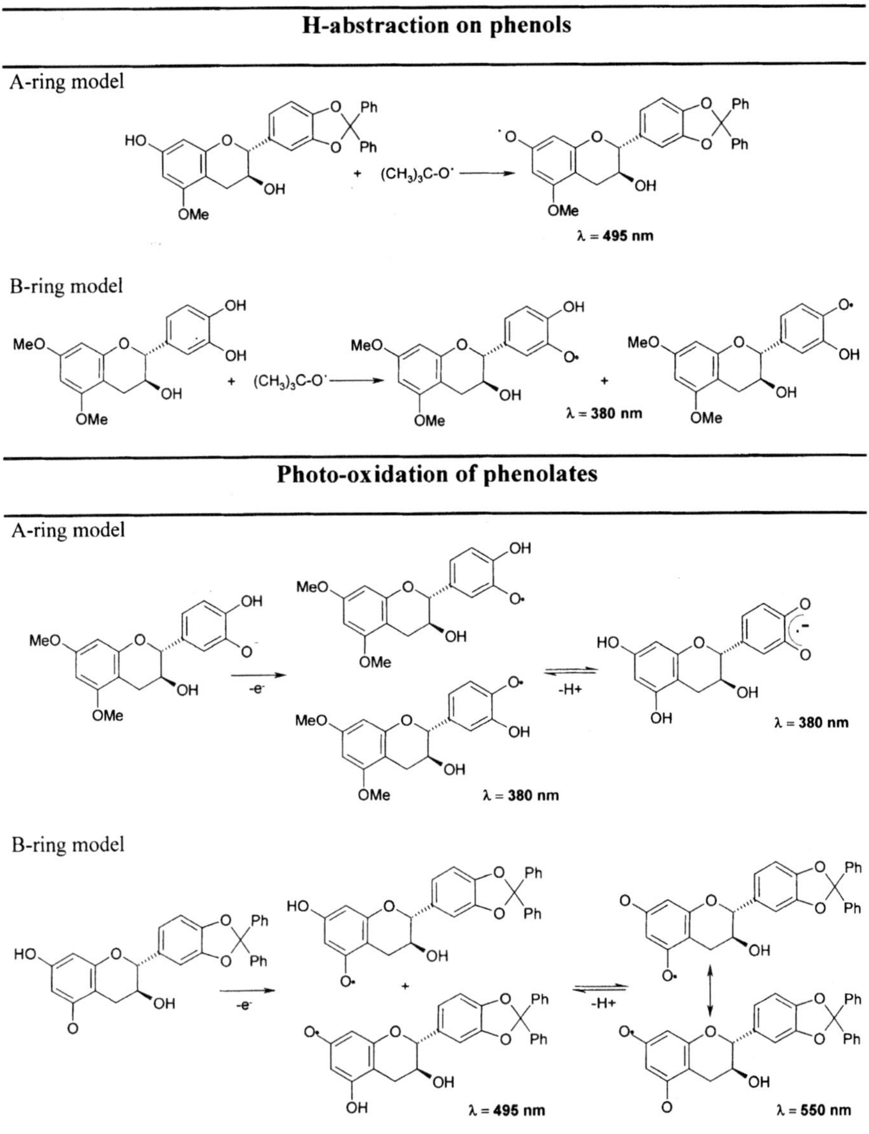

Scheme 5 *Results of photo-oxidation and H-abstraction on selectively protected flavan-3-ols*

On the one hand, H-abstraction by *tert*-butoxyl radical on catechin leads to two different radicals: the first one, a resorcinol-like phenoxyl radical absorbing at 495 nm and the second one a catechol-like phenoxyl radical characterized by an absorption band at 380 nm. The more precise examination of the absorbances shows nevertheless than the H-atom abstraction is a selective phenomenon since about 70% of the reactivity occurs on the A-ring (Scheme 6). On the other hand,

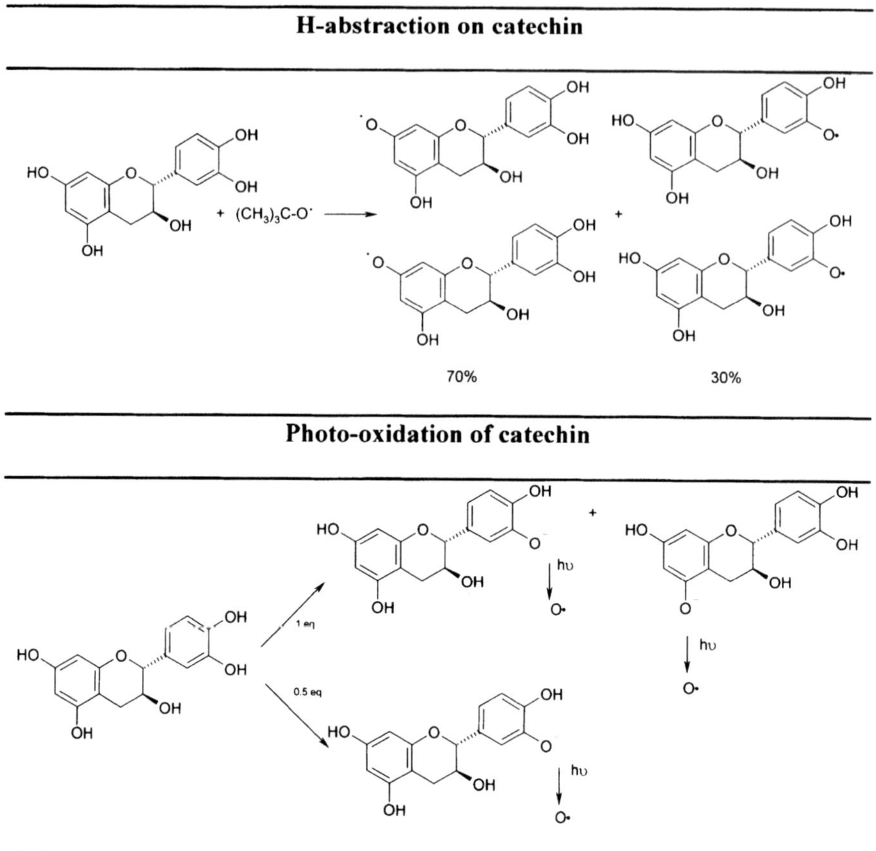

Scheme 6 *H-abstraction and photo-oxidation of catechin*

in the photooxidation experiments on catechin, in the presence of one half an equivalent of base, only the catechol-like radical absorbing at 380 nm appears, while in the presence of one equivalent of base both radicals are present (Scheme 6). This selectivity is related to the possibility of deprotonating a single or several phenolic functions as only phenolates absorb the laser light and are much more easily oxidized than neutral phenols. The 380 nm absorption is therefore related to the photo-oxidation of the first deprotonated phenolic group.

Also, during H-abstraction experiments,[15] it has been shown by studying the growth and decay kinetics obtained for both phenoxyl radicals, *i.e.* the resorcinol ($\lambda = 495$ nm) and catechol ($\lambda = 380$ nm) ones, that the resorcinol-like radical is a short-lived transient while the catechol-like one can be described as a long-lived transient, in agreement with a previous pulse-radiolytic study of Jovanovic *et al.*[8] But in this study, the decay of the resorcinol-like radical absorbing at $\lambda = 495$ nm is not correlated to any increase of the signal of the catechol-like radical characterized by an absorption band at $\lambda = 380$ nm. So we can conclude that the resorcinol-like radical transforms quickly into another transient which is not the catechol-like radical, contrary to the hypothesis made in the previous study.[8] In these conditions, the inter- or intra-molecular transformation of the resorcinol-like radicals into a B-ring radical is not observed.

In conclusion, this last study[15] using selectively protected catechin analogues allows to rationalization of the discrepancies regarding the structure of the phenoxyl radicals. Indeed, the two phenoxyl radicals formed upon oxidation of catechin have been characterized and identified unambiguously: the short-lived one is a resorcinol (A ring) like radical characterized by an absorption band at $\lambda = 495$ nm while the second one, a long-lived transient, can be ascribed as a catechol (B ring) like phenoxyl radical absorbing at $\lambda = 380$ nm. These results contradict the hypothesis concerning the structure of the phenoxyl radical of a recent pulse radiolysis study[14] but corroborate the one described by Jovanovic *et al.*[8] Yet, our experiments cannot confirm the latter hypothesis of the latter authors, *i.e.* an intra- or inter-molecular electron transfer from the A-ring phenoxyl radical leading to the transformation of the resorcinol-like phenoxyl radical into a catechol-like radical. This is supported by molecular modelling calculations which determine both the geometry and the stability of catechin in three states:[15] phenol, phenolate, phenoxyl radical. Examination of the charge and spin distribution of each phenolate and phenoxyl radical suggests that the two moieties of the molecule, *i.e.* chroman and the B ring, are independent so that delocalization of charge of spin over the whole molecular framework, as previously described, does not exist.

Moreover, these results allow, also for the first time, determination of the intrinsic reactivity of each ring: the A-ring is preferentially involved during H-atom transfer while the B ring is selectively involved during electron transfer. Finally, these photooxidation experiments indicate also that the first pK_as of the two rings are very close but that it is the B-ring that is the more acid, which establishes the protonation sequence of catechin, subject until now to discussion.

3 Acid-Base Properties of Flavan-3-ols

The knowledge of pK_a is essential since it is not only a prerequisite for calculating standard redox potentials but also for determining the precise form of phenol or phenolate present in biological medium. However, in the case of catechin and epicatechin, even if the four successive pK_a values have been well determined by various techniques using potentiometric and/or spectrophotometric measurements,[25,26] the deprotonation sequence is still under discussion. In fact, while

Table 3 *Dissociation constants of catechin and epicatechin*

	pK_{a1}	pK_{a2}	pK_{a3}	pK_{a4}	*Reference*
Catechin	8.79	9.44	11.18	13.25	25
	8.64	9.41	11.26	13.26	26
Epicatechin	8.72	9.49	11.23	13.40	26

Slabbert[25] proposes a sequence BAAB, the B-ring appearing, in this case, the more acid one, Kennedy *et al.*[26] prefer a deprotonation starting with the A-ring (ABAB).

However, the selectivity observed during our photo-oxidation experiments in the presence of one half an equivalent of base[15] (Scheme 6) indicates that the catechol B-ring is slightly more acidic in acetonitrile, allowing unambiguous establishment of the protonation sequence of catechin. Indeed, a similar order between the two rings must occur in water as the relative acidities of phenols are generally not affected by solvent effects,[28–30] supporting the protonation sequence suggested by Slabbert.[25]

However, the pK_a determined in these two studies characterize only the successive dissociation of catechin ($CatH_4$):

$$CatH_4 \rightarrow CatH_3^- + H^+ \qquad K_1$$

$$CatH_3^- \rightarrow CatH_2^{2-} + H^+ \qquad K_2$$

and similar equations for the third and fourth deprotonations, but the interesting data, above all at physiological pH where phenols are not completely ionized, are the microscopic pK_as of each phenolic function:

$$CatH^AH^BH^CH^D \rightarrow CatH^BH^CH^{D-} + H^+ \qquad k^A$$

$$CatH^AH^BH^CH^D \rightarrow CatH^AH^CH^{D-} + H^+ \qquad k^B$$

and similar equations for the C and D sites; the global constant is related to the four microconstants by the equation $K_1 = k^A + k^B + k^C + k^D$.

If the measurements in solution aqueous by the classical methods (UV spectra and/or pH measurement as a function of the number of equivalents of base added)[25,26] give access, as mentioned above, to the constant of successive dissociation, the only way to determine the microscopic constants, in the case of the polyphenols, is by ^{13}C NMR.[13,31] Indeed the successive deprotonations of the different phenolic functions present on flavan-3-ols induce great changes in the chemical environment of various carbon atoms of the skeleton, reflected in the ^{13}C-NMR chemical shifts.[32–36] More precisely, the deprotonation of a phenolic function induces deshielding for the *ipso* and *ortho* carbon atom and shielding for the *para* carbon atom. Such selective behaviour allows, with the unambiguous

Table 4 *Microscopic* pK_a *of each phenolic function of catechin*

Phenolic function	*Microscopic* pK_a ($\pm$ 0.05)
3′	9.02
4′	9.12
5	9.43
7	9.58

assignment of ^{13}C-NMR signals of flavan-3-ols, determination of the precise deprotonation sites of catechin and epicatechin.[31] The quantification of each existing species allows determination of the microscopic pK_a (Table 4).

Moreover, this study points out that the stereochemistry at C3 of flavan-3-ols appears to have no influence on the deprotonation of flavan-3-ols since the behaviour of epicatechin upon addition of base appears similar to that of catechin in the same medium.[31] So flavan-3-ols show a deprotonation behaviour very different from that of other types of flavonoids: while upon addition of successive equivalents of base, flavanones such as naringenin present successive dissociations which occur in the sequence 7-OH, 4′-OH, 5-OH, flavan-3-ols such as catechin and epicatechin exhibit competitive deprotonation on the B- and A-rings leading to a mixture of different monophenolates.[31]

4 Redox Properties of Flavan-3-ols

The interesting redox potential for understanding biological effects of flavan-3-ols is the one related to the couple phenoxyl radical/phenate as this potential is roughly one volt lower than the potential of the phenoxyl radical/phenol couple which furthermore may involve transiently the oxidation of the aromatic ring (Scheme 2).

Determination of Redox Potential of Phenol by Pulse Radiolysis

The redox potential can be generally determined by electrochemical methods, such as cyclic voltammetry. Nevertheless, the determination of thermodynamic constants requires first that the processes are reversible or quasi-reversible, in other words that the phenoxyl radicals formed upon this oxidation are stable enough during the time of the experiments. Unfortunately, this condition is usually not found in the case of phenolates since the phenoxyl radicals are known to be very reactive and very quickly undergo an oxidative coupling leading to the formation of dimers and even of polymers (Scheme 7), whose major consequence is the appearance of irreversible and badly defined waves during classic cyclic voltammetry.[37–42]

Thus pulse radiolysis appears particularly useful in the determination of the

$$\mathrm{PhOH} \rightleftharpoons \mathrm{PhO^-} + \mathrm{H^+}$$
$$\mathrm{PhO^-} \rightleftharpoons \mathrm{PhO^\bullet} + e^-$$
$$\mathrm{PhO^\bullet} + \mathrm{PhO^\bullet} \longrightarrow \mathrm{Dimer}$$

Scheme 7 *An example of polymerization reaction of phenol upon oxidation*

$$\mathrm{PhO^\bullet} + \mathrm{D^-} \underset{k_{-1}}{\overset{k_1}{\rightleftharpoons}} \mathrm{PhO^-} + \mathrm{D^\bullet}$$

Scheme 8 *Principle of determination of the redox potential of phenols by pulse radiolysis*

standard redox potential of a large number of phenoxyl radical/phenolate couples in basic aqueous medium[6,19,43–49] since it allows the study of the equilibrium shown in Scheme 8, where $PhO^{\bullet}$ is the phenoxyl radical formed by oxidation of the studied phenol PhO^- (or PhOH) by oxidants such as $N_3^{\bullet}$, $Br_2^{\bullet -}$, in the presence of D^-, a properly chosen electron donator presenting a known redox potential. The equilibrium constant K is therefore given, for this reaction, by

$$K = k_1/k_{-1} = [PhO^-][D^{\bullet}]/[PhO^{\bullet}][D^-]$$

K can therefore be determined by a kinetic study of these reactions (K_{Cin}), or by the determination of the concentrations at the equilibrium (K_{Abs}), *via*, for example, the study of the absorption spectra of the different compounds. The redox potential of the couple phenoxyl radical/phenolate can then be calculated using the Nernst equation: at 20 °C, $\Delta E = 0.059 \times \log K$, where K is the average of K_{Cin} and K_{Abs}.

The determination of the redox potential by this method requires the absence of all other reactions that could affect the equilibrium. This implies therefore that the equilibrium between phenolate and the phenoxyl radical has to be reached much more quickly than the dimerization or dismutation reactions. Also the experimental conditions have to be judiciously chosen in order to minimize the secondary reactions.

Simple Phenols

In the case of simple phenols, this technique has been used, with success, to determine the redox potential of many substituted phenols at different pHs. The results obtained for simple phenols or phenols able to be models of the A- or B-ring of flavonoids are summarized in Table 5.

Before discussing more precisely the effects of the various substituents, it is interesting to note that, in the case of these simple phenols, the values of E obtained by pulse radiolysis corroborate the one obtained by classical cyclic voltammetry.[50] In the latter cases, the values of the redox potential are obtained by the study of the peak potential E_p as a function of the scan rate v for a known dimerization rate k at a concentration C:

$$E_p = E_{red} + 0.902RT/F - (RT/3F)\ln(4kCRT/3Fv)$$

The redox potential summarized in Table 5 and the irreversible peak potential obtained by Bordwell *et al.*[22] appear to be very dependent of the substitution of the aromatic ring. More precisely, Hammett[19,22,51] correlations were obtained for the reduction potential determined reductions for the *para*-substituted phenoxyl radicals at pH = 7 (E_{pH7}) and at pH = 0 ($E°$) as a function of Brown constants σ^+:

$$E_{pH7} = 0.95 + 0.31 \times \Sigma\sigma^+$$

$$E° = 1.34 + 0.32 \times \Sigma\sigma^+$$

Table 5 *Redox potential, at different pH, of simple phenolic compounds (couple (phenoxyl radical /phenol or phenolate as a function of pH)*

	Pulse radiolysis		*Voltammetry*
Phenolic compound	E°_{pH7} *(V/ENH)*	$E^\circ_{pH11-12}$ *(V/ENH)*[e]	E_{pH12} *(V/ENH)*
Phenol	0.95[e,f]	0.79	0.86
4-Methylcatechol	0.52[d]/0.87[e]	0.68	0.71
4-Methoxyphenol	0.73[e,f]	0.54	0.58
4-Chlorophenol	0.94[e]	0.80	0.85
4-Fluorophenol	0.93[e]	0.76	0.80
4-Bromophenol	0.96[e]	0.82	0.84
4-Iodophenol	0.96[e]	0.82	0.81
4-Cyanophenol	1.17[e]	1.12	1.14
4-Nitrophenol	1.23[e]	1.22	1.28
4-Hydroxybenzoic acid	1.04[e]	0.90	0.97
Resorcinol	0.81[a]	–	–
Catechol	0.53[a]	–	–
3,5-Dihydroxyanisole	0.84[c]	–	–
2-Methoxy-4-methylphenol	0.68[d]	–	–
Methyl gallate	0.56[b]	–	–
3,4-Dihydroxybenzoic acid	0.60[b]	–	–
3,4-Dihydroxycinamic acid	0.54[b]	–	–

References: [a]6, [b]7, [c]8, [d]9, [e]19, [f]51.

The electron-donating substituents (more negative σ^+ values) reduce, as expected, the redox potentials of the phenol while the electron-withdrawing substituents increase it. These Hammett correlations allow excellent predictions in the case of simple phenolic compounds,[24,51] which do not have any substitution in the *ortho* position.[22] In contrast, they appear to be not very reliable in the case of the polyphenols, for three main reasons: (i) the majority of the polyphenolic compounds have an *ortho*-substituent, a situation in which Hammett correlations are not valid;[22,52] (ii) in water, the behaviour of the substituents of main interest in the case of flavan-3-ols (principally the hydroxy and methoxy groups) is abnormal and cannot be correlated to that observed in the gaseous phase; (iii) the data of the literature are often very scarce for this kind of compound, which does not allow reliable correlations to be established.

Polyphenols

The determination by pulse radiolysis of the redox potentials of polyphenols appears to be more delicate than for those of simple phenols. Simic *et al.*[6,9,24] were the only ones to perform this kind of study in the case of the flavan-3-ols and the results obtained for the couples *ortho*-semiquinonic radical/phenolate are summarized in Table 6.

The values determined by pulse radiolysis usually appear higher than the

Table 6 *Reduction potentials (V/NHE) of phenoxyl radicals at pH = 7 determined by different methods*

Phenoxyl radicals	E_{pH7}, *V, pulse radiolysis*	$E_{1/2}(pH = 7)$, *V, electrochemistry*
Catechin	0.57	0.40
Epigallocatechin	0.42	–
Epigallocatechin gallate	0.43	–

potentials determined by cyclic voltammetry,[53] and this observation is true for the whole family of flavonoids.[24] These discrepancies imply a different hypothesis on the effect of these polyphenolic compounds on human health. Indeed, if the redox potentials obtained either by pulse radiolysis or by classic cyclic voltammetry indicate that the flavan-3-ols ($0.4 < E_{pH7} < 0.6$ V) are able to inactivate alkylperoxyl radicals which are very toxic for the human body ($E_{pH7} = 1.05$ V)[47,54] as well as the superoxide radicals ($E_{pH7}(O_2^{-\bullet}/H_2O_2) = 0.94$ V)[18] [$O_2^{-\bullet}$ is the form present at pH = 7 since the pK_a of $HO_2^\bullet$ is 4.8[55]], the conclusions about the reactivity of these phenoxyl radical towards vitamin $E(E_{pH7} = 0.48$ V)[2] depend on the method used for the determination of the redox potential. While the redox potential obtained by electrochemistry leads to the conclusion that the phenoxyl radical of flavan-3-ols (E_{pH7}(elec) < 0.48 V) are able to reduce the oxidized vitamin E, allowing regeneration of the natural defence of the cell, the results obtained by pulse radiolysis indicate that only epigallocatechin and epigallocatechin gallate possess this ability. Moreover, these latter results indicate that the oxidation of vitamin E by catechin and epicatechin ($E_{pH7equi} > 0.48$ V), which is a very undesirable reaction, is thermodynamically feasible. So the consequences of the discrepancies of the literature allow us to doubt, not only the understanding of the role of flavan-3-ols in the cell, but also the validity of the results obtained by both techniques: classic cyclic voltammetry and pulse radiolysis in the case of polyphenolic compounds.

As we have already noted, classic cyclic voltammetry is not, in the case of polyphenolic compounds, the most reliable method to determine $E°$. However, pulse radiolysis seems not to be more appropriate in these conditions. Indeed, the determination of the redox potential by the study of the chemical equilibrium existing between the phenoxyl radical and a compound of appropriate and known potential requires, as mentioned above, the absence of any other reactions. However, in the case of flavan-3-ols, the existence of a second type of phenoxyl radical, a resorcinol-like radical, questions the validity of the measurement. Indeed, even if these resorcinol-like radicals constitute only 30% of the radicals formed upon oxidation, they could also react with the used reference D^-, modifying the chemical equilibrium studied by pulse radiolysis (Scheme 9).

The question is to know to what extent the equilibrium will be affected by the presence in the medium of resorcinol-like radicals. A recent study[56] showed that electron transfer between different cresols and inorganic radicals is diffusion limited when the potential difference (ΔE) between the two couples is greater than 500 mV, independent of the inorganic radical type, of the phenol or of pH.

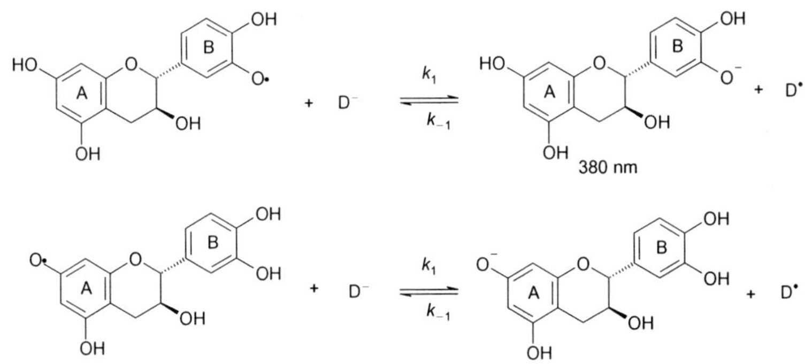

Scheme 9 *Reactions occurring during the pulse radiolysis experiment of catechin*

On the other hand, when this difference of potential (ΔE) is less than 500 mV the rate constants show a good correlation with the potential difference existing between the cresol and the inorganic radical. So if we suppose that the oxidation of phenolic compounds by organic radicals such as Trolox ($E_{pH7} = 0.48$ V) used by Simic *et al.*[8] can show similar behaviour, we can estimate the relative rate of reactions of the two phenoxyl radicals present in the solution, the resorcinol-like radicals and the catechol-like radicals, towards to the reference D^-. Considering the potentials of reduction of catechol, $E_{pH7} = 0.53$ V,[2] and resorcinol, $E_{pH7} = 0.81$ V,[8] one can reasonably think that the resorcinol-like radicals $[(E_{resorcinol} - E_{Trolox}) = 0.33$ V] will react more quickly than catechol-like radicals ($\Delta E = 0.20$ V). So the existence of the resorcinol-like radicals modifies the equilibrium studied in pulse radiolysis and cannot be neglected.

Determination of Redox Potential by Fast Cyclic Voltammetry on Selectively Protected Flavan-3-ols

In order to obtain reliable thermodynamic constants, we performed in a recent study[15] cyclic voltammetry on ultramicroelectrodes of selectively protected flavan-3-ols (Table 1), for example, two trimethylated flavan-3-ols offering only one free hydroxy group. By using ultramicroelectrodes (electrode with a micrometric diameter), it has been shown that a very high scan rate can be reached, up to one million volts per second, which implies that intermediates having a lifetime in the microsecond range can be characterized by direct electrochemical methods.[57] Even in these conditions, the only way to achieve a reversible process is to use protected flavan-3-ols. Indeed, while the use of fast cyclic voltammetry allows the determination of the $E°$ value of coniferyl alcohol,[41] the $E°$ of a simple catechol as caffeic acid cannot be determined.[39] The redox potential of each phenolic function of catechin in acetonitrile using fast cyclic voltammetry was investigated and is summarized in Table 7.[15] As expected, the catechol B-ring is more oxidizable than the resorcinol A-ring and these results are in good agreement with the acid–base properties discussed above: on each ring, the more basic site is the most oxidizable phenolic function. Besides, our study shows also

Table 7 *Standard potentials of flavan-3-ols in acetonitrile obtained by cyclic voltammetry at fast scan rate*

Catechin analogues	*Free phenolic position*	*E°, V (SCE)*
VII	7	–
VIII	5	0.285
V	4′	0.110
VI	3′	0.135

that theoretical calculation of the stability of the various catechin radicals (Table 8) confirms these trends:[15] the 4′-phenoxyl radical, corresponding to the more easily oxidizable phenolate is the most stable radical and the radicals can be ordered in regard to their values ($E_{PhO\bullet} - E_{PhO-}$) characterizing the electron affinity in the sequence 4′-OH, 3′-OH, 7-OH, 5-OH, which correlates well with the redox properties determined by cyclic voltammetry on ultramicroelectrodes.

The fact that the standard potential of each phenolic function follows the same trends as the pK_a implies that the B-ring, the more acid and the more oxidizable, would also present the lowest O–H dissociation energy (BDE). This is supported by the theoretical calculation of O–H bond strength reported in Table 8.[15] These results seem in contradiction with the selectivity observed on the A-ring during H-abstraction experiments of catechin. In fact they indicate that this selectivity is due to a kinetic control of the reaction.

5 Conclusion and Perspectives: A New Insight into Antioxidative Properties of Flavan-3-ols

The overall results on the physical chemistry of flavan-3-ols allowed a new approach of the mechanism implied in the antioxidative properties of flavan-3-ols. Indeed, a more differentiated concept of the protective effect of flavan-3-ols has just been proposed based, for the first time, on these physical chemical results: the mechanism implied in the scavenging properties of flavan-3-ols does not imply

Table 8 *Experimental and calculated formal potential E° and calculated bond dissociation enthalpy (BDE) for each phenolic function, variations with the radical positions in catechin*

Phenoxyl radical positions	*Experimental E° (V/SCE)*	*Experimental ΔE°*[a]	*Calculated ΔE°*[b] *(eV)*	*BDE (kcal mol⁻¹)*[c]
4′	0.110	0	0	75.1
3′	0.135	0.025	0.028	75.8
5	0.285	0.175	0.148	84.0
7	–		0.056	85.0

[a] $\Delta E° = E° - E°4'$.[15] [b] $\Delta E°$ estimated as $= (E^{elec}{}_{radical} - E^{elec}{}_{phenolate}) - (E^{elec}{}_{radical} - E^{elec}{}_{phenolate})_{4'}$.[15]
[c] BDE estimated as $E^{elec}{}_{phenoxylradical} + E^{elec}{}_{H\text{-}atom} - E^{elect}{}_{catechin} + RT$. Electronic.[15]

only hydrogen transfer as generally described but mainly electronic transfer from the phenolate. The major catechin metabolites in mammals, 3′- and 4′-methylcatechin, seem, moreover, to act as amphiphilic chain breaking antioxidants. Future research should therefore focus on mechanisms allowing the understanding of the protective effect of flavan-3-ols against cardiovascular disease other than simple antioxidant ones.

References

1. C. Cren-Olivé, S. Lebrun and C. Rolando, *J. Chem. Soc., Perkin Trans. 1*, 2002, 821.
2. G. E. Adams and B. D. Michael, *Trans. Faraday Soc.,* 1967, **63**, 1171.
3. E. J. Land and M. Ebert, *Trans. Faraday Soc.,* 1967, **63**, 1181.
4. R. H. Schuler, P. Neta, H. Zemel and R. W. Fessenden, *J. Am. Chem. Soc.*, 1976, **98**, 3825.
5. H. W. Richter, *J. Phys. Chem.*, 1979, **83**, 1123.
6. S. Steenken and P. Neta, *J. Phys. Chem.*, 1982, **86**, 3661.
7. S. V Jovanovic, S. Steenken, M. Tosic, B. Marjanovic and M. G. Simic, *J. Am. Chem. Soc.*, 1994, **116**, 4846.
8. S. V. Jovanovic, Y. Hara, S. Steenken and M. G. Simic, *J. Am. Chem. Soc.*, 1995, **117**, 9881.
9. S. V. Jovanovic, S. Steenken, Y. Hara and M. G. Simic, *J. Chem. Soc.*, *Perkin Trans 2,* 1996, 2497.
10. M. Erben-Russ, W. Bors and M. Saran, *Int. J. Radiat. Biol.*, 1987, **52**, 393.
11. W. Bors and M. Saran, *Free Radic. Res. Commun.*, 1987, **2**, 289.
12. W. Bors, C. Michel and S. Schikora, *Free Radic. Biol. Med.*, 1995, **19**, 45.
13. W. Bors, W. Heller and C. Michel, in *Flavonoids in Health and Disease*, eds C. Rice-Evans and L. Packer, Marcel Dekker, New York, 1997, p. 111.
14. W. Bors and C. Michel, *Free Radic. Biol. Med.*, 1999, **27**, 1413.
15. C. Cren-Olivé, P. Hapiot, J. Pinson and C. Rolando, *J. Am. Chem. Soc.*, 2002, in press.
16. T. A. Gadosy, D. Shukla and L. J. Johnston, *J. Phys. Chem.*, 1999, **103**, 8834.
17. M. Ye, K. Madden, R. W. Fessenden and R. H. Schuler, *J. Phys. Chem.*, 1986, **90**, 5397.
18. P. Wardman, *J. Phys. Chem. Ref. Data*, 1989, **18**, 1637.
19. J. Lind, X. Shen, T. E. Eriksen and G. Merenyi, *J. Am. Chem. Soc.*, 1990, **112**, 479.
20. M. Yi and S. Scheiner, *Chem. Phys. Lett.*, 1996, 262.
21. W. T. Dixon and D. Murphy, *J. Chem. Soc., Faraday Trans 2*, 1976, **72**, 1221.
22. F. Bordwell and J. P. Cheng, *J. Am. Chem. Soc.*, 1991, **113**, 1736.
23. J. Steadman and J. Syage, *J. Am. Chem. Soc.*, 1991, **113**, 1736.
24. S. V. Jovanovic, S. Steenken, M. G. Simic and Y. Hara, in *Flavonoids in Health and Disease*, eds C. Rice-Evans and L. Packer, Marcel Dekker, New York, 1997, p. 137.

25. N. P. Slabbert, *Tetrahedron*, 1977, **33**, 821.
26. J. A. Kennedy, M. H. G. Munro, H. K. J. Powell, L. J. Porter and Y. Foo, *Aust. J. Chem.*, 1984, **38**, 885.
27. C. Cren-Olivé, S. Lebrun, P. Hapiot, J. Pinson and C. Rolando, *Tetrahedron Lett.*, 2000, **41**, 5847.
28. J. Corse and L. L. Ingraham, *J. Am. Chem. Soc.*, 1951, **73**, 5706.
29. L. A. Cohen and W. M. Jones, *J. Am. Chem. Soc.*, 1963, **85**, 3397.
30. P. J. Pearce and J. J. Simkins, *Can. J. Chem.*, 1968, **46**, 241.
31. C. Cren-Olivé, J. M. Wierulesky, E. Maes and C. Rolando, *Tetrahedron Lett.*, 2002, **43**, 4545.
32. P. K. Agrawal and H. J. Schneider, *Tetrahedron Lett.*, 1983, **24**, 177.
33. J. W. Jarosszewski, L. Matzen, B. Frolund and P. Krogsgaard, *J. Med. Chem.*, 1996, **39**, 515.
34. S. Berger, *Tetrahedron*, 1981, **37**, 1607.
35. R. Haran, F. Nepveu-Juras and J. P. Laurent, *Org. Magn. Reson.*, 1979, **12**, 153.
36. G. E. Maciel and R. V. James, *J. Am. Chem. Soc.*, 1964, **86**, 3893.
37. P. J. Elving and A. F. Krivis, *Anal. Chem.*, 1958, **30**, 1645.
38. A. M. Yacynych and H. B. Mark, *J. Electrochem. Soc.*, 1976, **123**, 1346.
39. P. Hapiot, A. Neudeck, J. Pinson, H. Fulcrand, P. Neta and C. Rolando, *J. Electroanal. Chem.*, 1996, **405**, 169.
40. P. Hapiot, J. Pinson, P. Neta, C. Francesch, F. Mhamdi, C. Rolando and S. Schneider, *Phytochemistry*, 1994, **36**, 1013.
41. P. Hapiot, J. Pinson, C. Francesch, F. Mhamdi, C. Rolando and P. Neta, *J. Phys. Chem.*, 1994, **98**, 2641.
42. P. Hapiot, P. Neta, J. Pinson, C. Rolando and S. Schneider, *New J. Chem.*, 1993, **17**, 211.
43. S. V. Jovanovic, A. Harriman and M. G. Simic, *J. Phys. Chem.*, 1986, **90**, 1935.
44. P. S. Shudar and D. A. Armstrong, *J. Phys. Chem.*, 1987, **91**, 6532.
45. G. Merényi, J. Lind and X. Shen, *J. Phys. Chem.*, 1988, **92**, 134.
46. S. V. Jovanovic, S. Steenken and M. G. Simic, *J. Phys. Chem.*, 1990, **94**, 3583.
47. S. V. Jovanovic, I. Jankovic and L. Josimovic, *J. Am. Chem. Soc.*, 1992, **114**, 9018.
48. G. Merényi, J. Lind and M. Jonsson, *J. Am. Chem. Soc.*, 1993, **115**, 4945.
49. D. A. Armstrong, Q. Sun and R. H. Schuler, *J. Phys. Chem.*, 1996, **100**, 9892.
50. Li Cang and M. Z. Hoffman, *J. Phys. Chem. B*, 1999, **103**, 6653.
51. S. V. Jovanovic, M. Tosic and M. G. Simic, *J. Phys. Chem.*, 1991, **95**, 10824.
52. M. Charton, *Progr. in Phys. Org. Chem.*, 1971, **8**, 235.
53. W. F. Hodnick, E. B. Milosavljevic, J. H. Nelson and R. S. Pardini, *Biochem. Pharmacol.*, 1988, **37**, 2607.
54. N. Gotoh and E. Niki, *Biochem. Biophys. Acta*, 1992, **115**, 201.
55. K. J. A. Davies, *Biochem. Soc. Symp.*, 1996, **61**, 1.
56. M. Roder, G. Foldiak and L. Wojnarovits, *Radiation Phys. Chem.*, 1999, **55**, 515.
57. C. P. Andrieux, P. Hapiot and J. M. Saveant, *Chem. Rev.*, 1990, **90**, 723.

CHAPTER 8

Enzymatic Synthesis of Quercetin Glucosides and Glucuronides

G. W. PLUMB, K. O'LEARY, A. J. DAY AND G. WILLIAMSON

1 Introduction

The implication of flavonoids in the diet is currently receiving much interest and the biological activity of these polyphenolic components is of great importance in understanding their mode of action *in vivo*. Interest has been focused on the health benefits derived from a diet rich in flavonoids. These compounds, which are found in many fruits and vegetables,[1] may play a role in reducing the risk of developing long-term degenerative diseases such as cancer and cardio-vascular disease.[2,3] One class of flavonoids, the flavonols, exist in plants as conjugates of the aglycones, such as quercetin (see Figure 1) and kaempferol. They have been shown to exhibit high antioxidant activity[4,5] and inhibition of some enzyme activities.[6,7] In the majority of previous *in vitro* studies, the aglycones were used although, in the diet, flavonols are ingested as glycosides (onions, apples, tea and red wine being the main sources). Their biological activities have been shown to be dependent on the degree and nature of the substitution of the phenolic

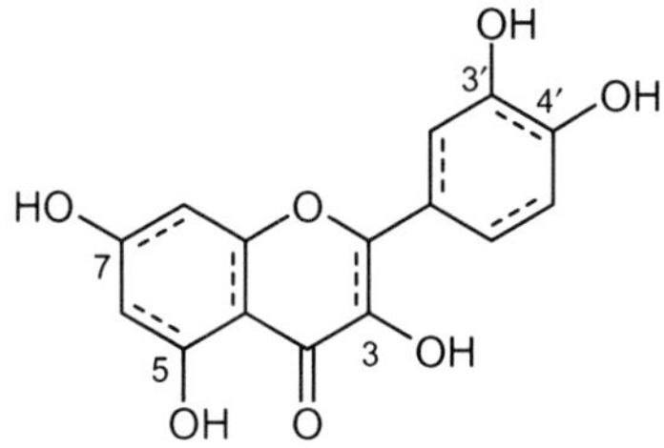

Figure 1 *Structure of quercetin showing the phenolic ring positions of the five hydroxyl groups*

hydroxyl groups. This has profound effects on both the radical scavenging properties and phase II enzyme induction.[8–10]

In order to study the effects of quercetin on the epithelial cells of the small intestine and its fate after absorption and metabolism, it is necessary to study the quercetin glycosides species actually present in food. In onion the most abundant forms of quercetin are the 4′-glucoside and the 3,4′-diglucoside,[8] whereas in other foods, such as tea and green bean, there are a variety of sugar substitutions at the 3-position.[9,11] In the case of onions, cooking has no effect on the level or nature of the quercetin glycosides present;[12] therefore these compounds will be the quercetin species presented to the cells in the small intestine.

Quercetin aglycone has not been detected in plasma using sensitive techniques such as LC–MS and so this molecule must be rapidly conjugated to glucuronide or sulfate moieties in the enterocyte.[13] Recently the major circulating forms of quercetin have been identified as glucuronides and sulfates of quercetin with diglucuronides and mixed sulfo-glucuronide conjugates as minor components.[14] Therefore, in order to study the cellular effects of quercetin on tissues such as the liver, kidney, brain *etc*., it is necessary to examine the biological activity of these circulating forms of quercetin in human plasma to fully understand the contribution of quercetin to human health. The study of the glucuronides of quercetin and not the aglycone itself is key to this research area.

There are a variety of enzymes present in *Allium* species involved in the metabolism of flavonols. For example, in onions there are glucosyltransferases and glucosidases present which are specific for the 3 and 4′-glucosides.[15] The optimum pHs are such that transglucosylation and glucosidation of quercetin 3-glucoside should be active under acidic conditions. In contrast, the same reactions for quercetin 4′-glucoside should be active under neutral or relatively alkaline conditions. Using this information, we enzymatically produced the three glucosides of quercetin, both labelled and unlabelled. Glucuronides were produced using UDP-glucuronyl transferase activity. The production of purified quercetin glycosides and glucuronides in milligram quantities is essential for studies of uptake and metabolism, as well as biological effects of these compounds in a variety of cellular and animal model systems. In addition we also describe the production of radio-labelled compounds for studies that require more sensitive detection methods.

2 Production of Quercetin Glycosides

Preparation of Crude Onion Enzyme Extract

Approximately 100 g of onion slices (the middle portion was used as it contains more of the glucosyltransferase activity and less of the glucosidase activity[15]) were homogenised with 10 g of polyvinylpyrrollidine and 300 ml of 25 mM McIlvaine's buffer (a solution of 0.1 M citric acid and 0.2 M disodium hydrogen orthophosphate) containing 10 mM β-mercaptoethanol. To ensure the stability of the enzyme, this process was performed on ice. The resulting homogenate was centrifuged at 10 000 g for 15 min at 4 °C. The supernatant was filtered under vacuum through Whatman 541 filter paper prior to a further centrifugation step at 10 000 g for 10 min at 4 °C. The supernatant was then subjected to a 95% ammonium sulfate cut to precipitate the protein fraction. The mixture was subsequently centrifuged at 15 000 g for 30 min at 4 °C. The pellets were redisolved in McIlvaine's buffer to a total volume of 4 ml and applied in four portions to a PD-10 desalting column. The eluate from the column was pooled and the volume made up to 6 ml. This constituted the crude onion enzyme preparation. For the large-scale production of quercetin glycosides, this extract was concentrated three-fold by evaporation under nitrogen.

Preparation of Crude Leek Enzyme Extract

The crude leek enzyme preparation was performed as described above for the onion with two modifications. The starting material was approximately 100 g of the green portion of the leek. Due to the efficiency of this crude enzyme preparation, the final step to concentrate the extract three-fold was not required.

Enzymatic Synthesis of Quercetin Glycosides

Quercetin (30 μl of 0.5 mg ml^{-1} stock) was incubated with 50 μl 2 mM UDP-glucose, 20 μl 10 × phosphate buffered saline (PBS) buffer at pH 6.5, 50 μl onion enzyme extract and 50 μl water at 37 °C for 2 h. The resulting HPLC chromatograms can be seen in Figure 2.

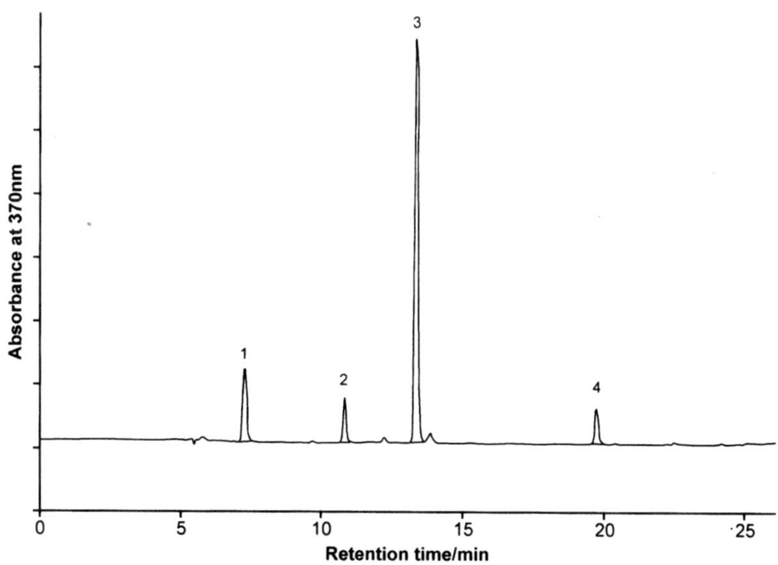

Figure 2 *HPLC chromatograph of quercetin glycosides from an incubation of quercetin with an onion extract (incubation for 2 hours at pH 6.5). For HPLC conditions, see text. Peak 1: Quercetin 3,4′-diglucoside; Peak 2: Quercetin 3-glucoside; Peak 3: Quercetin 4′-glucoside; Peak 4: Quercetin*

Ideally the production of the three quercetin glycosides from quercetin should be equimolar with a 100% yield (this is particularly important for subsequent production of radio-labelled compounds). From a previous study[15] it was thought this could be achieved by using the correct pH and incubation times. Figure 3 a, b and c show the enzymatic synthesis of glycosides at three pHs and at three incubation periods using the onion extract. After one hour incubation at pH 5.5, 6.5 and 7.7, the conversion from quercetin to glycosides was approximately 20–60%. After two and four hour incubations the conversion is much greater, typically 80–95%. At the three pHs chosen, the production of the quercetin 4′-glucoside predominated and the quercetin 3,4′- diglucoside appeared difficult to achieve. It was concluded that production of an equimolar mixture of the three glycosides with 100% efficiency would not be possible using these conditions. Substituting a leek extract for the onion extract, and using a one hour incubation in PBS at pH 7.2, the quercetin 3-glucoside was produced almost exclusively. These were the conditions selected for the final production of the 3-glucoside (see Figure 3d). To produce the other two glycosides, the onion extract was concentrated three-fold, and a long incubation time at pH 7.2 employed, to produce quercetin 3,4′-diglucoside in an almost equimolar mixture with the

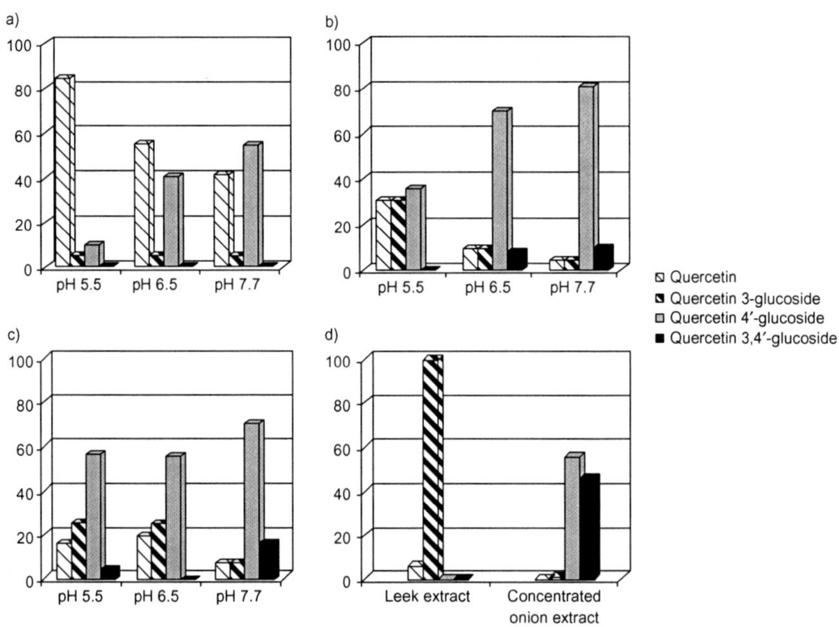

Figure 3 *Histogram showing the production of quercetin glucosides (in %) from (a) incubation of quercetin with onion extract for 1 h using pH 5.5, 6.5 and 7.7; (b) incubation of quercetin with onion extract for 2 h using pH 5.5, 6.5 and 7.7; (c) incubation of quercetin with onion extract for 4 h using pH 5.5, 6.5 and 7.7; (d) incubation of quercetin with leek extract for 1 h at pH 7.2 and incubation of quercetin with concentrated onion extract for 4 h at pH 7.2. For HPLC conditions, see text*

4′-glucoside. These conditions were used for the production of the 4′-glucoside and the 3,4′-diglucoside (see Figure 3d). By using these two methods, the three glycosides could be produced equally with a high yield (see Figure 3d).

Enzymatic Synthesis of Radio-labelled Quercetin Glycosides

The starting material used for this purpose was ^{14}C-labelled quercetin with a specific activity of 52.9 mCi mmol^{-1}. Quercetin 4′-glucoside and quercetin 3,4′-diglucoside were produced in roughly equal amounts by the incubation of 86.4 μl labelled quercetin (160 kBq), 50 μl 2 mM UDP-glucose, 50 μl 10 × PBS at pH 7.2, 300 μl concentrated onion enzyme extract and 13.6 μl water at 37 °C for 4 h. Quercetin 3-glucoside was produced by the incubation of 43.2 μl labelled quercetin (80 kBq), 50 μl 2 mM UDP-glucose, 50 μl 10 × PBS buffer at pH 7.2, 100 μl leek enzyme extract and 256.8 μl water at 37 °C for 2 h. This procedure yielded approximately 65 kBq of quercetin 3-glucoside after preparative HPLC purification.

HPLC Analysis for Flavonoids

Analytical HPLC

A modified version of a previously published analytical HPLC method was used.[16] The analysis was carried out by reversed-phase HPLC using a Hewlett Packard 100 model. Data were collected and analysed using Chemstation software. Solvents A (water:tetrahydrofuran:trifluoroacetic acid 98:2:0.1, v:v:v) and B (acetonitrile), were run at a rate of 1 ml min^{-1}, using a gradient of 17% B (2 min), increasing to 25% B (5 min), 35% B (8 min), 50% B (5 min) and then to 100% B (5 min). A column clean-up stage was performed at 100% B (5 min) followed by a re-equilbration at 17% B (15 min). The column was packed with Prodigy 5 μm ODS3 reversed-phase silica, 250 mm by 4.6 mm i.d. (Phenomenex, Macclesfield, UK). A guard column, 30 mm by 4.6 mm i.d. was also used to protect the column because the majority of samples passed through the column contained biological material. The eluent was monitored at 270 nm and 370 nm with a diode array detector. Internal standard (apigenin; 30 μM) was added to each sample to ensure repeatability and reproducibility of experiments. An external standard of rutin was run approximately every 6 runs.

Preparative HPLC

Preparative HPLC was performed using a Prodigy 5 μ ODS3 reversed-phase silica, 250 mm by 21.2 mm i.d. (Phenomenex Ltd, Macclesfield, UK) column with an isocratic solvent [(8:2) acetonitrile/0.1% triflouroacetic acid (TFA)] at a flow rate of 5 ml min^{-1}. The column effluent was monitored at 270 and 370 nm, and the fractions were collected using a Gilson fraction collector.

3 Production of Quercetin Glucuronides

Production of Pig Liver Extract

Hepatic tissue used for the pig liver extract was from a healthy pig liver obtained from a local abattoir and frozen immediately and stored at −70 °C until required. The pig liver extract was freshly prepared by the method of Lambert,[17] using 20 g of pig liver and a buffer of 20 mM BisTris propane, pH 7.0, containing 0.25 M sucrose and 0.15 M KCl. The final supernatant after ultra-centrifugation was free from lysosomal contamination as measured by an acid phosphatase assay.[17]

Enzymatic Synthesis of Flavonoid Glucuronides

Mixed quercetin glucuronides were produced enzymatically from pig liver. Quercetin (200 μM) was incubated for 180 min at 37 °C in HEPES buffer (25 mM, pH 7.2 containing 10 mM $MgCl_2$) a post-lysosomal fraction of pig liver, UDP-glucuronic acid (final concentration, 4 mM) UDP-glucosamine (final concentration, 2 mM). The extent of conjugation was greater than 95%. Quercetin glucuronides were identified by HPLC, UV spectra and LC–MS (see below). This method was particularly good for preparation of large amounts of quercetin 4′-glucuronide (see Figure 4).

An alternative method of producing quercetin glucuronides was to use a commercial preparation of bovine UDP glucuronyl transferase (UDPGT) obtained from Sigma-Aldrich (Poole, UK). Quercetin (typically 6.25 μl of a 16 μM stock) was added to 65 μl 2 mM UDP glucuronic acid (UDPGA), 50 μl, 5 mg ml^{-1} UDPGT and 378.75 μl 20 mM potassium phosphate buffer at pH 7. Aliqouts were taken from the mixture for HPLC analysis at 1, 2, 3 and 6 hours (see Figure 5). This method was particularly efficient at producing a mixture of the 3-, 3′- and 4′-glucuronides (incubation for one hour) for purification by HPLC.

Enzymatic Synthesis of Radio-labelled Quercetin Glucuronides

The starting material used for this purpose was ^{14}C-labelled quercetin with a specific activity of 52.9 mCi mmol^{-1}. Quercetin 4′-glucuronide was produced by the incubation of radio-labelled quercetin (80 kBq) in 25 mM Hepes buffer at pH 7.2 containing 10 mM $MgCl_2$, a post-lysosomal fraction of pig liver, 4 mM UDP-glucuronic acid and 2 mM UDP-glucosamine for 3 hours at 37 °C. HPLC conditions are described above. This procedure yielded approximately 60 kBq of quercetin 4′-glucuronide after HPLC purification.

Quercetin 3-, 3′- and 4′-glucuronides were produced by incubating quercetin (60 kBq) with UDPGA, UDPGT in 20 mM potassium phosphate buffer at pH 7, as described above, to yield approximately 15 kBq of each glucuronide after HPLC purification (see above for conditions).

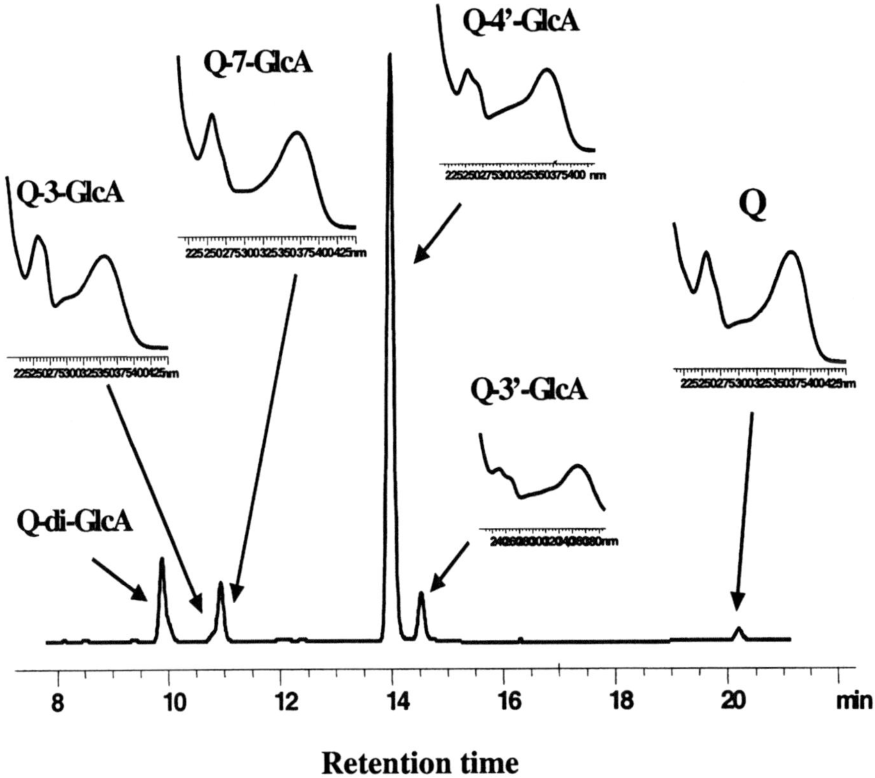

Figure 4 *HPLC chromatogram (with UV spectra for each peak) of quercetin glucuronides from a post-lysosomal fraction of pig liver (quercetin-3-glucuronide, quercetin-7-glucuronide, quercetin 4'-glucuronide, quercetin-3'-glucuronide). The position of conjugation was determined previously.*[20,21] *For HPLC conditions, see text*

Liquid Chromatography–Mass Spectrometry (LC–MS)

Positive ion electrospray LC–MS measurements were performed using a Micromass Quattro II (Micromass, Manchester, UK), equipped with a Z-spray™ source. The Micromass Quattro II triple quadrupole tandem mass spectrometer incorporates an improved electrospray and APCI LC/MS interface and is also equipped with nanospray (a form of electrospray useful for small samples), GC–MS, EI and CI. The instrument has a mass range of 1–4000 Da and a low-resolution accurate mass measurement capability in APCI mode. Applications will include LC–MS and LC–MS–MS determination of stable isotope labelled nutrient and non-nutrient metabolites, LC–MS analysis of complex mixtures of natural products, accurate determination of the molecular weights and purity of proteins and polysaccharides, and peptide sequencing by MS–MS.

Electrospray Ionisation (ES or ESI) ionises samples directly from solution. Samples are sprayed at flow rates of 10–1000 $\mu l\,min^{-1}$ through a capillary needle held at a potential of 2–3.5 kV: older style ES sources can only accommodate lower flow rates (typically 10–50 $\mu l\,min^{-1}$). This process generates a spray of

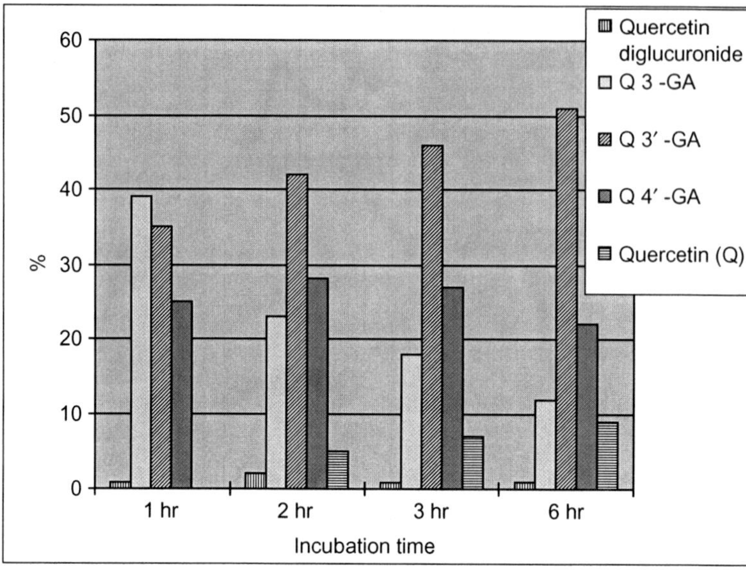

Figure 5 *Histogram showing the production of quercetin glucuronides (in %) from the incubation of quercetin (6.25 μl of a 16 μM stock) with 65 μl 2 mM UDP glucuronic acid (UDPGA), 50 μl 5 mg ml^{-1} UDPGT and 378.75 μl 20 mM potassium phosphate buffer at pH 7.0 for incubations of 1, 2, 3 and 6 hours*

charged droplets at atmospheric pressure. The solvent spray passes through a staged pumping system and vaporises. Charged molecules are expelled from the shrinking droplets and are mass analysed by the mass spectrometer (this can be of any type but quadrupole and magnetic sector instruments are the most widely used). Large molecules, proteins for example, acquire multiple charges, bringing their mass/charge ratio into the measurement range of conventional mass spectrometers that only have a mass range of a few thousand Da. ES can be combined with HPLC: 2.1 mm bore HPLC columns are preferred, but not essential, for this purpose. ES is particularly useful for the analysis of large molecules (*e.g.* proteins of up to 100 000 Da molecular mass) or smaller, highly polar, thermally labile compounds.

Samples were introduced using a Hewlett Packard 1050 HPLC (Agilent, Stockport, UK) equipped with a diode array detector. Eluent flow (1 ml min^{-1}) was split between the diode array detector and the mass spectrometer ion source in the approximate ratio 8:1. The electrospray capillary voltage was set to 3.5 kV and the cone voltage to 28 V. Source block temperature was 140 °C and desolvation temperature 350 °C. Nitrogen was used as the drying and nebulising gas at flows of 400 l h^{-1} and 20 l h^{-1} respectively. Full scan spectra were obtained in positive ion mode from *m/z* 50–1000 at a scan rate of 1.5 s and a cone voltage of 27. Selected ion monitoring was conducted on mass channels 303, 479 and 655 with a scan window of 0.2, dwell times of 0.1 s/channel and an interchannel delay of cycle time of 0.03 s. Diode array spectra were scanned from 190 and 450 nm, with an interval of 2 nm. Instrument control, data acquisition and processing were performed using Micromass MassLynx™ version 3.4 data system and software.

HPLC Analysis for Flavonoids

Analytical and preparative HPLC was performed as described in the previous section.

4 Concluding Remarks

We have described the synthesis of a range of quercetin glycosides and glucuronides by enzymatic means. The synthesis and subsequent purification offers a straightforward method of producing these compounds in milligram amounts. The method can also be modified to produce radio-labelled compounds, and, if a further modification is applied (to use radio-labelled glucose or glucuronic acid), to produce double-labelled compounds. The radio-labelled quercetin glycosides have already been successfully used in uptake and metabolism studies[18,19] and will be invaluable in the future for a range of studies related to absorption, metabolism and cellular effects.

Acknowledgements

The work in this chapter was funded by the Biotechnology and Biological Research Council and the European Union (POLYBIND–QLK1–1999-00505). The author would also like to thank Unilever, Vlaardingen, The Netherlands for the gift of radio-labelled quercetin and Fred Mellon and Mhorag Campbell for their experimental expertise.

References

1. J. B. Harborne, *'Encyclopedia of Plant Physiology Secondary Plant Products'*, Springer, New York, USA, 1980.
2. G. Block, *Nutr. Rev.*, 1992, **50**, 207.
3. P. Knekt, R. Jarvinen, A. Reunanen and J. Maatela, *Br. Med. J.*, 1996, **312**, 478.
4. C. A. Rice-Evans, N. J. Miller, G. P. Bolwell, P. M. Bramley and J. B. Pridham, *Free Rad. Res.*, 1995, **22**, 375.
5. W. Bor, W. Heller, C. Michel and M. Saran, *Methods Enzymol.*, 1990, **186**, 343.
6. T. Walle, E. A. Eaton, and U. K. Walle, *Biochem. Pharmacol.*, 1995, **50**, 731.
7. E. L. daSilva, T. Tsushida and J. Terao, *Arch. Biochem. Biophys.*, 1998, **349**, 313.
8. G. Williamson, G. W. Plumb, Y. Uda, K. R. Price and M. J. C. Rhodes, *Carcinogenesis*, 1997, **17**, 2385.
9. G. W. Plumb, K. R. Price and G. Willamson, *Redox Rep.*, 1998, **4**, 13.
10. G. W. Plumb, K. R. Price, M. J. C. Rhodes and G. Willamson, *Free Rad. Res.*, 1997, **27**, 429.
11. G. W. Plumb, K. R. Price and G. Williamson, *Redox Rep.*, 1999, **4**, 123

12. K. R. Price, J. R. Bacon and M. J. C. Rhodes, *J. Sci. Food Agric.*, 1997, **45**, 938.
13. J.-H. Moon, R. Nakata, S. Oshima, T. Inakuma and J. Terao, *Am. J. Physiol.*, 2000, **279**, R461.
14. A. J. Day, F. Mellon, D. Barron, G. Sarrazin, M. R. A. Morgan and G. Williamson, *Free Rad. Res.*, 2001, **35,** 941.
15. T. Tsushida and M. Suzuki, *Nippon Shokuhin Kagaku Kaishi*, 1996, **43**, 642.
16. K. R. Price, J. R. Bacon and M. J. C. Rhodes, *J. Agric. Food Chem.*, 1997, **45**, 938.
17. N. Lambert, *'Protein Purification Methods: A Practical Approach'*, IRL Press, Oxford, UK, 1989.
18. J. M. Gee, M. S. DuPont, A. J. Day, G. W. Plumb, G. Williamson and I. T. Johnson, *J. Nutr.*, 2000, **130**, 2765.
19. G. W. Plumb, W. R. McLauchlan, P. A. Kroon, A. J. Day C. B. Faulds, J. M. Gee, M. S. DuPont and G. Williamson, *Polyphenols Commun.*, 2000, **2**, 401.
20. A. J. Day, Y.-P. Bao, M. R. A. Morgan and G. Williamson, *Free Rad. Biol. Med.*, 2000, **26**, 1234
21. K. A. O'Leary, A. J. Day, P. Needs, W. S. Sly, N. M. O'Brien and G. Williamson, *FEBS Lett.*, 2000, **503**, 103.

CHAPTER 9

Chemical Synthesis of Flavonoid Conjugates

DENIS BARRON, CÉCILE CREN-OLIVÉ AND PAUL W. NEEDS

1 Introduction

Flavonoid compounds are widely found in food and so they represent an integral part of the diet.[1] More particularly, four classes of compounds are of special interest: flavonols like quercetin (onions), isoflavones like genistein (soya), flavanones like naringenin or hesperetin (Citrus) and finally flavan-3-ols like catechin and their polymerised forms, the proanthocyanidins (wine, tea, chocolate) (Figure 1).

All these compounds are quickly metabolised in the body by methylation, glucuronidation, sulfation[1] and/or ring cleavage.[2] Thus, *in vivo*, it is the conjugate forms of flavonoid compounds that exert biological activity and not their native

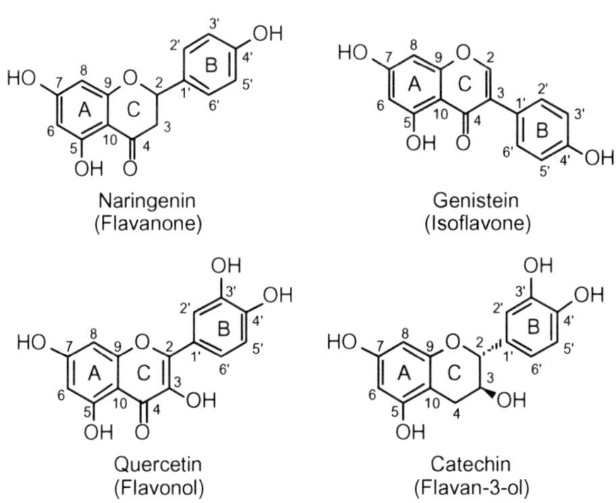

Figure 1 *Structures of some of the flavonoids present in the diet*

forms. While the general reactions involved in the conjugation of flavonoids are well established, the exact structural identification of the conjugates often remains unclear, mostly due to the lack of appropriate reference compounds. Surprisingly, it is only recently that regioselective methods have been designed for the preparation of flavonoid conjugates. The purpose of this chapter is to review the different chemical methods useful in the preparation of flavonoid conjugates. The survey is restricted to the preparation of methylated, glucuronidated and sulfated conjugates of quercetin, catechin, naringenin and genistein.

2 Strategies for the Preparation of Flavonoid Conjugates

The Total Synthesis of Basic Structures

Introduction

Naringenin, quercetin and (+)-catechin can be purchased at relatively low prices. Most monomethylated derivatives of quercetin, such as isorhamnetin (3′-methyl), tamarixetin (4′-methyl) and rhamnetin (7-methyl), as well as the isoflavones daidzein, genistein and their 4′-methyl ethers (formononetin and biochanin A, respectively) are also commercially available. However, the latter compounds are sold in mg amounts and the cost of the quantity required for conjugate preparation may turn out to be prohibitive. In this case it may be advantageous to prepare those compounds in bulk quantities in the laboratory, especially since most procedures are quite easily designed. In this section only the preparation of basic skeletons by total synthesis is described. The compounds obtained by an hemisynthetic transformation of a natural precursors are covered in the following section.

Total synthesis of isoflavones

This is a two-step process (Scheme 1), involving the synthesis of a phenyl benzyl ketone, followed by the formation of the pyran ring using dimethylformamide as a carbon donor. The preparation of the phenyl benzyl ketone follows two different routes depending on the degree of hydroxylation of the A-ring of the desired isoflavone. Phenyl benzyl ketones involved in the synthesis of resorcinol-type isoflavones, like daidzein or formononetin, are prepared by direct acylation of resorcinol with 4-hydroxy- or 4-methoxy-phenylacetic acids. The synthesis of the phenyl benzyl ketones, intermediates in the obtention of phloroglucinol type isoflavones like genistein or biochanin A, involves an Hoesch condensation of a phenylacetonitrile on phloroglucinol, followed by hydrolysis of the resulting ketimine. The Hoesch condensation involves saturation with HCl gas, which is no problem if the proper equipment is used (Figure 2).

The formation of the pyran ring is a common process to both routes. A very elegant method, making use of microwave-assisted synthesis has been reported.[3]

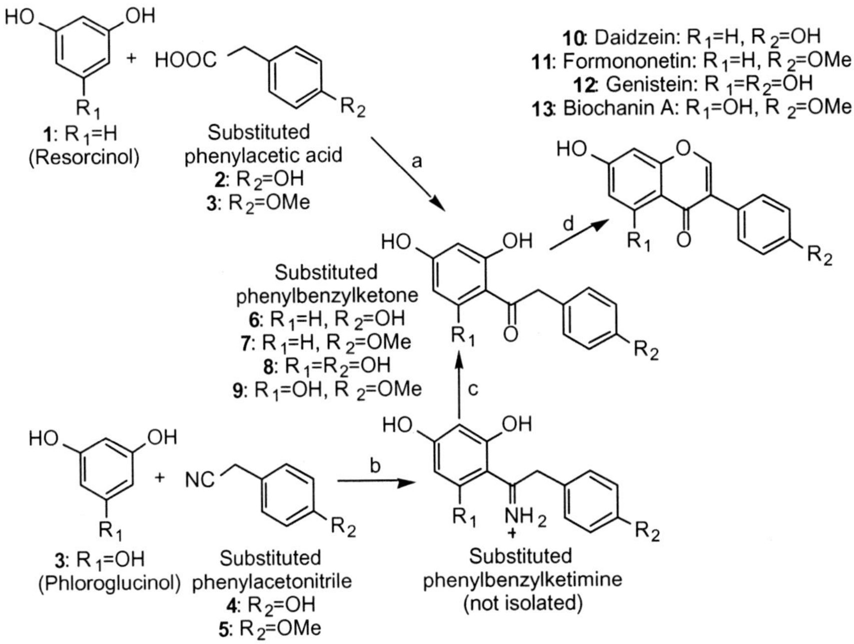

Scheme 1 *Total synthesis of common isoflavones:*[3] (a) *BF_3 etherate, DMF, 70–80 °C;* (b) *HCl gas, Et_2O;* (c) *H_2O;* (d) *BF_3 etherate, $MeSO_2Cl$, DMF, 60–70 °C*

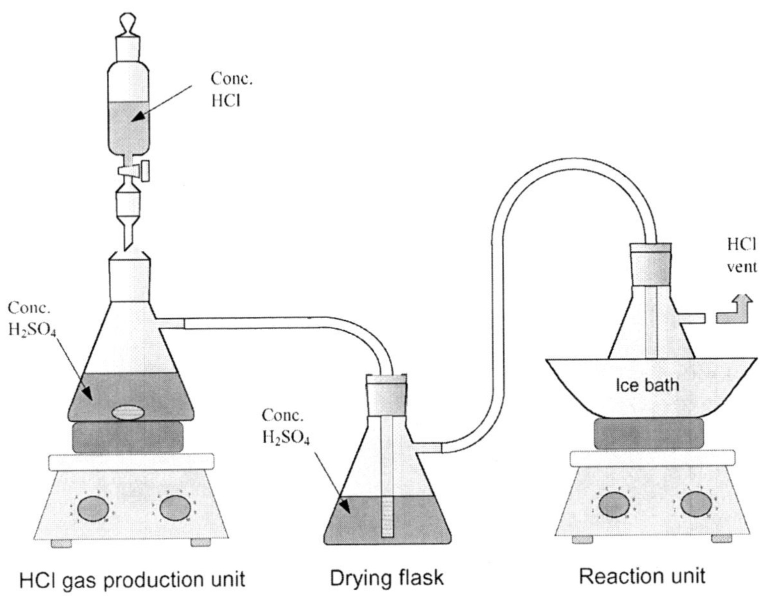

Figure 2 *Example of an apparatus for Hoesch condensation*

In the absence of proper microwave equipment, however, the reaction can be easily carried out by conventional heating, although in more reduced yield. The very common isoflavones daidzein, genistein and their 4′-monomethyl ethers formononetin and biochanin A are obtained in good 70–90% yield using microwave-mediated synthesis.[3]

Total synthesis of 3-methoxyflavones

As we shall discuss later in Section 2, selective methylation of quercetin at position 3 is difficult to achieve. On the contrary, 3-methylquercetin (**21**) is of easy access by total synthesis (Scheme 2).[4] The key intermediate is ω-methoxyphloroacetophenone (**15**), prepared in excellent yield by Hoesch condensation of methoxyacetonitrile (**14**) on phloroglucinol (**3**). Acetophenone (**15**) is the precursor of rings A and C of 3-methylquercetin. The precursor of ring B is a 3,4-dihydroxybenzoic acid, activated as its chloride. The preparation of this precursor is by itself a three-step process. Direct benzylation of 3,4-dihydroxybenzoic acid would lead to benzyl esterification of the acid function, in addition to protection of the phenolic groups. Thus it is more advantageous to benzylate 3,4-dihydroxybenzaldehyde (**16**). The benzyl aldehyde (**17**) is oxidised in excellent yield by potassium permanganate to its corresponding benzoic acid derivative (**18**), finally converted to the chloride (**19**). The protected flavone (**20**) is obtained in one step from acetophenone (**15**) and benzoyl chloride (**19**) by a modified Baker–Venkataraman transformation. Since the reaction is performed in the presence of an excess of benzoyl chloride, compound **20** is first obtained in partially esterified form. Thus a saponification step is included in the work-up of

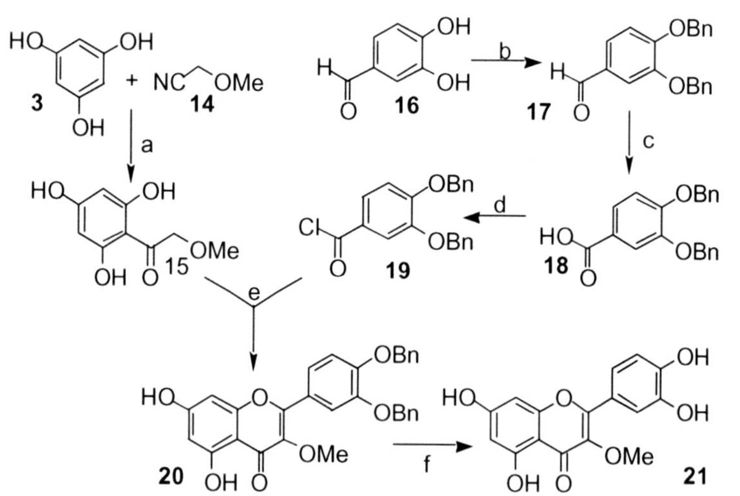

Scheme 2 *Total synthesis of 3-methylquercetin* (**21**):[4] (a) *HCl/$ZnCl_2$/Et_2O, 80%;* (b) *Bn–Br/K_2CO_3/DMF;* (c) *$KMnO_4$/Me_2CO–H_2O, 93%;* (d) *thionyl chloride/DMF, 98%;* (e) *K_2CO_3/TBAHS/toluene, 69%;* (f) *$Pd(OH)_2$/cyclohexene/EtOH, 99%*

compound **20** (see Section 3). Debenzylation of **20** yields 3-methylquercetin (**21**) in excellent yield.

The Hemisynthesis of Conjugates

Introduction

When the native flavonoid compound is commercially available in bulk quantities, the hemisynthetic preparation of conjugates is obviously more advantageous than the total synthesis. In this case, however, the regioselectivity of the synthesis relies on the differences in acidity of the various phenolic groups of the molecule. Examination of the ionisation constants listed in Table 1 demonstrates that very different situations prevail for the various classes of flavonoid compounds. Isoflavones and flavanones display a marked difference in reactivity between rings A and B, related to a lack of conjugation between ring B and the 4-carbonyl. This is due to the reduction of the 2,3-double bond in flavanones, and to the migration of ring B at position 3 in the case of isoflavones. In both cases, the A-ring remains conjugated to the 4-carbonyl, enhancing the acidity of the 7-phenolic hydroxyl, as compared to the 4′ one. On the contrary, the 5-hydroxyl is by far the least acidic, due to its hydrogen bonding with the 4-carbonyl. The case of flavonols like quercetin is more complex. Still the 5-hydroxyl remains the least acidic; however, both the A- and B-rings are conjugated to the 4-carbonyl. The reduced acidity of the 3′-hydroxyl is easily explained by its *meta* orientation as compared to the conjugated chain. The reactivity of the 3-hydroxyl is similar to that of the 3′-hydroxyl, due at least in part to a possibility of weak hydrogen bonding with the 4-carbonyl. In addition this position is clearly more hindered than the others. The *para* position of the 7- and 4′-hydroxyls is responsible for their enhanced acidity. However, the difference of reactivity of the latter positions

Table 1 *Experimental* pK_a *values for the four classes of flavonoids covered in this chapter*

	Isoflavones	*Flavanones*	*Flavonols*	*Flavan-3-ols*
3-OH			9.60[d]	
5-OH	11.76[a]	Not available	11.56[d]	9.43[e]
7-OH	7.48[b]	6.70[c]	7.39[d]	9.58[e]
3′-OH			9.75[d]	9.02[e]
4′-0H	9.53[b]	9.10[c]	8.28[d]	9.12[e]
Usual order of reactivity[f]	$7 \gg 4' \gg 5$	$7 \gg 4' \gg 5$	$7 > 4' > 3 \sim 3' \gg 5$	$3' \sim 4' \sim 5 \sim 7$

[a] Determined spectrophotometrically for 5-hydroxyisoflavone.[5]
[b] Tentative assignment from the data of reference 6.
[c] Tentative assignment from the data of reference 7.
[d] Taken from reference 8.
[e] Microscopic pK_a; taken from reference 9.
[f] See text pg 192.

is much less important than in the case of flavanones and isoflavones. Both pK_a determinations (Table 1) and experimental results plead for higher acidity of the 7-hydroxyl, although the subject has been a matter of debate in the past.

The more acidic hydroxyls of flavonols, made more nucleophilic by deprotonation under the mildly basic conditions commonly employed for conjugation, are often more reactive; here acidity equates with reactivity. But in some cases, a small equilibrium concentration of a more nucleophilic phenoxide formed from a less acidic hydroxyl can react preferentially in some circumstances. Treatment with a strong base in an aprotic solvent to give multiple deprotonation can also favour reaction at the most nucleophilic phenoxide ion (formed from the least acidic hydroxyl). Under acidic or neutral conditions, the greater intrinsic nucleophilicity of the less acidic hydroxyls can make them the more reactive. These factors, in particular cases, may lead to regioselective conjugation; in others, such as glucuronidations, mixtures of conjugates are obtained and a strategy of selective protection must be used to direct regioselectivity.

The case of flavan-3-ols like catechin is by far the most difficult. Both the 4-carbonyl and the 2,3-double bond have been lost, which results in similar acidities for all the phenolic hydroxyls (Table 1). Thus very different strategies based upon selective and successive protection of the different functions of catechol and resorcinol of the skeleton must be employed for the regioselective synthesis of catechin conjugates.

Therefore isoflavones and flavanones are expected to offer the best regioselectivity, followed by flavonols and finally by flavan-3-ols. The validity of this sequence will be mainly illustrated through the preparation of sulfated and methylated conjugates. At present, the preparation of glucuronidated conjugates still mostly relies on enzymatic synthesis and this aspect is covered in Chapter 8. However, a few long standing and some more recent developments in the chemical synthesis of flavonoid glucuronides are included in this chapter.

Hemisynthesis of sulfated conjugates. Flavonoid 7-monosulfates

The method (Scheme 3) is based on the dicyclohexylcarbodiimide (DCC) mediated esterification of phenolic hydroxyls by tetrabutylammonium hydrogen sulfate (TBAHS).[10] The phenolic hydroxyls are sulfated in a stepwise manner, in order of decreasing acidity. All the sulfated conjugates so far identified as metabolites are monosulfates. Thus the scope of the DCC method is mainly restricted to the preparation of 7-monosulfates. As expected, isoflavones like daidzein or genistein, and flavanones like naringenin, gave their 7-sulfated analogues as sole products,[11] due to a large difference in acidity between the 7- and 4′-hydroxyls (Table 1). In the case of quercetin, the 7-monosulfate was still major but lower regioselectivity was obtained since appreciable amounts of quercetin-7,4′-disulfate were produced as well.[10] Finally sulfation of (+)-catechin afforded a very complex mixture of conjugates (Barron, unpublished results), illustrating the weak differences in acidity of its phenolic hydroxyls (Table 1). In the latter case, no identification of the reaction products has been attempted yet

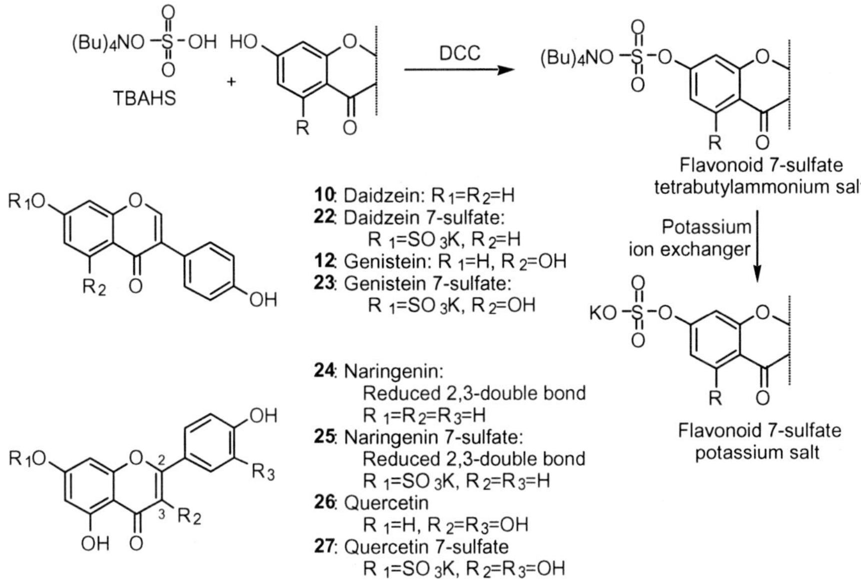

Scheme 3 *The DCC-mediated synthesis of flavonoid 7-monosulfates*[10,11]

since the separation of sulfated conjugates of similar sulfation level is usually difficult.

Gel filtration on Sephadex LH-20®, however, is very efficient in the separation of sulfates of different sulfation level, the disulfates being eluted before the monosulfates. One of the main disadvantages of the DCC method is the formation of a huge precipitate of dicyclohexylurea (DCU). Most of it can be removed by filtration of the reaction medium. Unfortunately, appreciable amounts of DCU remain soluble and tend to crystallise in the next steps of the purification procedure. When mg quantities of sulfate conjugates are prepared, the gel filtration step on Sephadex LH-20® achieves the elimination of DCU. Unfortunately, when more than 100 mg of conjugates are to be purified, the monosulfate band tends to co-elute with DCU during filtration.

It is of the utmost importance to remove this contaminant completely since it greatly disturbs the following ion exchange step. Complete removal of DCU, however, can be achieved as follows: the monosulfate fraction from Sephadex LH-20® is taken in an aqueous solution of tetramethylammonium hydroxide. Due to the presence of the sulfate, together with the ionisation of the phenolic groups under strong alkaline conditions, the flavonoid monosulfate immediately dissolves. On the contrary, DCU, of basic character, remains totally insoluble and can be completely eliminated by centrifugation. Subsequent acidification of the aqueous filtrate with an aqueous solution of TBAHS, followed by extraction with ethyl acetate, yields the monosulfate tetrabutylammonium salt, free of contamination. This method is applicable with no problems to poorly hydroxylated isoflavone- and flavone-sulfated conjugates that are stable in strong alkaline conditions. Naringenin 7-monosulfate opens to its corresponding chalcone forms in alkaline conditions. However, the acidification step with TBAHS usually leads to recyclisation to flavanone. Quercetin conjugates may partly degrade in alkaline conditions and in this case gel filtration of small amounts of conjugates remains the preferred method of elimination of DCU.

The last step of the synthetic scheme is the conversion of the tetrabutylammonium salt to its corresponding potassium salt. The usual technique involves the addition of a saturated solution of potassium acetate in methanol to a methanolic solution of the sulfate conjugate tetrabutylammonium salt. The less soluble potassium salt precipitates and can be isolated by centrifugation. While conversion is very efficient in the case of flavonoid disulfates, due to total insolubility of the potassium salt in methanol, on the contrary the conversion of flavonoid monosulfates is more problematic. In fact the partial solubility of the monosulfate potassium salts in methanol may cause partial conversion. In addition, as mentioned earlier, if the monosulfate tetrabutylammonium salt fraction is contaminated by DCU, addition of saturated potassium acetate creates a salting out effect. Thus DCU precipitates, together with the tetrabutylammonium salt that strongly binds to it. This results at the best in no conversion, and at the worst in total degradation of the sample. Efficient conversion of monosulfate esters can be carried out in all cases by filtration on a Dowex K ion exchange resin. The general strategy for the purification of flavonoid sulfate conjugates is summarised in Figure 3.

Flavonoid 3-monosulfates

Position 3 of the flavonoid ring is quite hindered. Therefore the DCC/TBAHS method, making use of bulky reagents, does not provide a direct access to, for example, quercetin 3-sulfate. The latter compound can be obtained in two steps, involving both chemical and enzymatic syntheses (Scheme 4).

Sulfation of quercetin by TBAHS under forced conditions yields quercetin 3,7,4′-trisulfate as sole product.[10] Quercetin 3-sulfate is generated from the trisulfate ester by enzymatic hydrolysis with aryl sulfatase.[12] At this point it is important to note that positive, or negative, hydrolysis of an unknown compound with aryl sulfatase is by no means an unequivocal demonstration of the presence or the absence of a sulfate group. In fact, the 3-sulfate of flavonols is resistant to aryl sulfatase hydrolysis,[12] probably due again to the hindered nature of this position. In addition, it has been recently demonstrated that flavonoid phosphates can be cleaved by this enzyme.[13]

Quercetin 3′-monosulfate

Sulfation of quercetin with sulfamic acid[14] (Scheme 5) gives rise to a complex mixture of sulfate esters among which quercetin 3′-sulfate is major. Thus this compound usually crystallises at one point or another of the working-up procedure. The mechanism of this reaction has never been studied in detail, but it seems likely that the formation of the 3′-conjugate is favoured, under these acidic conditions, for the reasons outlined elsewhere in this section. Quercetin 3′-sulfate

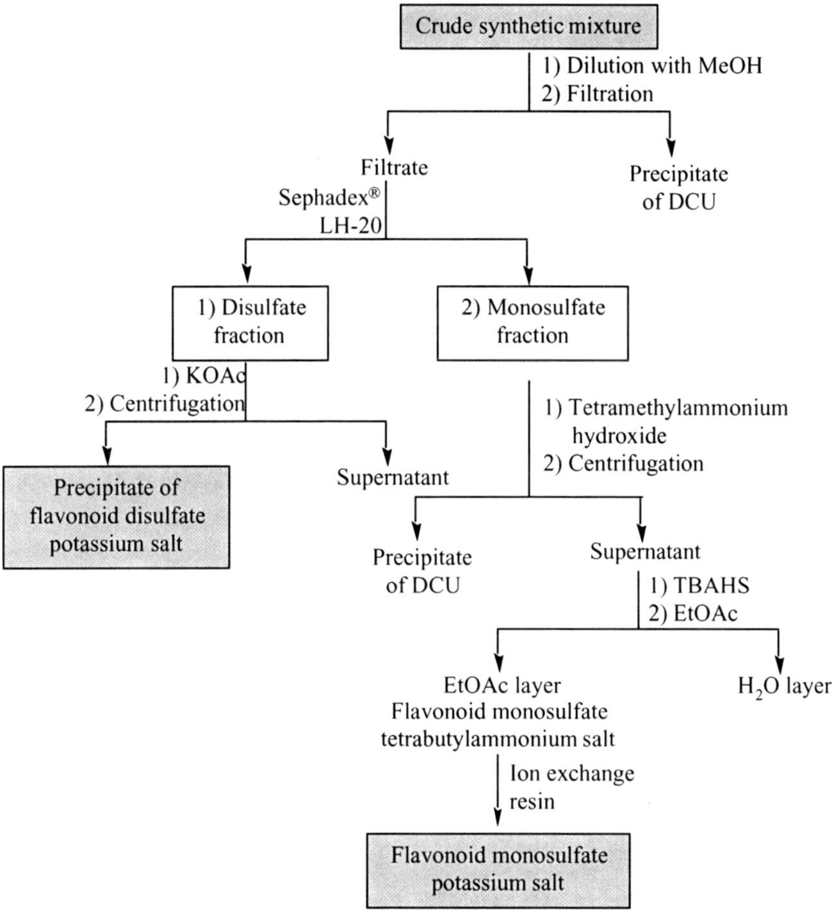

Figure 3 *A general strategy for flavonoid sulfate purification*

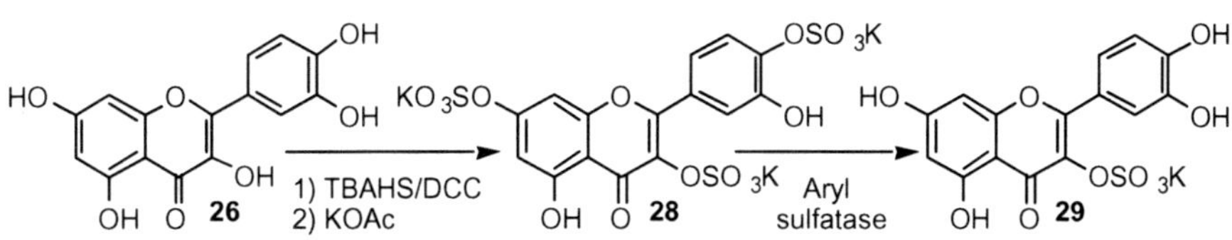

Scheme 4 *Preparation of quercetin 3-sulfate* (**29**)[10,12]

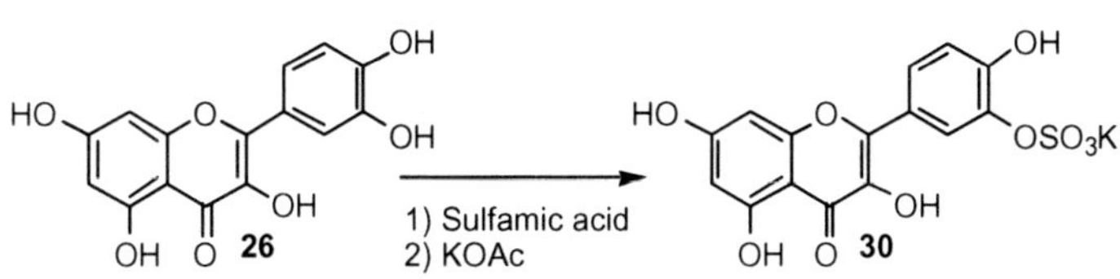

Scheme 5 *Access to quercetin 3′-sulfate* (**30**)[14]

has been identified as one of the circulating form of quercetin in the human body.[14]

Hemisynthesis of glucuronidated conjugates. Alkylation by glucuronic acid methyl ester

In contrast to sulfation, direct glycosylation[15] of the flavonoid nucleus generally results in poor yields and/or complex mixtures of products, and glucuronidation is no exception. Treatment of quercetin with acetobromoglucuronic acid methyl ester and potassium carbonate in DMF gave, after deprotection, a mixture from which quercetin 4′-*β*-D glucuronide and quercetin 3-*β*-D-glucuronide were isolated as major products, but in yields of only 2.8% and 5.1% respectively.[16] Usually selective protection of the phenolic functions is required, only leaving the desired position available for conjugation. Kaempferol 3-*β*-D-glucuronide[17] and quercetin 3-*O*-*β*-D-glucuronide[18] were made by treatment of 4′,7-di-*O*-benzylkaempferol and 4′,7-di-*O*-benzylquercetin respectively with acetobromoglucuronic acid methyl ester and silver carbonate. More recently, quercetin 7-*O*-*β*-D-glucuronide (**35**)[19] has been prepared (Scheme 6) using a similar approach starting from 3,3′,4′,5-tetra-*O*-benzoylquercetin (33).[20]

Thus quercetin (**26**) was perbenzoylated and treated with two equivalents of Ag^+ [as silver(I) carbonate] and 2.1 equivalents of phenol in pyridine. Under these conditions phenol displaced both the 7- and 4′-benzoyl groups to give phenyl benzoate and 3,3′,5-tri-*O*-benzoylquercetin (**32**). Treatment of the latter with a slight excess of penta-*O*-benzoylquercetin and 3.3 equivalents of Ag^+ led to a disproportionation reaction to give 3,3′,4′,5-tetra-*O*-benzoylquercetin (**33**) as

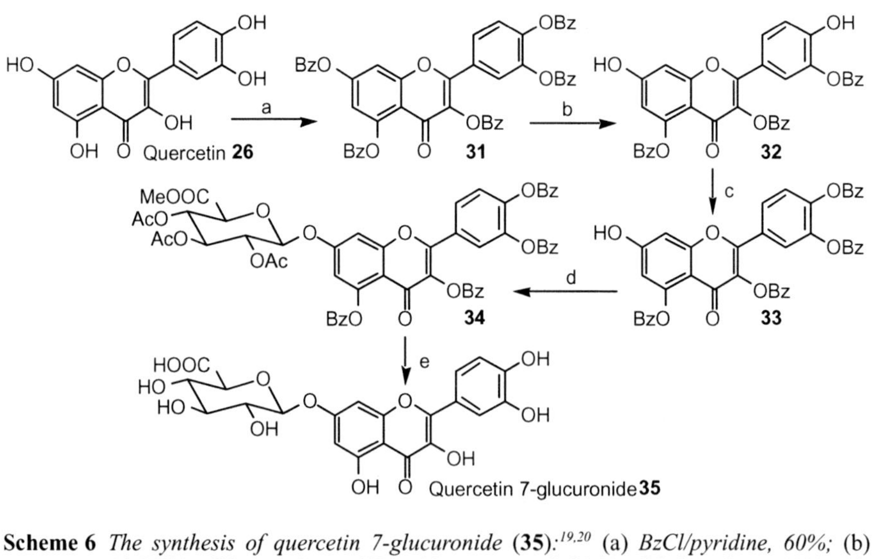

Scheme 6 *The synthesis of quercetin 7-glucuronide* (**35**):[19,20] (a) *BzCl/pyridine, 60%;* (b) *PhOH/Ag_2CO_3/pyridine, 60%;* (c) *compound* **31**/*Ag_2CO_3/pyridine, 58%;* (d) *acetobromoglucuronic acid methyl ester/Ag_2CO_3/collidine/CH_2Cl_2;* (e) *1 M NaOH, 8% from* **33**

the major product. Glucuronidation of **33** yielded the 7-conjugate (**34**). Quercetin 4′-*β*-D-glucuronide was obtained as a by-product (presumably due to partial migration of the 4′-benzoyl group to the 7-position under the reaction conditions). Yields in all cases, after deprotection, are typically only ⩽10% however. Treatment with methyl 2,3,4-tri-*O*-(trichloroacetimidyl-*β*-D-glucopyranuronate[21] and boron trifluoride etherate gave an improved yield, after deprotection, of 20%, but the reaction could not be repeated. Treatment of 4′,7-di-*O*-benzyl-3′-*O*-methylquercetin[22] with the imidate gave, this time reproducibly, isorhamnetin 3-*β*-D-glucuronide (Needs, unpublished results).

Daidzein 7-*β*-D-glucuronide has been recently synthesised (Scheme 7).[23] The method takes advantage of the differences in acidity between the 4′- and the 7-hydroxyls. Attempts to directly submit daidzein to glucuronidation failed, and protection of the 4′-position was necessary. Thus polyacetylated daidzein (**36**) was first selectively deacetylated at position 7, reacted with acetobromoglucuronic acid methyl ester and silver carbonate, and the conjugate (**38**) deprotected by saponification. Genistein 7-*β*-D-glucuronide has since been prepared from 4′,5-di-*O*-acetylgenistein in the same way.[24]

Obviously more work is required to optimise the synthesis of flavonoids, protected in suitable positions, that can be used for subsequent glucuronidation. Total synthesis of partially protected flavonols *via* ring annulation of suitable A- and B-ring precursors should bring the advantages of greater versatility and larger amounts of material to glucuronidate, afforded by the convergent approach. Moreover, recent findings that glucosides can be easily oxidised to glucuronides[25,26] open new avenues.

Oxidation of flavonoid glucoside to flavonoid glucuronide

Glucosylation of phenols can, indeed, be performed in mild conditions using phase-transfer catalyst such as tetrabutylammonium bromide[27] or tris[2-(2-meth-

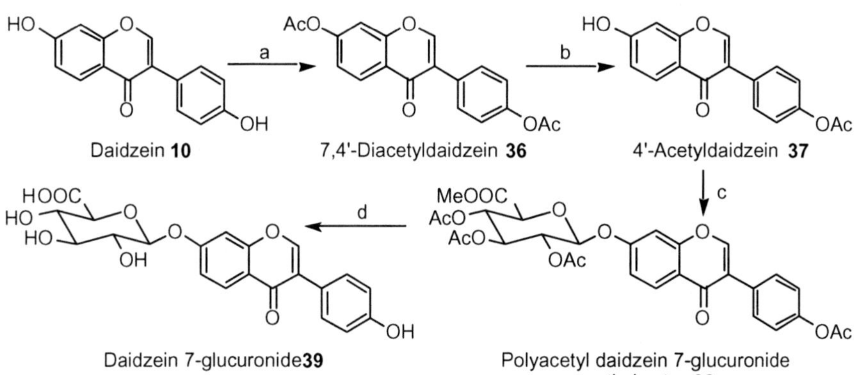

Scheme 7 *Preparation of daidzein 7-glucuronide* (**39**):[23] (a) *Ac_2O/H_2SO_4, 87%;* (b) *imidazole/THF–H_2O, 94%;* (c) *acetobromoglucuronic acid methyl ester/ Ag_2CO_3/collidine/CH_2Cl_2, 30%;* (d) *Na_2CO_3/MeOH-H_2O, 78%*

oxyethoxy)ethyl]amine (TDA-1) in good to fair yields.[28] These are generally higher than the yields of conjugation with acetobromoglucuronic acid methyl ester, except in particular cases such as the synthesis of quercetin 7-*O*-*β*-D-glucoside from treatment of 3,3′,4′,5-tetra-*O*-benzoyl quercetin with acetobromoglucose, where the yield is only slightly higher than that for the corresponding glucuronidation.[29] Thus the protected flavonoid compound could be first glucosylated, and the flavonoid glucoside selectively oxidised to flavonoid glucuronide by TEMPO in catalytic amount in the presence of sodium hypochlorite, which regenerates the free radical.[30,31] A recent work has just proposed the synthesis of quercetin 3-*O*-*β*-D-glucuronide using this approach.[26]

More precisely, selective protection of the catechol B-ring of quercetin (**26**) is achieved by dichlorodiphenylmethane leading to the ketal (**40**) (Scheme 8), which gives a selective access in the series substituted on the 3-position.[32] Under these conditions, position 3 of compound **40** was found to be the one of higher reactivity, although the equilibrium concentration of the 3-phenoxide will be lower than the 7-phenoxide, it is more nucleophilic – allowing the selective glucosylation at this position, and leading to **41** in 54% yield. Diglucosylated quercetin is formed in trace amounts. At this point, cleavage of the ketal (**41**) by hydrogenolysis catalysed by palladium hydroxide on charcoal leads to quercetin-3-*O*-*β*-D-glucoside (isoquercetin) in 86% isolated yield, while oxidation of **41** gives the glucuronide form.

The oxidation by TEMPO of the glucoside to the glucuronide requires the protection of the free hydroxyl functions in positions 5 and 7 since phenols are reactive under such conditions. Benzylation of these two phenolic positions with benzyl bromide in excess gives the fully protected quercetin-3-glucoside, and classical deprotection of the acetoxy protecting group on the sugar residue (sodium methylate then neutralisation by anionic resin in protonated form) leads

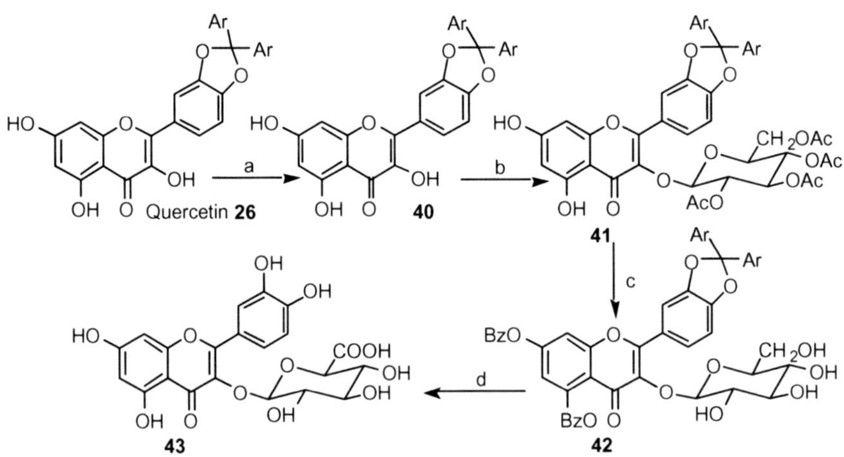

Scheme 8 *Synthesis of quercetin-3-O-β-D-glucuronide* (**43**):[26] (a) *Ph_2CCl_2, 180 °C, 37%;* (b) *2,3,4,6-tetra-O-acetyl-α-D-glucopyranosyl bromide, K_2CO_3, DMF, 54%;* (c) (i) *BnBr/K_2CO_3/DMF, 92%,* (ii) *MeONa, MeOH then Amberlite 120 H^+, 88%;* (d) *NaOCl, TEMPO then H_2/Pd$(OH)_2$, 59%*

to the unacetylated glucoside (**42**). The oxidation of the protected quercetin-3-*O*-*β*-D-glucoside (**42**) to the corresponding glucuronic acid by TEMPO in presence of NaBr and catalysed by NaOCl can now occur safely. The difficulty arises, at this step, from the low solubility of protected glucoside (**42**) but is circumvented using phase-transfer catalysis by tetrabutylammonium between CH_2Cl_2 and saturated sodium hydrogen carbonate.[30] After catalytic hydrogenation, quercetin-3-*O*-*β*-D-glucuronide (**43**) is obtained in 27% isolated yield from **40**.

So this new and original strategy provides an easy access not only to glucuronide quercetin, but also to glucuronide flavonoids more sensitive to drastic conditions such as catechin since this route proposes the glucuronidation of phenols in mild conditions. Moreover, this approach also opens a route to labelled molecules, needed in metabolic studies, starting from commercially available labelled glucose.

Hemisynthesis of methylated conjugates. Methylated quercetin

Despite the differences in the pK_a values of the five phenolic groups of quercetin, direct methylation of quercetin results in the formation of four different products (*i.e.* the 3-methyl, the 4′-methyl, the 7-methyl and the 3,7-dimethyl conjugates), even when one equivalent of methyl iodide is used.[32] Among the monomethylated compounds, 7-methylquercetin was present in 55% amount, which is in accordance with the highly acidic character of the 7-hydroxyl. However, most of the quercetin remained unreacted. When two equivalents of methyl iodide are used, most quercetin is consumed but extensive dimethylation takes place (formation of 3,7- and 3,4′-dimethylquercetin), while the major monomethylated product is now 3-methylquercetin. So obviously methylation of quercetin is less regiospecific than sulfation (see p. 192). However, a comparison between the two cases is interesting at this point. Monosulfation of quercetin yields, similarly to methylation, the 7-conjugate as major monoconjugated product.[10] But in contrast to methylation, the amount of unreacted quercetin is low, and the diconjugated contaminant is sulfated at positions 7 and 4′, and not at positions 3 and 7. Increasing the amount of reagents results in the formation of quercetin 7,4′-disulfate as sole product in the case of sulfation, and of a mixture of 3,7- and 3,4′-dimethylquercetin in the case of methylation. Clearly methylation involves the reaction of quercetin phenolate with methyl iodide. This reactivity is governed by two parameters with opposite effect. Obviously the rate is proportional to the phenolate concentration. On the other hand the more basic phenolate gave the higher rates. So in a polyphenolic compound with close acidity constant the alkylation site is difficult to predict. The sulfation involves the reaction of neutral quercetin, as it is not dissociated in pyridine, with the activated form of hydrogen sulfate formed by addition on DCC. This reaction is sensitive to the nucleophilicity of the oxygen atom rather than to their basicity. Furthermore the activated form of hydrogen sulfate is a bulky reagent. So sulfation at position 3 may be prevented due to the difficult approach of TBAHS at this rather hindered position, resulting in selective formation of quercetin 7,4′-disulfate. Another notable difference between methylation and sulfation exists when large excesses of reagents are employed. In fact, while the formation of 3,7,3′,4′-tetramethylquercetin is possible, sulfation stops at quercetin 3,7,4′-trisulfate and no tetrasulfate is formed. This, again may be related to a difficult approach of TBAHS at position 3′ when a sulfate group preexists at position 4′.

So, the regioselective methylation of quercetin implies protection of hydroxyl groups. Thus benzylation of quercetin with 3.5 equivalents of benzyl bromide mainly gives rise to 3,7,3′,4′-tetrabenzylquercetin (**44**), together with 3,7,4′-tribenzylquercetin (**47**) (Scheme 9).[32] Access to azaleatin (**46**) is possible from **44** by methylation of the 5-hydroxyl, yielding **45**, followed by debenzylation by catalytic hydrogenation. The 5-hydroxyl, however, is less acidic due to hydrogen bonding with the 4-carbonyl. Therefore an excess of methyl iodide must be used and the reaction allowed to take place for 12 h. If methylation of tribenzylquercetin (**47**) is carried out with only one equivalent of methyl iodide, selective methylation at position 3′ takes place and isorhamnetin (**49**) is obtained after debenzylation of **48**.

The hemisyntheses of 7- and 4′-methylquercetins (rhamnetin and tamarixetin, respectively) imply protection of the catechol group of quercetin (Scheme 10). This is achieved by using dichorodiphenylmethane as protective group. From the central intermediate **40**, rhamnetin (**52**) and tamarixetin (**56**) are obtained through two different routes. Controlled benzylation of **40** with one equivalent of benzyl bromide gives rise to the 3-benzyl derivative (**50**) in 64% yield. The latter is then easily methylated at position 7 to give **51**. The final reaction to rhamnetin (**52**) is

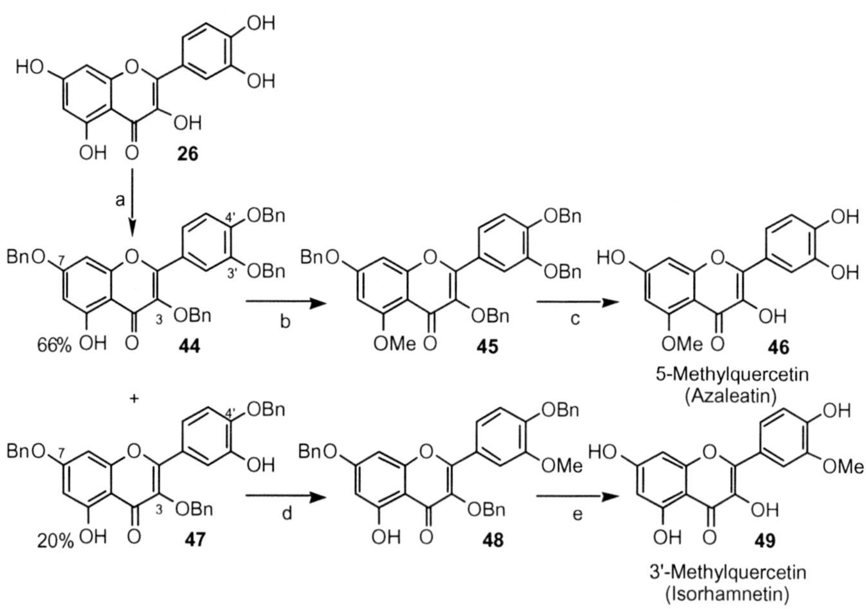

Scheme 9 *The syntheses of azaleatin* (**46**) *and isorhamnetin* (**49**)*:*[32] (a) *BnBr/K_2CO_3/DMF;* (b) *Excess MeI/K_2CO_3/DMF, 93%;* (c) *H_2/Pd$(OH)_2$, 88%;* (d) *1 eq. MeI/K_2CO_3/DMF, 90%;* (e) *H_2/Pd$(OH)_2$, 85%*

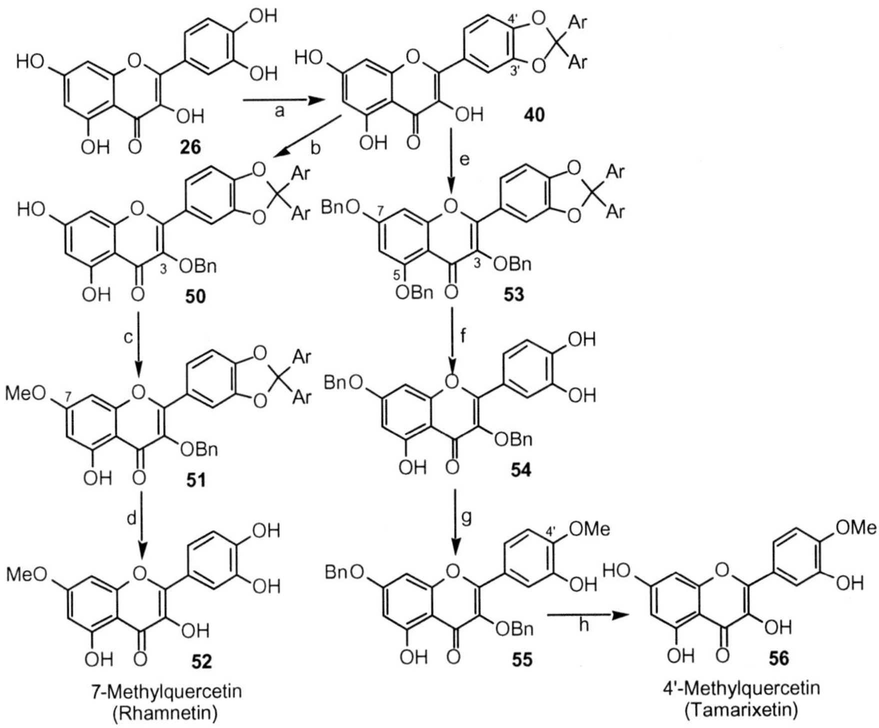

Scheme 10 *Hemisyntheses of rhamnetin* (**52**) *and tamarixetin* (**56**):[32] (a) *Dichlorodiphenylmethane, 180 °C, 37%;* (b) *1 eq. BnBr/K_2CO_3/DMF, 64%;* (c) *1 eq. MeI/K_2CO_3/DMF, 60%;* (d) *H_2/Pd(OH)$_2$/EtOH and 80% aq. AcOH, 80%;* (e) *4 eq. BnBr/K_2CO_3/DMF, 95%;* (f) *80% aq. AcOH, 60%;* (g) *1 eq. MeI/K_2CO_3/DMF, 90%;* (h) *H_2/Pd(OH)$_2$/EtOH, 80%*

deprotection, which requires two steps. Catalytic hydrogenation eliminates the 3-benzyl group, while removal of the diphenylmethane group is better accomplished by acidic hydrolysis. In the hemisynthesis of tamarixetin, compound **40** is polybenzylated using an excess of benzyl bromide, yielding **53**. The next step is again removal of the diphenylmethane by acidic hydrolysis, which results in the elimination of the 5-benzyl group as well. In this case the instability of the 5-benzyl group is due to strong stabilisation of the liberated 5-hydroxyl by hydrogen bonding with the 4-carbonyl. The resulting compound (**54**) is then submitted to controlled methylation with one equivalent of methyl iodide, which results in excellent regioselectivity of the reaction for the most acidic 4′-hydroxyl, yielding compound **55** in 90% yield. Final debenzylation of **55** by catalytic hydrogenation yields tamarixetin (**56**) in 80% yield.

Methylated catechin

As pointed out earlier, the case of catechin is by far the most difficult, due to no significant differences in the acidities of the four phenolic groups. The only structural feature we can play with is the presence of the catechol group on ring B of catechin. Therefore the approach for the hemisynthesis of methylated catechin is very quickly directed towards strategies based upon selective protection–deprotection of the A and B rings,[33] strategies quite similar to that used for quercetin.

However, since catechin is known to undergo quite readily a base-catalysed epimerisation at C-2 to form *ent*-epicatechin through reversible opening of the C ring *via* a B ring quinone methide intermediate,[34] the choice of the reagent is rather limited. Indeed, among the various protecting groups used usually for catechol moities, only dichlorodiphenylmethane,[33] di-*tert*-butyldichlorosilane[33] and cyclic borate[35,36] are reported to lead to the formation of permanent or transient dioxolane rings. However, cyclic borate is rather delicate to use since the protected compound is not isolated.[35,36] The use of dichlorodiphenylmethane,[33] or di-*tert*-butyldichlorosilane[33] opens access to two synthetic pathways: the first one leading to the A ring monomethylated compound, the second one to the B ring monomethylated compounds by the means of specific protection of the A ring and deprotection of the B ring.

Thus reaction of catechin (**57**) with dichlorodiphenylmethane gives compound **58** in 22% yield (Scheme 11). Methylation of **58** with one equivalent of dimethyl sulfate affords a mixture of the 5- and 7-methyl conjugates (**59** and **60**), in 70% overall yield. The 5-methyl ether, however, represents 90% of the mixture and no selective access to compound **60** is possible by this approach. Debenzylations of **59** and **60** give rise respectively to 5-methylcatechin (**61**) and to 7-methylcatechin (**62**), which are separated by column chromatography using ethyl acetate–hexane (30:70) as eluent.

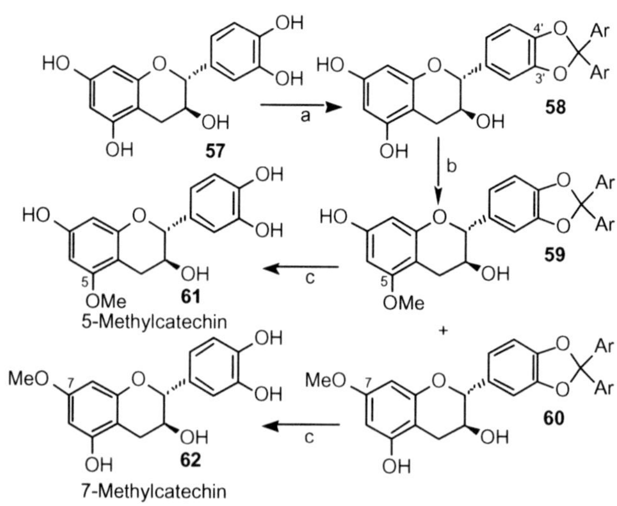

Scheme 11 *Hemisyntheses of 5-methylcatechin* (**61**) *and 7-methylcatechin* (**62**):[33] (a) *Dichlorodiphenylmethane/Et_3N/MeCN, 22%;* (b) *1 eq. Dimethyl sulfate/K_2CO_3/Me_2CO, 70% of a mixture of* **59**/**60** *9/1;* (c) *H_2/Pd$(OH)_2$/EtOH–THF* (**61** *92%;* **62** *86%)*

The hemisyntheses of 3′- and 4′-methyl catechin (Scheme 12) involve the use of a different protector for the catechol ring, *i.e.* the di-*tert*-butyldichlorosilane. This choice was guided by the observation that, upon chromatography on silica gel, this group opens and yields a 1:1 mixture of 3′- and 4′-monoprotected isomers, leaving either positions 4′ or 3′ for further conjugation. The crucial step of this synthesis consists in the selective and successive protection of B and A rings. Indeed, after protecting selectively the B-ring catechol moiety, a protection of the A ring whose deprotection conditions are different from those of the B-ring

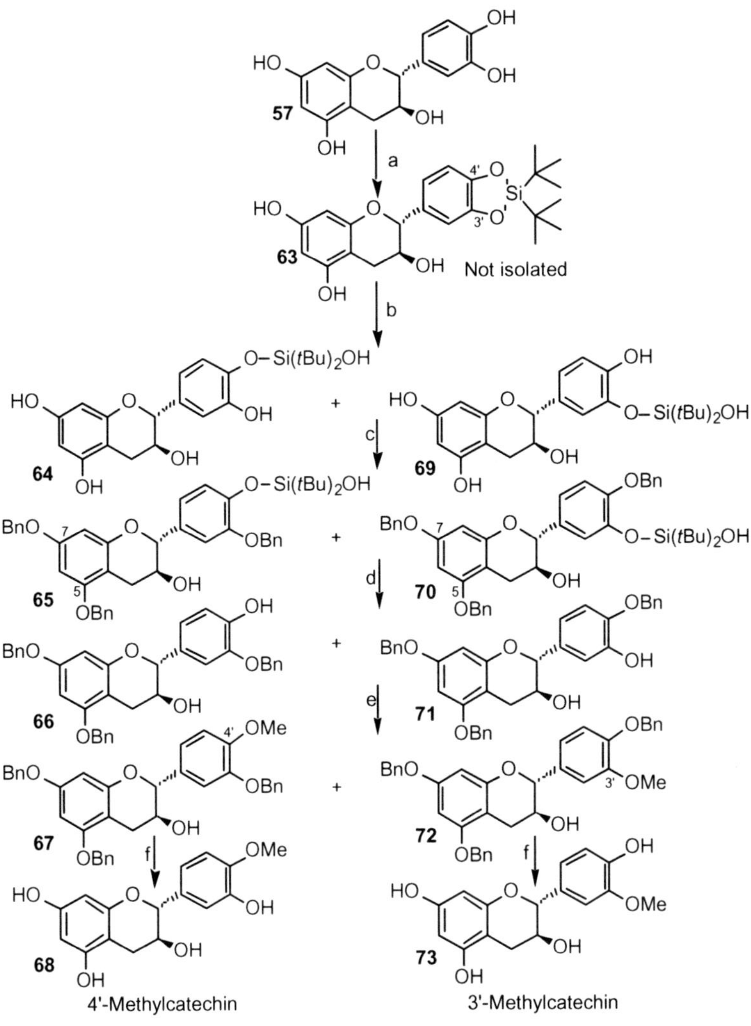

Scheme 12 *Hemisyntheses of 4′-methylcatechin* (**68**) *and 3′-methylcatechin* (**73**):[33] (a) *$Cl_2Si(Bu_t)_2/Et_3N/MeCN$;* (b) *Chromatography on silica gel, 75% from* **57**; (c) *$BnBr/K_2CO_3/DMF$, 70%;* (d) *Bu_4NF/THF, 95%;* (e) *$Me_2SO_4/K_2CO_3/Me_2CO$* (**67** *30% and* **72** *35%);* (f) *H_2/10% Pd–C/THF* (**68** *85% and* **73** *87 %)*

protection (to allow selective deprotection of the B-ring and so selective methylation on B ring) has to be found.

More precisely catechin (**57**), after reaction with di-*tert*-butyldichlorosilane in the presence of triethylamine, gives the catechol protected catechin (**63**). This compound is directly submitted to a column chromatography on silica gel, yielding a 1:1 mixture of compounds **64** and **69** [75% yield from **57**]. This 1:1 mixture is successively submitted to benzylation affording **65** plus **70** (70%), deprotection of the silyl ether by tetrabutylammonium fluoride [**66** + **71**, 95%], and finally methylation by dimethyl sulfate [**67** + **72**]. After purification of the latter mixture of compounds by flash chromatography and successive recrystallisations from petroleum ether–AcOEt, 30% of compound **67** and 35% of compound **72** are isolated. Finally debenzylation of **67** and **72** by catalytic hydrogenation gives rise respectively to 4′-methylcatechin (**68**) (85%) and 3′-methylcatechin (**73**) (87%).

So these new methodologies based on differentiation of the catechol B ring of catechin from the resorcinol-like A ring by the formation of permanent or transient dioxolanes open access to the first chemical synthesis of the 3′- and 4′-methylcatechin, the major metabolites of catechin in human and rats.

3 Typical Experimental Procedures

All the compounds with no literature reference have been prepared according to our own laboratory procedures. Other procedures have been reproduced with permission from the cited reference.

2,4-Dihydroxyphenyl 4-Hydroxybenzyl Ketone (6)

To a mixture of 5 g of resorcinol (**1**) (45 mmol) and 7 g of 4-hydroxyphenylacetic acid **2** (46 mmol) were added 116 ml (0.9 mol) of BF_3-etherate, dropwise, under stirring and inert argon atmosphere. The medium was allowed to react for 3.5 h at 70–80 °C under stirring, which led to the formation of a yellow precipitate. The latter was isolated by filtration, washed with 3 × 50 ml of 12% aqueous sodium acetate, and finally rinsed with water. This led to the isolation of 8.8 g of 2,4-dihydroxyphenyl-4-hydroxybenzylketone **6** (36 mmol, 80%).

2,4,6-Trihydroxyphenyl 4-Hydroxybenzyl Ketone (8)

4.4 g (33 mmol) of 4-hydroxyphenylacetonitrile (**4**) were dissolved in 200 ml of boiling dry diethyl ether. After cooling of the ether solution, 4.7 g (37 mmol) of phloroglucinol (**3**) and 1.3 g of powdered freshly fused zinc chloride were added. The mixture was placed in a vacuum flask, cooled to 0 °C (ice bath) and saturated with dry HCl gas (Figure 2, prepared by careful addition of 300 ml of concentrated H_2SO_4 in 300 ml concentrated HCl). Leaving the reaction medium for 24 h at 0 °C resulted in the formation of a bulky precipitate. The ether supernatant was eliminated by decantation (CAUTION: HCl vapours). The precipitate was directly dissolved in the flask by 250 ml of 50% aqueous methanol (warm if necessary). The solution was transferred to a 1 l round flask and refluxed for 18 h. After cooling, the solution was concentrated by half under reduced pressure and placed at 0 °C for 1 h. This resulted in the formation of a yellow precipitate which was isolated by filtration and washed with H_2O to give 2.8 g (11 mmol, 33%) of 2,4,6-trihydroxyphenyl 4-hydroxybenzyl ketone (**8**).

Daidzein (10)

8.5 g (35 mmol) of 2,4-dihydroxyphenyl 4-hydroxybenzyl ketone (**6**), dissolved in 100 ml of dry DMF were placed in a three-necks round flask. 115 ml (0.9 mol) of BF_3-etherate were added dropwise with stirring under argon. The solution was warmed to 50 °C and 8 ml of methanesulfonyl chloride (105 mmol) in 12 ml of dry DMF were added dropwise under argon. The medium was warmed at 70 °C for 3.5 h under inert atmosphere. After cooling, the medium was diluted with a mixture of crushed ice and 12% aqueous sodium acetate. The resulting precipitate was isolated by filtration and washed with H_2O. The precipitate was refluxed for 5 min in 2 × 50 ml of methanol. Evaporation of the mixed methanolic solutions yielded 4.5 g (18 mmol, 51%) of daidzein (**10**). The yield in daidzein can be increased to 71% by microwave-assisted synthesis. Microwave oven owners are invited to consult reference 3 for a typical procedure.

ω-Methoxyphloroacetophenone (15)

To a solution of 25 g of anhydrous phloroglucinol (**3**) (0.2 mmol) in 200 ml of dry ether were added 5 g of freshly fused and finely powdered $ZnCl_2$ and 25 ml of methoxyacetonitrile (**14**) (0.33 mmol). The flask was placed in an ice bath and a stream of dry hydrogen chloride (Figure 2, prepared by addition of 100 ml of conc. HCl to 100 ml of conc. H_2SO_4) was passed through the solution, under stirring. The medium was left at 4 °C for 48 h after which the ether supernatant was eliminated (CAUTION: HCl vapours). The bulky precipitate formed was suspended in 500 ml H_2O and the solution was refluxed for 2 h. The hot solution was clarified by filtration, allowed to cool down and left overnight at 4 °C. The crystals of the acetophenone (**15**) were isolated by filtration and washed with H_2O. Yield: 39 g (0.18 mmol, 90%).

3,4-Dibenzyloxybenzoyl Chloride (19)[4]

A mixture of 10 g (31 mmol) of 3,4-dibenzyloxybenzaldehyde (**17**) and 7.4 g (47 mmol) of $KMnO_4$ in 800 ml acetone plus 80 ml H_2O was refluxed for 6.5 h under stirring. The mixture was evaporated to dryness and the residue was warmed in 1.6 l of 1% aqueous Na_2CO_3. The brown precipitate was eliminated by filtration and washed with 600 ml of warmed H_2O. The washing and the filtrate were mixed and acidified (5N HCl), which resulted in the precipitation of 3,4-

dibenzyloxybenzoic acid (**18**). The precipitate was isolated by filtration, washed with H_2O and recrystallised from EtOH. Yield: 9.6 g (28.7 mmol, 93%).

17 g of (**18**) (51 mmol) were dissolved in 200 ml of thionyl chloride and, after addition of 2.1 ml of DMF (28 mmol), the mixture was refluxed under stirring for 6 h. Evaporation of the solvent under reduced pressure yielded 17.6 g (50 mmol, 98%) of 3,4-dibenzyloxybenzoyl chloride (**19**).

3-Methylquercetin (21)[4]

400 mg (2 mmol) of acetophenone (**15**), 1.4 g (4 mmol) of chloride (**19**), 1.68 g (12 mmol) of finely powdered anhydrous K_2CO_3, 1 g of tetrabutylammonium hydrogen sulfate (3 mmol), and 25 ml of toluene were refluxed for 6 h under stirring and N_2 atmosphere. Toluene was decanted from the solid and 15 ml of H_2O were added to the solid. The aqueous extract was extracted with CH_2Cl_2 and the later extract was mixed with the toluene solution. After evaporation of the solvents, the residue was treated with 20 ml of 5% aqueous K_2CO_3 and refluxed for 4 h (saponification of esters of **20**). Saturation of the solution with solid CO_2 induced the formation of a semi-solid which was collected and purified by column chromatography on silica gel using 2% MeOH in CH_2Cl_2 as solvent. The isolated compound (**20**) was crystallised in anhydrous EtOH, to give 685 mg (1.38 mmol, 69%) of pure (**20**).

220 mg of compound **20** (0.44 mmol) and 0.23 g of 20% $Pd(OH)_2$ on carbon were suspended in 10 ml of EtOH. 5 ml of cyclohexene were added and the mixture was refluxed for 30 minutes. The mixture was filtered and the solid was washed with EtOH. Evaporation of the filtrate plus the washings yielded 137 mg (0.43 mmol, 98%) of 3-methylquercetin (**21**).

Daidzein 7-Sulfate Potassium Salt (22)

To 1 g of daidzein (10) (4 mmol), dissolved in 20 ml of dry pyridine were added 10 g (48 mmol) of dicyclohexycarbodiimide (DCC). After 15 min, this solution was poured to a suspension of 2.7 g (8 mmol) of TBAHS in 10 ml of dry pyridine. After complete homogenisation, the medium was left at 4 °C for 5 days. The precipitate of dicyclohexylurea which was formed during the reaction was eliminated by filtration and rinsed with methanol. The filtrate was purified by MPLC on Sephadex® LH-20 using methanol as solvent and UV detection. First UV-absorbing fraction (A) contained a disulfate ester while the following UV-absorbing fraction (B) contained a monosulfate ester.

Fraction B from the column of LH-20 was still appreciably contaminated with dicyclohexylurea. Thus the fraction was taken in 100 ml of 10% aqueous tetramethylammonium hydroxide and the insoluble dicyclohexylurea was eliminated by centrifugation. The supernatant was neutralised with a solution of 15% TBAHS in H_2O and the medium was extracted with EtOAc. After evaporation of the ethyl acetate, the residue of crude daidzein monosulfate tetrabutylammonium salt was taken in the minimum volume of MeOH, and an equal volume of a saturated solution of KOAc in MeOH was added. The precipitate formed was isolated by centrifugation. The precipitate (after solubilisation in 50% aqueous MeOH) and the supernatant were purified separately by MPLC on Sephadex® G-10 using 0.1 M aqueous KH_2PO_4 (to ensure total conversion to potassium salt), H_2O (elimination of the buffer), and finally a gradient of MeOH in H_2O (elution of daidzein monosulfate) as eluents. Again, elution of daidzein 7-monosulfate was followed by UV detection. Total recovery of crude daidzein 7-monosulfate: 1.44 g.

Quercetin 3-Sulfate (29)

To 1 g of quercetin (**26**) (3 mmol), dissolved in 30 ml of dry pyridine were added 18.3 g (88.7 mmol, 30 eq.)) of DCC. After 15 min, this solution was poured to a suspension of 20 g (59 mmol, 20 eq.) of TBAHS in 20 ml of dry pyridine. After complete homogenisation, the medium was left at room temperature for 12 days. The reaction medium was filtrated to eliminate the precipitate of DCU and the filtrate containing the reaction products was set aside. The precipitate of DCU was rinsed with 70 ml of 10% aqueous tetramethylammonium hydroxide, followed by 20 ml of H_2O. The two last solutions were added to the first filtrate which led to further precipitation of DCU. The solution was clarified by filtration and acidified by addition of a solution of 30 g of TBAHS in 50 ml H_2O. Quercetin 3,7,4′-trisulfate tetrabutylammonium salt was extracted from the solution with EtOAc and the extract was rinsed with H_2O and evaporated to dryness. The residue was taken in the minimum volume of MeOH, 10 ml of a saturated solution of KOAc in MeOH were added and the precipitate of quercetin 3,7,4′-trisulfate potassium salt (**28**) was isolated by centrifugation. The pellet was resuspended in 20 ml of MeOH and centrifuged again. Yield of crude quercetin trisulfate: 2.5 g.

0.45 g of crude quercetin trisulfate (**28**) were placed in a centrifuge tube and dissolved in 30 ml of citrate buffer[37] pH 4.5. 520 mg of aryl sulfatase (16 400 units g^{-1} of solid) were gradually added and the medium incubated at 30 °C for 16 h. Note that the required quantity of aryl sulfatase and the reaction time might need adjustment depending on the activity of the enzymatic preparation. The medium was centrifuged. The supernatant was set aside and the pellet was washed three times with MeOH (centrifugation after every washing). The three methanol washings were combined with the supernatant of the first centrifugation. The mixed solution was directly submitted to gel filtration on Sephadex G-10®. The column was eluted with H_2O until complete elimination of the protein (UV detection). Subsequent elution of quercetin 3-sulfate (**29**) was achieved by a gradient from 100% H_2O to 100% MeOH in 20 minutes. Yield of **29**: 30 mg.

Quercetin 3′-Sulfate (30)

Quercetin (1 g, 3 mmol) and finely powdered sulfamic acid (0.58 g, 6 mmol) were heated in pyridine (5 ml; 80 °C, 1 h) with continuous stirring. After cooling the reaction medium was diluted with 5% aqueous KOAc (100 ml, pH 7.2) and extracted with ethyl acetate. On standing, a precipitate appeared in the ethyl acetate extract, which was isolated by filtration and purified by gel filtration on Sephadex® G-10 using a gradient of MeOH in H_2O as solvent, yielding 162 mg of quercetin-3′-sulfate.

Quercetin 7-*β*-D-glucuronide (35) (and Quercetin 4′-*β*-D-glucuronide)

3,5,3′,4′-Tetrabenzoylquercetin (**33**)[20] (300 mg, 0.42 mmol), methyl 2,3,4-tri-*O*-acetyl-*α*-D-glucopyranosyluronate bromide, (185 mg, 1.1 eq.), Ag_2CO_3 (127 mg, 1.1 eq.) 3 Å sieves (500 mg) and dry CH_2Cl_2 (5 ml) were stirred, in the dark, under Ar (5 min). Collidine (60 μL, 1.1 eq.) was added. After 3 h, the mixture was filtered (celite) The filtrate and celite rinses (5% MeOH/CH_2Cl_2, CH_2Cl_2 2 × 50 ml) were combined and washed with 1 M HCl (50 ml), H_2O (50 ml), 0.1 M $Na_2S_2O_3$ (50 ml), H_2O (50 ml), satd $NaHCO_3$ (50 ml), and H_2O (50 ml), and dried ($MgSO_4$). The evaporated solid was suspended in 1 M NaOH (20 ml), stirred under Ar (0 °C, 60 min), warmed to room temperature, left for 1 h, and heated at reflux (10 min). After cooling, Dowex 50 W resin (H^+ form, 40 ml) was added, and stirred (10 min); the mixture was filtered, and filtrate and resin washes (50% aq. MeOH: 50 ml; MeOH: 50 ml) were combined and evaporated. The resultant solid was dissolved in 5% aq. MeOH (100 ml), extracted with CH_2Cl_2 (3 × 75 ml); the aqueous phase was evaporated. The residue was dissolved in 50% aq. MeOH (8 ml), and filtered. The products were isolated by preparative reverse-phase HPLC, on a ODS column using a gradient of acetonitrile in 0.1% aqueous TFA, and were pure by analytical HPLC. Yield of quercetin 7-*β*-D-glucuronide: 17 mg, 8%. Yield of quercetin 4′-*β*-D-glucuronide: 6 mg, 3%.

Daidzein 7-Glucuronide (39)

To a solution of 6.4 g (19.0 mmol) of 4′-7-di-*O*-acetyldaidzein (**36**) in 750 ml of THF (gentle warming) was added after cooling 750 ml of a 1 M solution of imidazole in H_2O (adjusted to pH 7 with HCl). The mixture was stirred for 15 min at 25–30 °C, after which time an initial slight precipitate had redissolved. The mixture was acidified to pH 4.5 (AcOH), diluted with 1 l H_2O and extracted with CH_2Cl_2. The extract was washed with H_2O, satd. Na_2CO_3 and again H_2O. After drying ($MgSO_4$) and evaporation of the organic phase to dryness, the residue was recrystallised from MeOH, to give 5.28 g (17.8 mmol, 94%) of 4′-*O*-acetyldaidzein (**37**). To a stirred suspension of 3 g (10.1 mmol) of compound **37**, 4.43 g (11.1 mmol) methyl 2,3,4-tri-*O*-acetyl-*β*-D-glucopyranosyluronate, 3.07 g (11.1 mmol) $AgCO_3$ and 2 g of ground 3 Å molecular sieves in 120 ml of CH_2Cl_2 were added 1.47 ml (11.1 mmol) of collidine, over 10 min at room temperature under argon. The mixture was stirred in the dark at room temperature for 7 days. The medium was filtered through celite and the celite washed with 50 ml of 10% MeOH in CH_2Cl_2. The filtrates were combined and washed with aqueous AcOH, H_2O, 0.1 M $Na_2S_2O_3$, H_2O, $NaHCO_3$ and H_2O, successively. After drying

($MgSO_4$) and evaporation to dryness, the residue was taken in 50 ml of 1% MeOH in CH_2Cl_2 and left overnight at −20 °C. The precipitate of unreacted (**37**) was removed by filtration and the filtrate was purified by column chromatography on silica using a gradient of MeOH in CH_2Cl_2 (from 1% to 10%) as solvent. This resulted in the isolation of 3.7 g of crude compound **38**. 0.5 g of crude (**38**) were further purified by MPLC on silica using 1% MeOH in CH_2Cl_2 as solvent, yielding 254 mg (0.41 mmol) of pure (**38**). 0.5 g of crude **38** (containing 254 mg of pure **38**) were suspended in 250 ml of MeOH (sonication) at room temperature. 7 ml of 0.5 M aqueous Na_2CO_3 were added and the mixture was stirred for 2 h under Ar, after which time 20 ml of Dowex 50 W resin (H^+ form) were added. The medium was stirred for 10 min. The eluate was collected from the resin and the latter washed with MeOH. The combined eluates were purified by preparative HPLC on ODS using a gradient of 0.1% aqueous TFA in 80% aqueous MeCN. This yielded 139 mg (0.32 mmol, 78%) of daidzein 7-glucuronide (**39**). The material was stored in solution in 50% aqueous MeOH at −20 °C (the material, in its acid form, slowly esterifies in MeOH alone).

Quercetin 3-O-β-D-Glucuronide (43)

A mixture of quercetin (**26**) (2.0 g, 5.92 mmol) and dichlorodiphenylmethane (3 equiv., 3.39 ml, 17.76 mmol) was intimately mixed then heated at 180 °C for 10 min. The crude resulting products mixture was then purified by flash column chromatography on silica gel using CH_2Cl_2/EtOAc (85/15) as eluent and was recrystallised from $CHCl_3$ to afford **40** (1.0 g, 37% yield). A solution of 2,3,4,6-tetra-*O*-acetyl-α-D-glucopyranosyl bromide (470 mg, 1.13 mmol) in 5 ml of DMF and potassium carbonate (156 mg, 1.13 mmol) were added under argon to a solution of **40** (510 mg, 1.1 mmol) in 20 ml of DMF. The mixture was stirred at room temperature for 6 h, then diluted with 50 ml of water, extracted with 70 ml of EtOAc. The organic layer was washed with 50 ml of water and dried over $MgSO_4$ and the solvent was evaporated. The residue was purified by flash column chromatography using EtOAc/petroluem ether (15/85) as eluent. This yielded 490 mg of **41** (yield 56%). Benzyl bromide (0.187 ml, 1.57 mmol) and potassium carbonate (217 mg, 1.057 mmol) were added to a solution of **41** (500 mg, 0.63 mmol) in 30 ml of DMF under an argon flow and stirring was maintained overnight. The resulting mixture was diluted with 70 ml of water, extracted with 100 ml of EtOAc, then the organic layer was washed with 70 ml of water and finally dried over $MgSO_4$. Solvent was removed under vacuum to leave a white solid (crude acetylated **42**), which was purified by flash column chromatography on silica gel using CH_2Cl_2/EtOAc (95/5) as eluent (565 mg, 92% yield). This white solid was dissolved in a mixture of MeOH (20 ml) and THF (20 ml) under argon. A solution of sodium methoxide in methanol (15 mg 10 ml^{-1}) was added and the resulting mixture left at ambient temperature for 3 h. At this point TLC (EtOAc/methanol, 9/1) indicated the complete disappearance of the reactive. The solution was neutralised with Amberlite® resin 120 (H^+ form) and the resin filtered off; then the solution was evaporated and dried under vacuum. Compound

42 was purified by flash column chromatography using EtOAc/methanol (90/10) as eluent (291 mg; yield 88%). A solution of (**42**) (100 mg, 0.124 mmol), TEMPO (2 mg, 0.013 mmol), tetrabutylammonium bromide (3 mg, 0.009 mmol) and KBr (2 mg, 0.017 mmol) in 5 ml of dichloromethane and 1 ml of a saturated aqueous solution of $NaHCO_3$ were vigorously stirred at room temperature. Then sodium hypochlorite (15 ml of 5.25% solution) was added over a period of 30 min. The resulting mixture was stirred for an additional period of 2 h. The reaction mixture was diluted with water (25 ml) and extracted with EtOAc (80 ml); then the organic layer was washed with water (25 ml) and dried over $MgSO_4$ and evaporated. A suspension of the residue in a mixture of EtOH (10 ml) and THF (10 ml) was treated at room temperature with 30% palladium hydroxide (30 mg) under a flow of hydrogen for 10 h, filtered on Celite® and eluted with EtOH (20 ml). After concentration of the filtrate under vacuum, the resulting compound was purified on a LC-18 SPE cartridge (12 ml model, 2 g reverse phase) using a methanol, water gradient as eluent and afforded analytically pure quercetin-3-*O*-*β*-D-glucuronide (**43**) (35 mg; yield 59%).

Isorhamnetin (49)

To a solution of 5 g (14.8 mmol) of quercetin (**26**) in 100 ml of DMF were added 7.14 g (51.8 mmol) of anhydrous K_2CO_3, and 6.2 ml (52 mmol) of benzyl bromide, under stirring at 0 °C. Stirring was continued for 2 h at 0 °C, and at room temperature for 12 h. After dilution of the medium with H_2O and extraction with EtOAc, the extract was purified by flash chromatography on silica gel using CH_2Cl_2 as eluent, to give 6.5 g of 3,7,3′,4′-tetrabenzylquercetin (**44**) (9.8 mmol, 66%) and 1.7 g of 3,7,4′-tribenzylquercetin (**47**) (3 mmol, 20%). To a stirred solution of 1.2 g (2.1 mmol) of tribenzylquercetin (**47**) in 30 ml of DMF, were added 0.38 g (2.75 mmol) of anhydrous K_2CO_3 and 0.13 ml (2.1 mmol) of methyl iodide. The medium was stirred overnight under argon, diluted with H_2O and extracted with EtOAc. Purification of the extract by flash chromatography on silica using 5% EtOAc in CH_2Cl_2 as eluent yielded 1.1 g of 3′-methyl-3,7,4′-tribenzylquercetin (**48**) (1.9 mmol, 90%). A suspension of 200 mg (0.34 mmol) of **48** plus 10 mg of 10% palladium hydroxide on carbon, in a mixture of 30 ml of EtOH and 30 ml of THF, was treated with a hydrogen flow for 6 h. The medium was filtered on Celite® and eluted with EtOH. After evaporation of the solvent the residue was crystallised in MeOH to give 91 mg of isorhamnetin (**49**) (0.29 mmol, 85%).

Rhamnetin (52)

To a solution of 1 g of compound **40** (2.14 mmol) in 30 ml of DMF were added 0.38 g of K_2CO_3 (2.74 mmol) and 0.24 ml (2 mmol) of benzyl bromide. The mixture was stirred at 0 °C for 2 h and at room temperature for 12 h. The medium was diluted with H_2O, extracted by EtOAc and the extract was purified by flash chromatography on silica gel using 5% EtOAc in CH_2Cl_2 as solvent, yielding 0.76 g (1.37 mmol, 64%) of compound (**50**). To a solution of 0.5 g of compound

(**50**) (0.9 mmol) in 30 ml of DMF were added 0.25 g of K_2CO_3 (1.8 mmol) and 0.075 ml (1.2 mmol) of methyl iodide. The mixture was stirred at room temperature for 6 h, diluted with H_2O and extracted with EtOAc. The organic extract was purified by flash chromatography on silica gel using a gradient of 20% to 50% EtOAc in petroleum ether as solvent, to give 307 mg (0.54 mmol, 60%) of compound **51**. A stirred suspension of 0.7 g (1.23 mmol) of **51** and 10 mg of 10% $Pd(OH)_2$ on carbon, in 20 ml of THF/EtOH 2:1 was exposed to a hydrogen flow. Stirring was maintained overnight, after which the medium was filtered on Celite®, which was washed with 20 ml of EtOH. The filtrate and washing were mixed and evaporated to dryness. The residue was taken in 30 ml of 80% aqueous AcOH and the solution was refluxed overnight under stirring. After H_2O dilution, the medium was extracted with EtOAc. After evaporation of the organic layer, the residue was recrystallised in CH_2Cl_2, to give 311 mg (0.98 mmol, 80%) of rhamnetin (**52**).

4′-Methylcatechin (68) and 3′-Methylcatechin (73)

To a stirred solution of 1 g (3.44 mmol) of catechin hydrate (**57**) in 50 ml of MeCN were added 2.4 ml (17.2 mmol) of triethylamine, and 1.46 ml (6.9 mmol) of di-*tert*-butyldichlorosilane at 0 °C under argon. The mixture was stirred at room temperature for 12 h, diluted with H_2O and extracted with EtOAc. The EtOAc extract was purified by flash chromatography on silica gel 60 using 40% petroleum ether in Me_2CO as solvent, yielding 1.16 g (2.58 mmol, 75%) of a mixture of compounds **64** and **69**. To a stirred solution of 0.5 g (1.11 mmol) of **64** plus **69** in 20 ml of dry DMF were added 0.47 g (3.4 mmol) of anhydrous K_2CO_3 and 0.4 ml (3.4 mmol) of benzyl bromide. The mixture was stirred at room temperature for 12 h. After dilution with H_2O and extraction with EtOAc, the organic extract was purified by flash chromatography on silica gel 60 using 40% EtOAc in hexane as eluent. This yielded 0.56 g (0.78 mmol, 70%) of a mixture of compounds **65** and **70**. To a solution of 1.6 g (2.2 mmol) of **65** plus **70** in 30 ml of THF were added 11 ml (11 mmol) of 1 M terabutylammonium fluoride in THF under argon. The mixture was stirred at room temperature under argon for 30 min, diluted with H_2O and extracted with CH_2Cl_2. The extract was purified by flash chromatography on silica gel 60 using 40% EtOAc in hexane as solvent, yielding 1.16 g (2.1 mmol, 95%) of a mixture of compounds **66** and **71**.

A solution of this mixture of compounds **66** and **71** (0.5 g, 0.91 mmol), potassium carbonate (0.63 g, 4.52 mmol) and dimethyl sulfate (0.44 ml, 4.52 mmol) in acetone (30 ml) was refluxed for 5 h. After removal of the solvent, water (30 ml) and AcOEt (40 ml) were added to the reaction mixture. The organic layer was dried ($MgSO_4$). The separation of the two isomeric compounds **67** and **72** was performed by flash column chromatography on silica gel using AcOEt/hexane (30:70) and successive recrystallisations from petroleum ether/AcOEt. This afforded 156 mg of **67** (0.27 mmol, 30%) and 183 mg of **72** (0.32 mmol, 35%). A suspension of compound **67** or **72** (150 mg, 0.26 mmol) in THF (15 ml) was stirred with activated Pd/C (10%, 10 mg) under an atmosphere of hydrogen for 10 h, filtered on Celite® and eluted with EtOH (30 ml) and CH_2Cl_2 (30 ml). The filtrate was concentrated under vacuum and the resulting product **68** or **73** recrystallised from $CHCl_3$/MeOH to give white solids (**68**:**67**, 2 mg, 85% and **73**:**68**, 7 mg, 87%).

References

1. A. Scalbert and G. Williamson, *J. Nutr.*, 2000, **130**, 2073S.
2. A. M. Aura, K. A. O'Leary, G. Williamson, M. Ojala, M. Bailey, R. Puupponen-Pimiä, A. M. Nuutila, K. M. Oksman-Caldentey and K. Poutanen, *J. Agric. Food Chem.*, 2002, **50**, 1725.
3. Y. C. Chan, M. G. Nair, R. C. Santell and W. G. Helferich, *J. Agric. Food Chem.*, 1994, **42**, 1869.
4. F. Boers, B. L. Deng, G. Lemiere, J. Lepoivre, A. De Groot, R. Dommisse and A. J. Vlietinck, *Arch. Pharm. Pharm. Med. Chem.*, 1997, **330**, 313.
5. M. Akira, T. Masanori, I. Hisamitsu and S. Tetsuo, *Bunseki Kagaku*, 1973, **22**, 179 (*Chem. Abstr.*, **78**: 164954).
6. W. Guoqing and Z. Runya, *Shenyang Yaoxueyuan Xuebao* 1992, **9**, 63 (*Chem. Abstr.*, **117**: 258064).
7. S. V. Jovanovic, S. Steenken, M. Tosic, B. Marjanovic and M. G. Simic, *J. Am. Chem. Soc.*, 1994, **116**, 4846.
8. K. Lemanska, H. Szymusiak, B. Tyrakowska, R. Zielinski, A. E. M. F. Soffers and I. M. C. M. Rietjens, *Free Radical Bio. Med.*, 2001, **31**, 869.
9. C. Cren-Olivé, J. M. Wieruszeski, E. Maes and C. Rolando, *Tetrahedron Lett.*, 2002, **43**, 4545.
10. D. Barron and R. K. Ibrahim, *Tetrahedron*, 1987, **43**, 5197.
11. D. Barron, G. Sarrazin, C. Bayet and G. Comte, *J. Agric. Food Chem.*, 2003, submitted.
12. D. Barron and R. K. Ibrahim, *Z. Naturforsch.*, 1988, **43c**, 625.
13. A. Kanakubo, K. Koga, M. Isobe, T. Fushimi, T. Saitoh, Y. Ohshima and Y. Tsukamoto, *Tetrahedron*, 2001, **57**, 8801.
14. A. J. Day, F. Mellon, D. Barron, G. Sarrazin, M. R. A. Morgan and G. Williamson, *Free Radical Res.*, 2001, **35**, 941.
15. T. Tsushida and M. Suzuki, *Jpn. Food Sci. Technol.*, 1995, **42**, 100.
16. J. K. Moon, T. Tsushida, K. Nakahara and J. Terao, *Free Radical Bio. Med.*, 2001, **30**, 1274.
17. H. Wagner, H. Danninger, O. Seligmann, M. Nogradi, L. Farkas and N. Farnsworth, *Chem. Ber.*, 1970, **103**, 3678.
18. H. Wagner, H. Danninger, O. Seligmann and L. Farkas, *Chem. Ber.*, 1970, **103**, 3674.
19. K. O'Leary, A. J. Day, P. W. Needs, W. S. Sly, N. M. O'Brien and G. Williamson, *FEBS Lett.*, 2001, **503**, 103.
20. L. Farkas, M. Nógràdi, B. Vermes, A. Wolfner, H. Wagner, L. Hörhammer and H. Krämer, *Chem. Ber.*, 1969, **102**, 2583.
21. A. V. Stachulski and G. N. Jenkins, *Nat. Prod. Rep.*, 1998, 173.

22. L. Jurd, *J. Org. Chem.*, 1962, **27**, 1294.
23. P. W. Needs and G. Williamson, *Carbohydr. Res.*, 2001, **330**, 511.
24. D. B. Clarke, A. S. Lloyd, N. P. Botting, M. F. Oldfield , P. W. Needs and H. Wiseman, *Anal. Biochem.*, 2002, **309**, 158.
25. F. Melvin, A. McNeill, P. J. F. Henderson and R. B. Herbert, *Tetrahedron Lett.*, 1999, **40**, 1201.
26. M. Bouktaib, A. Atmani, C. Cren-Olivé, S. Lebrun, K. Wähälä and C. Rolando, *Tetrahedron Lett.*, 2002, **43**, 6263.
27. P. Lewis, S. Kaltia and K. Wähälä, *J. Chem. Soc., Perkin Trans. 1*, 1998, 2481.
28. N. Daubresse, C. Francesch, F. Mhamdi and C. Rolando, *Synthesis*, 1998, 157.
29. J. G. Berrin, W. R. McLauchlan, P. Needs, G. Williamson, A. Puigserver, P. A. Kroon and N. Juge, *Biochem., J.*, 2002, **269**, 249.
30. L. P. Anelli, C. Biffi, F. Montanari and S. Quici, *J. Org. Chem.*, 1987, **52**, 2559.
31. N. J. Davis and S. L. Flitsch, *Tetrahedron Lett.*, 1993, **34**, 1181.
32. M. Bouktaib, S. Lebrun, A. Atmani and C. Rolando, *Tetrahedron*, 2002, **58**, 10001.
33. C. Cren-Olivé, S. Lebrun and C. Rolando, *J. Chem. Soc., Perkin Trans. 1*, 2002, 821.
34. K. D. Sears, R. L. Casebier and H. L. Hergert, *J. Org. Chem.*, 1974, **39**, 3244.
35. D. E. Hathway and J. W. T. Seakins, *J. Chem. Soc.*, 1957, 1562.
36. M. S. Van Dyk, J. P. Steynberg and D. Ferreira, *Tetrahedron Lett.*, 1990; **31**, 2643.
37. G. Gomori, in *Methods in Enzymology*, eds S. P. Colowick and N. O. Kaplan, Academic Press, New York, 1955, **1**, 138.

CHAPTER 10

Quantitative Determination of Flavonols in Plant Foods and Biological Fluids

I. C. W. ARTS, D. P. VENEMA AND P. C. H. HOLLMAN

1 Introduction

Flavonols constitute one of the major subclasses of flavonoids. Flavonols have a C-ring structure with a double bond at the 2–3 position (Figure 1). Major flavonols are quercetin (3,5,7,3′,4′-pentahydroxyflavone), kaempferol (3,5,7,4′-tetrahydroxyflavone), myricetin (3,5,7,3′,4′,5′-hexahydroxyflavone) and isorhamnetin (3,5,7,4′-tetrahydroxy-3′-methoxyflavone). The average intake of flavonols in the Netherlands estimated in 1987, using data from a nationwide survey among a representative sample of the population, was 21 mg day^{-1}, of which quercetin contributed 16 mg day^{-1}, kaempferol 3.9 mg day^{-1}, and myricetin 1.4 mg day^{-1}.[1] Average intakes of flavonols in a number of epidemiological prospective cohort studies published so far are 20 mg day^{-1} in middle-aged to older American males,[2] 14 mg day^{-1} in postmenopausal American women,[3] 26 mg day^{-1} in elderly Dutch men,[4] 4 mg day^{-1} in Finnish middle-aged men and women[5] and 26 mg day^{-1} in Welsh middle-aged men.[6] Major sources of flavonols are onions,

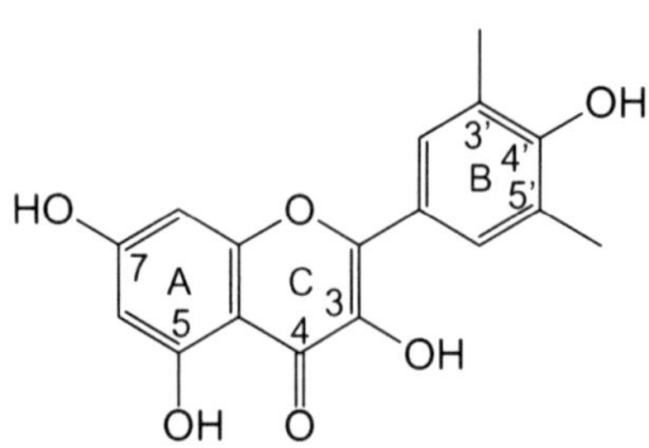

Figure 1 *General structure of flavonols*

apples and tea in tea-drinking countries like the United Kingdom and the Netherlands.[7]

2 Flavonols in Plant Foods

Flavonols are usually found in plants bound to sugars as *O*-glycosides. This is a general feature of flavonoids, with one notable exception, the catechins. Free flavonoids, flavonoids without their attached sugars, are called aglycones. Aglycones of flavonols are not present in fresh plants, but may be present as a result of food processing. Sugars are predominantly bound to the flavonoid nucleus *via* a β-glycosidic bond. Sugar molecules can bind to various positions in the parent flavonoid, although there is a preference for the 3-position. More than 80 different sugars, monosaccharides (10 species), disaccharides (39), trisaccharides (30), and even tetrasaccharides (1), have been found bound to flavonoids in plants. As a result, 179 different quercetin glycosides have been described in nature.[8] A multitude of glycosides also occurs in common foods such as tea[9] and apples.[10,11]

Total quercetin levels in vegetables are generally below 10 mg kg^{-1}, except for onions (280–490 mg kg^{-1}), kale (110 mg kg^{-1}), broccoli (30 mg kg^{-1}), and beans (45–60 mg kg^{-1}). Kaempferol is present only in kale (210–470 mg kg^{-1}), endive (15–90 mg kg^{-1}), broccoli (60 mg kg^{-1}) and leek (10–60 mg kg^{-1}). In most fruits the quercetin content averages 15 mg kg^{-1}, except for apples (20–70 mg kg^{-1}), apricot (25 mg kg^{-1}), and blackcurrants (37 mg kg^{-1}). Quercetin levels in red wine are between 5 and 15 mg l^{-1}, whereas in fruit juices they are generally below 5 mg l^{-1}. Black tea infusions contain 10–25 mg l^{-1} quercetin, 7–17 mg l^{-1} kaempferol and 2–5 mg l^{-1} myricetin.[12–15]

UV Detection of Total Flavonols after Acid Hydrolysis

We have developed and optimized a method to determine flavonol contents of foods after acid hydrolysis of the individual flavonol glycosides.[16] This is a practical approach because no pure reference compounds are available for many of the glycosides that occur in nature.

Extraction and Hydrolysis

Flavonol glycosides were simultaneously extracted from freeze-dried plant material and hydrolysed to their aglycones by adding to 0.5 g of the sample material 40 ml of 62.5% aqueous methanol (to which 2 g l^{-1} of antioxidant *tert*-butylhydroxyquinone (TBHQ) was added). To this extract was added 10 ml of 6 M HCl and the extract was carefully mixed. This resulted in a final concentration of 1.2 M HCl in 50% aqueous methanol (v/v). After refluxing at 90 °C for 2 h with regular swirling, the extract was allowed to cool, was subsequently made up to 100 ml with methanol and sonicated for 5 min. Approximately 2 ml was filtered through a 0.45 μm filter (Acrodisc CR PTFE, Gelman) prior to injection.

Chromatography

For chromatographic separation 10 μl of sample was injected onto a Nova-Pak C_{18} (Waters Associates, Milford, MA) column (3.9 × 150 mm, 4 μm) protected by a Perisorb RP-18 guard column (3.9 × 40 mm, 30–40 μm). Both columns were placed in a 30 °C column oven. The HPLC system consisted of a Kratos (Kratos Analytical Systems, Ramsey, NJ) Spectroflow 400 pump controlled by a Kratos Spectroflow 450 solvent programmer and a Marathon (Spark Holland, Emmen, The Netherlands) autoinjector with a fixed 10 μl loop. The compounds were detected using a Linear (Linear Instrument Corp., Reno, NV) Model 204 UV-vis detector set at 370 nm wavelength. Two isocratic mobile phases were used to achieve baseline separation of all compounds of interest. The first mobile phase consisted of 25% acetonitrile in 0.025 M phosphate buffer (pH 2.4) with a flow rate of 0.9 ml min^{-1}. Because of inadequate separation of quercetin and luteolin (a flavone), a second eluent consisting of 45% methanol in the same buffer was used. The latter mobile phase was used only for peak identification and for quantification if both quercetin and luteolin were present in one sample. Peaks were identified by comparing their retention times with the retention times of the pure standards. A diode array detector (HP 1040A, Hewlett Packard, Palo Alto, CA) was used to confirm peak identity and to check peak purity.

Precision, Recovery and Detection Limits

The within-run relative standard deviations for quercetin, kaempferol and myricetin in control samples composed of onion/leek, cranberry, and endive/lettuce ranged from 2.5 to 5.6%. The between-run reproducibility of the assay, as determined by five duplicate analyses of the same control samples on separate days within a period of one month, was slightly higher. It ranged from 4.1 to 11.8%.

Recoveries were determined by spiking pure standards to the extraction solutions of the control samples at 50 and 100% of their measured flavonol content. The recoveries approximated 100% for all compounds and varied between 77 and 110%.

The limits of detection of flavonols in foods, defined as the concentration producing a peak height three times the standard deviation of the baseline noise, were 1 mg kg^{-1} fresh weight for quercetin, 2 mg kg^{-1} for kaempferol and 1 mg kg^{-1} for myricetin.[13] In beverages, the detection limits were 0.5, 1 and 0.5 mg l^{-1} for quercetin, kaempferol and myricetin, respectively.[12]

Applicability

The method presented here allows fast, quantitative and reproducible determination of the flavonols quercetin, kaempferol and myricetin in a variety of plant foods without the need for extensive sample clean-up or expensive equipment. Since this method was developed, more efficient columns with higher plate numbers have become available. As a consequence, the use of two different mobile phases is no longer necessary. Baseline separation of all compounds of interest can be achieved with acetonitrile eluents only. A drawback of this method using fixed hydrolysis/extraction conditions is that flavonol yield and degradation occur simultaneously during acid hydrolysis and optimum conditions are not the same for each food. Completeness of glycoside hydrolysis varies with reaction time and acid concentration and is dependent on the type of glycoside and the sample matrix. For a number of foods optimal hydrolysis conditions have been published.[13,16] Suboptimal hydrolysis conditions can lead to an underestimation of up to 50% of the true flavonol content. Prolonged hydrolysis can lead to degradation of the aglycone. Under the conditions described here, possible degradation of quercetin-3-glucoside, the most easily hydrolysable flavonol, is limited to 10% and recoveries of reference compounds (aglycones) spiked to different foods at different levels were satisfactory.

UV Detection of Individual Flavonol Glycosides

Because the type of attached sugar is an important determinant of the bioavailability of flavonoids, it is of interest to measure quantities of individual glycosides in foods. The bioavailability of various glycosides may differ three-fold.[17,18]

Extraction

For extraction, 25 ml methanol/water (1/1) was added to 0.5 g of freeze-dried sample and the mixture was refluxed for 15 min at 90 °C under nitrogen. Then the sample extract was allowed to cool, it was transferred into a flask and the volume was made up to 50 ml with methanol/water (1/1). After 5 min sonication the mixture was filtered through a 0.45 μm filter (Acrodisc CR PTFE, Gelman) and injected onto the HPLC system without further clean-up. Pure reference standards were available for the following flavonols: quercetin-3-rhamnoside, quercetin-3-galactoside, quercetin-3-glucoside, quercetin-3-arabinoside, quercetin-3-arabinoglucoside, quercetin-4′-glucoside, kaempferol-3-glucoside and kaempferol-3-robino-7-rhamnoside (all Roth, Germany), quercetin-3-rhamnoglucoside (Sigma, St. Louis, MO), kaempferol-3-rhamnoglucoside, myricetin-3-rhamnoside and quercetin-3-*O*-sulfate (Extrasynthese, France) and quercetin-3,4′-diglucoside (kindly provided by M. Rhodes, IFR, Norwich, UK). Stock solutions of these compounds were stable at 4 °C for a period up to three months.

Chromatography

The HPLC system consisted of a Perkin-Elmer ISS-100 automatic injector, a Gynkotek 480 pump, and an Inertsil ODS-3 (GL Sciences Inc.) column (4.6 × 150 mm, 5 μm) protected by a Newguard RP-18 guardcolumn (3.2 × 15 mm, 7 μm) both placed in a thermostated column oven at 30 °C, and a UV detector (Milton Roy SM4000) with the wavelength set at 370 nm. The detector output was sampled using a Nelson (PE Nelson, Cupertino, CA) series 900 interface and Nelson integrator software (Model 2600 rev. 5). Of several columns tested, the Inertsil ODS-3 gave the most symmetrical peaks and high plate numbers. For chromatographic separation, 50 μl of the sample extract was injected onto the column and separated isocratically with a mobile phase consisting of 18% acetonitrile in 0.025 M phosphate buffer (pH 2.4) at a flow rate of 1 ml min^{-1}. Mobile phases containing methanol as the organic modifier or acetic acid instead of phosphate buffer gave poorer separation and were therefore rejected. Peaks were identified by comparing their retention times with those of the pure standards. A diode array detector (HP 1040A, Hewlett Packard, Palo Alto, CA) was used to confirm peak identity. For identification of unknown flavonol glycosides, a directly coupled HPLC–NMR–MS method has been developed and applied for the identification and confirmation of quercetin and phloretin glycosides in apple peel.[19]

Glucuronides of quercetin and of kaempferol, for which no standards are available, occur frequently in foods. It is hard to obtain baseline separation of the flavonol glucosides and their corresponding glucuronides. Thus, to allow quantification of glucosides in glucuronide-containing foods and to indicate the presence of unknown glucuronides, an ion-pairing system with gradient elution was developed. By increasing the pH of the eluent from 2.4 to 4.0, the glucuronides but not the glycosides are ionized, which causes a downward shift in their retention time. The mobile phases for glucuronide identification were 13% acetonitrile (eluent A) and 50% acetonitrile (eluent B) in 0.025 M phosphate buffer (pH 4.0) with 0.005 M tetrapentylammonium bromide. Gradient elution was as follows: in 10 min from 5 to 20% B, then in 20 min from 20 to 80% B, 2 min 80%, and in 3 min back to 5%. A diode array detector (HP 1040A, Hewlett Packard, Palo Alto, CA) was used to confirm that the glucuronides detected were indeed of the flavonol type.

Optimization

Extraction optimization experiments were performed in duplicate with three model foods: onions, apples and french beans. The percentage of methanol in the aqueous extraction solvent was varied from 30 to 100% in 10% steps. The extraction time was varied from 5 to 60 min at 90 °C and compared to extraction for 1 min at room temperature in a Waring Blendor. Yields of the main flavonol conjugates in each product were comparable at all methanol concentrations (Figure 2). A lower concentration of methanol in the sample extract permits injection of larger volumes onto the HPLC column; therefore 50% methanol in water was chosen as the extraction solvent. Other glycosides were found to have similar extractability characteristics. Reflux extraction gave somewhat higher yields than cold extraction (Figure 3), but the refluxing time only moderately affected the yield. As an optimum, 15 minutes of refluxing was chosen.

Precision, Recovery and Detection Limits

The within-run repeatability of the method was determined for quercetin-3-galactoside from apple, and for quercetin-3,4′-diglucoside and quercetin-4′-gluco-

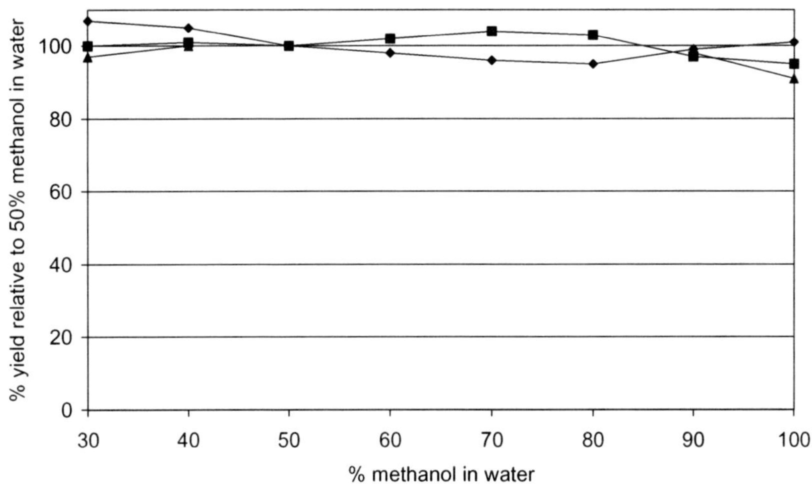

Figure 2 *Influence of percentage methanol in the extraction solvent on the yield of quercetin-3-galactoside from apple (◆), quercetin-4′-glucoside from onions (■) and quercetin-3-glucuronide from french beans (▲) (value for 50% methanol set at 100%; each datapoint represents the average of duplicate analyses)*

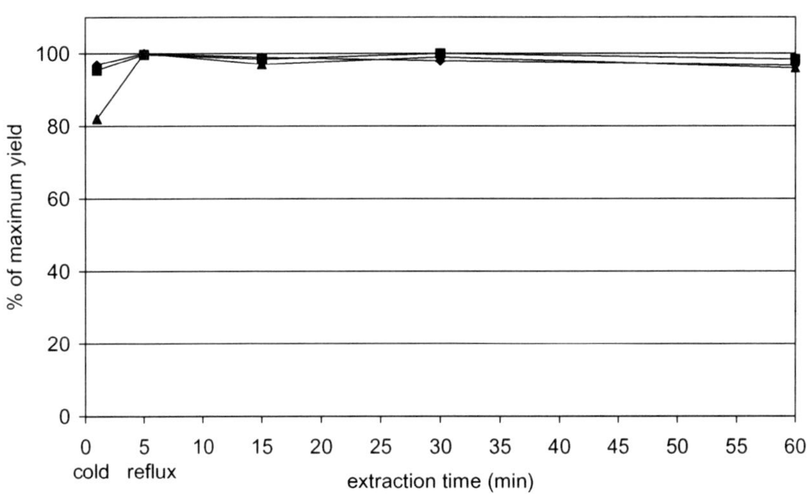

Figure 3 *Influence of extraction time on the yield of quercetin-3-galactoside from apple (◆), quercetin-4′-glucoside from onions (■) and quercetin-3-glucuronide from french beans (▲) (each datapoint represents the average of duplicate analyses)*

side from onion. The relative standard deviations were 8.4% for the galactoside, 2.7% for the diglucoside and 3.6% for the 4′-glucoside.

Recoveries of a number of glycoside standards that were added to apple or onion prior to extraction at 50 and 100% of their measured levels ranged from 77 to 125% (Table 1).

Table 1 *Recovery of flavonol glycosides spiked at 50 and 100% of their measured level to apple or onion*[a]

	Recovery of spike (%)	
Type of glycoside	*50% addition*	*100% addition*
Quercetin-3-rhamnosylglucoside	123	125
Quercetin-3-galactoside	76	82
Quercetin-3-glucoside	95	95
Quercetin-3-arabinoside	77	89
Quercetin-3-rhamnoside	94	101
Quercetin-4′-glucoside	96	95

[a] Average of duplicate analyses.

The limit of detection in food samples was approximately 10 mg kg^{-1} fresh food for most of the conjugated flavonols.

Applicability

The method presented here provides a simple way of quantifying a variety of flavonol glycosides in plant foods. Of course, quantification depends upon the availability of pure reference standards. Because no standards are available, the flavonol glucuronides, compounds that occur in commonly consumed foods like french beans, endive and red wine, could not be quantified and no recovery data are given. Unknown flavonol glycosides can be identified with the HPLC–NMR–MS method described by Lommen and co-workers.[19] Quantification of these unknown compounds may be possible against other flavonol glycosides with similar UV absorption characteristics. The detection limit of the method permits the detection of small amounts of many individual flavonol glycosides occurring in foods, but trace quantities in foods or biological fluids cannot be measured.

3 Flavonols in Biological Fluids

After ingestion with the diet, flavonol structures are modified by metabolism inside the body. In order to be able to evaluate the effects of flavonols, it is important to measure these metabolites in body fluids and tissues. For the metabolism of flavonoids, two compartments are important. The first consists of tissues in the body, where biotransformation enzymes located in the gut wall, the liver and kidney act upon absorbed flavonoids. As a result the polar hydroxyl groups of flavonols are conjugated with glucuronic acid and sulfate. In addition, *O*-methylation by the enzyme catechol-*O*-methyltransferase plays an important role in the inactivation of the catechol moiety, *i.e.* the two adjacent (ortho) aromatic hydroxyl groups, of flavonols.[7] *O*-Methylation of quercetin gives rise to the common metabolite isorhamnetin. The second metabolically active compartment is the colon, where microorganisms degrade unabsorbed flavonoids and flavonoids that have been absorbed and subsequently secreted with bile. Degradation by microorganisms involves splitting of the heterocyclic oxygen containing ring, to form a variety of phenyl carboxylic acids, which can be absorbed. In this chapter only methods for metabolites with an intact flavonol structure will be given.

HPLC methods with UV detection[16] do not meet the requirements of sensitivity and specificity to quantify flavonols in biological fluids. Thus, extensive sample enrichment and clean-up procedures would have to be considered. However, the enrichment attainable would probably not be sufficient for analyses in plasma. Consequently, detection techniques with enhanced sensitivity are needed. Fluorometry is a detection technique that could fulfil the requirements of sensitivity and specificity needed for body fluids. We have developed a post-column derivatization technique for fluorescence detection of flavonols after acid hydrolysis with HPLC that does not require expensive equipment.[20] This very sensitive method has been applied in a number of pharmacokinetic and absorption studies of flavonols in humans.[17,18,21,22] A second method was developed to determine flavonol conjugates and total flavonol aglycones after enzymatic hydrolysis followed by electrochemical detection. Electrochemical detection (coulometry) had shown to offer sufficient sensitivity (limit of detection 0.5 ng ml^{-1} for the detection of quercetin in plasma.[23]

Fluorescence Detection of Total Flavonols after Acid Hydrolysis

Extraction of Plasma

Flavonols and their conjugates (glucuronides and sulfates) were extracted from plasma and simultaneously hydrolysed to the aglycone form using 2 M HCl in aqueous methanol. For plasma 1 ml of methanol containing 2 g l^{-1} TBHQ and 0.4 ml of 10 M HCl were added to 0.6 ml of plasma in a 4 ml vial (Waters, Milford, MA) and mixed. The vial was sealed tightly with a cap and septum (both Waters, Milford, MA), inserted into a preheated aluminium block (Reacti-Block C-1, Pierce Europe, Oud-Beijerland, The Netherlands), heated in an oven at 90 °C for 5 hours, allowed to cool, and centrifuged at 1000 *g* for 15 min after addition of 2 ml of methanol containing 2 g l^{-1} TBHQ. A volume of 1.5 ml of the upper phase was transferred into an HPLC vial and 15 μl 100 g l^{-1} ascorbic acid was added.

Extraction of Urine

Flavonol conjugates (glucuronides and sulfates) were extracted from urine and simultaneously hydrolysed to the aglycone form using 2 M HCl in aqueous methanol. For urine 15 ml methanol containing 2 g l^{-1} TBHQ and 5 ml 10 M HCl were added to 5 g urine followed by mixing, the extract was refluxed at 90 °C for 8 hours with regular swirling, allowed to cool and subsequently brought to a final volume of 50 ml with methanol containing 2 g l^{-1} ascorbic acid. Urine extracts were sonicated for 5 min and filtered through a 0.45 μm filter for organic solvents (Acrodisc CR PTFE; Gelman Sciences, Ann Arbor, MI) before HPLC analysis.

Chromatography

For HPLC analysis 20 μl of the sample was injected onto an Inertsil ODS-2 column (4.6 × 150 mm, 5 μm; GL Sciences Inc., Tokyo, Japan) protected by a NewGuard RP-18 column (3.2 × 15 mm, 7 μm; Brownlee, Applied Biosystems Inc., Foster City, CA) using acetonitrile/0.025 M phosphate buffer pH 2.4, 31:69 (v/v) as mobile phase, at a flow rate of 1 ml min^{-1} using a Merck Hitachi L-6200 A pump (Hitachi Ltd., Tokyo, Japan). The columns were placed in a column oven set at 30 °C. The column effluent was mixed with 0.4 ml min^{-1} 1.5 M $Al(NO_3)_3$ in methanol containing 7.2% (v/v) acetic acid in a post-column reactor placed in the column oven. The post-column reactor consisted of a 15 m (0.25 mm internal diameter) Teflon PTFE tubing (Upchurch Scientific Inc., Oak Harbor, WA) coiled to a diameter of 3 cm and connected to the HPLC column with a low dead volume tee (Upchurch Scientific Inc., Oak Harbor, WA). Column effluent and aluminium reagent entered the tee counter-currently at an angle of 180°. A Merck Hitachi L-6000 A pump (Hitachi Ltd., Tokyo, Japan) was used to generate the aluminium reagent. The fluorescence of the ensuing quercetin-metal complex was measured at 485 nm using a Jasco FP 920 (Jasco Corporation, Japan) fluorescence detector with excitation wavelength set at 422 nm. The detector output was sampled using a Nelson (PE Nelson, Cupertino, CA) series 900 interface and Nelson integrator software (model 2600, rev.5) and peak area of the quercetin-metal complex was determined.

Precision, Recovery and Detection Limits

The relative standard deviation of duplicate analyses of quercetin was 4% for plasma and 6% for urine. The reproducibility of the quercetin assay was determined by duplicate analyses of 15 identical samples of plasma and of urine on 15 different days. The between series relative standard deviation of these quercetin determinations was 10% for plasma and 9% for urine.

Recovery of 100 ng quercetin aglycone added to 1 ml plasma was 88 ± 3% (six additions). Addition of 250 ng quercetin aglycone per gram of urine yielded a recovery of 99 ± 7% (3 additions).

The limit of detection, *i.e.* the concentration producing a peak height three times the standard deviation of the baseline noise, was 2 ng ml^{-1} for quercetin in plasma and 3 ng ml^{-1} for quercetin in urine. The limits of detection of standard solutions were 0.15 ng ml^{-1} for quercetin, 0.05 ng ml^{-1} for kaempferol, 0.45 ng ml^{-1} for myricetin, and 0.05 ng ml^{-1} for isorhamnetin. The calibration graph for quercetin was linear up to 5000 ng ml^{-1}, showing an extended dynamic range.

Applicability

Metal ions may bind to the 3-hydroxyl and 4-keto oxygen and to the 5-hydroxyl and 4-keto oxygen (Figure 1). Two adjacent hydroxyls, such as the 3′- and 4′-hydroxyls in quercetin, may also be involved in chelation, but at low pH this complex is labile.[24] It was shown[20] that in this post-column system only flavonoids that contain both a 3-hydroxyl and a 4-keto oxygen had sufficient fluorescence intensity. Flavonoids without a 3-hydroxyl or a 4-keto oxygen group, such as flavones and catechins, escape fluorescence detection. A free 3-hydroxyl group is essential. Rutin, which contains a sugar bound to the 3-hydroxyl group, does not form a fluorescent chelate. Thus, this method is suitable to quantify all flavonols as aglycones in biological fluids.

Electrochemical Detection of Conjugated Flavonols and of Total Flavonols after Enzymatic Hydrolysis

There is still debate whether intact flavonoid glycosides, the forms in which most flavonoids occur in plants, can be found circulating in plasma.[25] Certainly, glucuronidation, sulfation and methylation of absorbed flavonoids has been shown to occur extensively in humans.[26,27] A sensitive and selective method using electrochemical detection has been developed for the determination of conjugated flavonols in biological fluids.[25] We here describe a slight modification of this method, with which not only conjugated flavonols but also total flavonols after enzymatic hydrolysis can be measured. Quantification of individual flavonol conjugates is dependent upon the availability of pure reference compounds.

Extraction

For the determination of individual quercetin and isorhamnetin glycosides, glucuronides and sulfates, plasma samples were mixed with two volumes of acetonitrile and thoroughly vortexed. Then, one volume of 20% orthophosphoric acid, containing ascorbic acid ($3\ g\,l^{-1}$) was added. Samples were centrifuged for 10 min at 10 000 g (4 °C), and the supernatant was analysed with HPLC. The method has been used to determine whether intact quercetin-3-glucoside and quercetin-4′-glucoside can cross the intestinal barrier and may thus be expected to be found in plasma.[25] Quercetin, quercetin-3-*O*-sulfate and isorhamnetin-3-glucoside were obtained from Extrasynthese (Genay, France), quercetin-3-glucoside was obtained from Apin Chemicals Ltd. (Oxfordshire, England) and quercetin-4′-glucoside was obtained from Carl Roth GmbH (Karlsruhe, Germany). A mix of quercetin glucuronides used in this study was kindly provided by Karen O'Leary (IFR, Norwich, UK).

Hydrolysis

For the quantitative determination of total flavonols 100 μl plasma or bile was incubated with 0.5 mg β-glucuronidase/sulfatase from Helix Pomatia equivalent to 300 units β-glucuronidase and 20 units sulfatase activity (G1512, Sigma,

St. Louis, MO) in 40 μl 0.6 M acetate buffer pH 4.9 with ascorbic acid (6 g l^{-1}) for 0.5 h (bile) or 2 h (plasma) at 37 °C. Samples were extracted as described above.

Chromatography

A Merck Hitachi L-6000A pump (Hitachi Ltd., Tokyo, Japan), equipped with a Gilson 234 autosampler (Gilson, Villiers-le-Bel, France) and a coulometric detector (Coularray detector model 5600, ESA, Chelmsford, MA) set at 100, 300 and 500 mV was used (Pd as reference). Quercetin and isorhamnetin were quantified at 100 mV because at this potential the oxidation is extremely selective (Figure 4). Separation of the 20 μL of sample that was injected, was performed on a Chromolith RP-18e column (100 mm × 4.6 mm, Merck, Darmstadt, Germany) protected by a NewGuard RP18 guard column (15 mm × 3.2 mm, 7 μm, Perkin Elmer, Norwalk, CT) maintained at 30 °C with an ESA column heater (ESA, Chelmsford, MA). The solvents for the gradient elution were 5% acetonitrile (v/v) (solvent A) and 50% acetonitrile (v/v) (solvent B) in citrate buffer (25 mmol l^{-1}, pH 3.7). The following gradient, at a flow rate of 2.5 ml min^{-1}, was used: 0–12 min, linear gradient to 60% B; 12–12.5 min, linear gradient to 100% B; 12.5–14.5 min, 100% B; 14.5–15 min, linear gradient to 0% B. The total run time was 16 min.

Optimization

The amount of enzyme and the period of hydrolysis were varied to find optimum conditions for enzymatic hydrolysis of quercetin and isorhamnetin glucuronides and sulfates into their aglycones. All analyses were performed in duplicate.

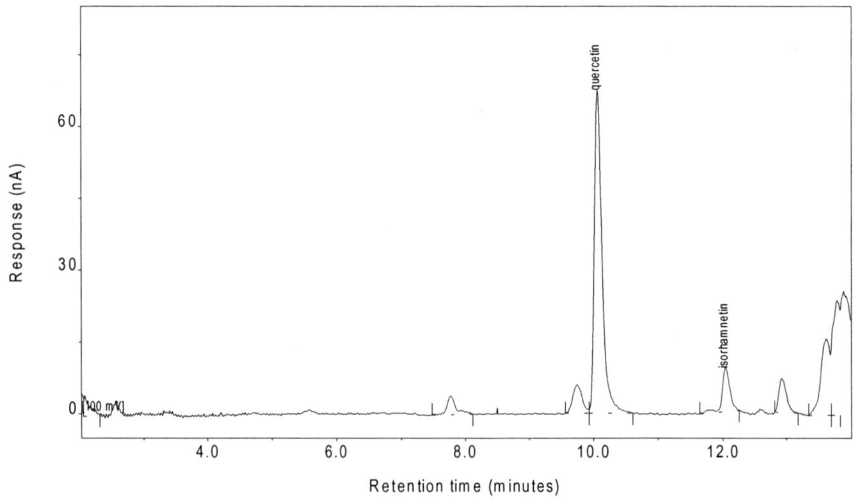

Figure 4 *Typical chromatogram of enzymatically hydrolysed rat plasma with coulometric detection at 100 mV*

To determine the amount of enzyme needed for complete deconjugation, rat plasma that contained quercetin and isorhamnetin conjugates was spiked with a mix of quercetin glucuronides to increase the amount of conjugates present. To 100 μl plasma, 0, 0.125, 0.25, 0.5, 1, or 2 mg of β-glucuronidase/sulfatase from Helix Pomatia (G1512, Sigma, St. Louis, MO) was added. The relative yields are presented in Table 2. Adequate yield of quercetin and isorhamnetin, combined with complete disappearance of the added flavonol glucuronides was attained with $\geqslant$0.5 mg enzyme.

To assess the time needed for complete hydrolysis of flavonol conjugates, 100 μl of bile or plasma was incubated with 0.5 mg β-glucuronidase/sulfatase from Helix Pomatia for 0.5, 1, or 2 h. Maximum yield for bile was attained with 0.5 h incubation, whereas for plasma 2 h of incubation was necessary to achieve maximum values (Table 3). Incubation of the sample for a period longer than

Table 2 *Relative recovery of quercetin and isorhamnetin from quercetin glucuronide enriched rat plasma with varying concentrations of β-glucuronidase/sulfatase from Helix pomatia*[a]

	Relative recovery (%)	
Enzyme (mg)	*Quercetin*	*Isorhamnetin*
0	12	21
0.125	83	81
0.25	84	97
0.5	88	103
1	94	104
2	100[b]	100[b]

[a] Average of duplicate analyses, 100 μl plasma incubated with enzyme in 40 μl 0.6 M acetate buffer pH 4.9 with ascorbic acid (6 g l^{-1}) for 2 h at 37 °C.
[b] Value set at 100%.

Table 3 *Relative yield of quercetin and isorhamnetin from human bile and rat plasma and recovery of quercetin and isorhamnetin aglycones spiked to human bile with increasing incubation time*[a]

	Relative yield (%)				*Recovery of spike (%)*	
	Bile		*Plasma*			
Incubation time (h)	*Quercetin*	*Isorhamnetin*	*Quercetin*	*Isorhamnetin*	*Quercetin*	*Isorhamnetin*
0.5	100[b]	100[b]	100[b]	100[b]	100	100
1	95	99	159	138	89	88
2	91	88	153	170	86	83

[a] Average of duplicate analyses, 100 μl sample incubated with 0.5 mg enzyme in 40 μl 0.6 M acetate buffer pH 4.9 with ascorbic acid (6 g l^{-1}) at 37 °C.
[b] Value set at 100%.

0.5 h caused a decrease in the recovery of quercetin and isorhamnetin aglycones added before extraction at a final concentration of 0.24 μM (Table 3). Similar recoveries were found when quercetin and isorhamnetin aglycones were spiked to saline with or without addition of enzyme (data not shown). Also, no difference in recoveries was found when a high (17.5 $g\,l^{-1}$) or low (6 $g\,l^{-1}$) concentration of ascorbic acid was added as an antioxidant during incubation. Apparently the incubation time as such, at relatively high temperature and pH, causes the loss of flavonols independent from enzyme or sample matrix.

Precision, Recovery and Detection Limits

Relative standard deviations were determined for quercetin-3-glucoside (by injecting eight duplicate samples), quercetin aglycone (17 duplicate analyses), and isorhamnetin aglycone (13 duplicate analyses). They were 0.4% for quercetin-3-glucoside, 0.06% for quercetin, and 0.2% for isorhamnetin.

Recovery of conjugated standards (quercetin-3-glucoside, quercetin-4′-glucoside, quercetin-3-sulfate and the mixture of quercetin glucuronides) added to plasma ranged from 83 to 115%.[25] Recoveries for quercetin and isorhamnetin aglycones added to bile and incubated for 0.5 h were both 100%. With increasing incubation time the recoveries deteriorated; for plasma the longer incubation time was necessary to achieve complete deconjugation of the flavonol glucuronides and sulfates. For plasma incubated at 37 °C for 2 h, recoveries ranged between 76 and 85% for quercetin and between 71 and 78% for isorhamnetin. This is comparable to the 70% recovery for quercetin in plasma reported previously by Erlund and coworkers using a similar method.[23] All standards and metabolites extracted from the plasma were stable during storage overnight in the autosampler.

The limits of detection of standard solutions were 0.6 $ng\,ml^{-1}$ for quercetin and 0.8 $ng\,ml^{-1}$ for isorhamnetin. Detection limits may be improved by detecting at higher Coularray potentials (250 mV for quercetin), but then the specificity is lower. No detection limits have been determined for conjugated flavonols.

Applicability

Coulometric detection combined with the use of a Chromolith column offers an extremely fast, easy, sensitive and selective method for the determination of free and conjugated flavonols in biological fluids. Contrary to the above described method using post-column derivatization and fluorescence detection, this method does not require a free 3-hydroxyl and a 4-keto oxygen group. However, with fewer electrochemically active groups (such as hydroxyl groups) the sensitivity and specificity of the assay decreases. For example apigenin (a trihydroxyflavone) has a substantially higher detection limit and can be detected only at higher potentials. A free hydroxyl group at the 3-positive appears to improve detection and also appears to shift the maximum response towards lower potentials. To quantify flavonols in biological fluids no sample clean-up is necessary and recoveries of samples spiked with aglycones were in general satisfactory, but may be somewhat low with longer enzymatic incubation. Enzymatic hydrolysis of conjugated flavonols into aglycones offers a relatively fast alternative to acid hydrolysis.

References

1. M. G. L. Hertog, P. C. H. Hollman, M. B. Katan and D. Kromhout, *Nutr Cancer*, 1993, **20**, 21.
2. E. B. Rimm, M. B. Katan, A. Ascherio, M. J. Stampfer and W. C. Willett, *Ann. Intern. Med.*, 1996, **125**, 384.
3. L. Yochum, L. H. Kushi, K. Meyer and A. R. Folsom, *Am. J. Epidemiol.*, 1999, **149**, 943.
4. M. G. L. Hertog, E. J. M. Feskens, P. C. H. Hollman, M. B. Katan and D. Kromhout, *Lancet*, 1993, **342**, 1007.
5. P. Knekt, R. Järvinen, A. Reunanen and J. Maatela, *Br. Med. J.*, 1996, **312**, 478.
6. M. G. L. Hertog, P. M. Sweetnam, A. M. Fehily, P. C. Elwood and D. Kromhout, *Am. J. Clin. Nutr.*, 1997, **65**, 1489.
7. P. C. H. Hollman and I. C. W. Arts, *J. Sci. Food Agr.*, 2000, **80**, 1081.
8. C. A. Williams and J. B. Harborne, in *Flavone and Flavonol Glycosides*, ed. J. B. Harborne, Chapman & Hall, London, 1994.
9. A. Finger, U. H. Engelhardt and V. Wray, *J. Sci. Food Agric.*, 1991, **55**, 313.
10. C. E. Lister, J. E. Lancaster, K. H. Sutton and J. R. L. Walker, *J. Sci. Food Agric.*, 1994, **64**, 155.
11. A. J. Dick, P. R. Redden, A. C. DeMarco, P. D. Lidster and T. B. Grindley, *J. Agric. Food Chem.*, 1987, **35**, 529.
12. M. G. L. Hertog, P. C. H. Hollman and B. van de Putte, *J. Agric. Food Chem.*, 1993, **41**, 1242.
13. M. G. L. Hertog, P. C. H. Hollman and M. B. Katan, *J. Agric. Food Chem.*, 1992, **40**, 2379.
14. A. Crozier, M. E. J. Lean, M. S. McDonald and C. Black, *J. Agric. Food Chem.*, 1997, **45**, 590.
15. U. Justesen, P. Knuthsen and T. Leth, *J. Chromatogr. A*, 1998, **799**, 101.
16. M. G. L. Hertog, P. C. H. Hollman and D. P. Venema, *J. Agric. Food Chem.*, 1992, **40**, 1591.
17. P. C. H. Hollman, M. N. C. P. Buijsman, Y. van Gameren, P. J. Cnossen, J. H. M. de Vries and M. B. Katan, *Free Radic. Res.*, 1999, **31**, 569.
18. P. C. H. Hollman, J. M. P. van Trijp, M. N. C. P. Buysman, M. S. van der Gaag, M. J. B. Mengelers, J. H. M. de Vries and M. B. Katan, *FEBS Lett.*, 1997, **418**, 152.
19. A. Lommen, M. Godejohann, D. P. Venema, P. C. H. Hollman and M. Spraul, *Anal. Chem.*, 2000, **72**, 1793.
20. P. C. H. Hollman, J. M. P. van Trijp and M. N. C. P. Buysman, *Anal. Chem.*, 1996, **68**, 3511.

21. P. C. H. Hollman, J. H. M. de Vries, S. D. van Leeuwen, M. J. B. Mengelers and M. B. Katan, *Am. J. Clin. Nutr.*, 1995, **62**, 1276.
22. J. H. M. de Vries, P. C. H. Hollman, S. Meyboom, M. N. C. P. Buysman, P. L. Zock, W. A. van Staveren and M. B. Katan, *Am. J. Clin. Nutr.*, 1998, **68**, 60.
23. I. Erlund, G. Alfthan, H. Siren, K. Ariniemi and A. Aro, *J. Chromatogr. B*, 1999, **727**, 179.
24. K. R. Markham, *Techniques of Flavonoid Identification*, Academic Press, London, 1982.
25. A. L. A. Sesink, K. A. O'Leary and P. C. H. Hollman, *J. Nutr.*, 2001, **131**, 1938.
26. U. Fuhr and A. L. Kummert, *Clin. Pharmacol. Ther.*, 1995, **58**, 365.
27. J. H. Moon, R. Nakata, S. Oshima, T. Inakuma and J. Terao, *Am. J. Physiol. Regulatory Integrative Comp. Physiol.*, 2000, **279**, R461.

CHAPTER 11

Analysis and Purification of Catechins and Their Transformation Products

TIMOTHY J. BOND,* JOHN R. LEWIS, ADRIENNE DAVIS AND ALAN P. DAVIES

1 Introduction

This chapter has been written with practising scientists in mind and is meant to be not only informative, but useful, and an easy to follow guide in identifying sources of the desired polyphenol(s). The information presented here also shows how to further isolate and purify the said compound and most importantly, how to reassure yourself that the compound isolated is the desired compound in the required purity!

This chapter covers the major catechins and their oxidation products, the theaflavins (TFs) and thearubigins (TRs). For information on the less abundant proanthocyanidins the authors recommend you see Chapters 12 and 13 on this subject in this volume. For those not familiar with tea chemistry the first two texts referenced and the references therein give an excellent entry point to the literature.[1,2]

Where possible protocols are included that have been carried out in our laboratories, including scale and an idea of yield/losses that can be expected. Hints, tips and advantages of protocols are given where applicable. Happy isolation!!!

2 Catechin and Related Compound Isolation

Catechins are flavan-3-ols and their corresponding 3-*O*-gallate esters. There are various sources in the natural world from the tea bush (*Camelia sinensis* var.

*Corresponding author

assamica and var. *sinensis*), and willow (*Salix caprea*), to the now common garden plant *Bergenia crassiflora*. The tea plant, and in particular the young shoots (flush), are weight for weight probably the richest source of catechins. The major catechins present in tea are epicatechin, epigallocatechin, epicatechin-3-*O*-gallate, epigallocatechin-3-*O*-gallate, catechin and gallocatechin (Figure 1). In the variety *sinensis* the flavonoid level (% dry wt) is in the range 10–20%, whereas in *assamica* it is 10–25%. Additionally the ratio of flavan-3-*O*-gallates to flavan-3-ols is higher in *assamica* compared with the variety *sinensis*. This should be borne in mind when deciding on the raw materials for extraction.

Extraction

Extraction is performed by reducing the size of the plant material (hence the diffusion pathways) and using polar solvents (water, methanol, ethanol). When choosing a solvent, considering the solubility of the desired compounds and extraction conditions is essential. Extraction of catechins should be carried out preferably at low temperatures to minimise oxidative reactions and most importantly epimerisation.[3] Epimerisation is the inversion of a single stereocentre in a molecule. This can have the effect of changing both physical parameters (optical rotation, chromatographic characteristics) and (potentially) biochemical behaviour/actions (absorption, enzyme recognition).

Exposure to high temperatures (>80 °C) causes inversion of the stereocentre at C-2 of the catechin structure.[4,5] In the authors' considered opinion the low levels

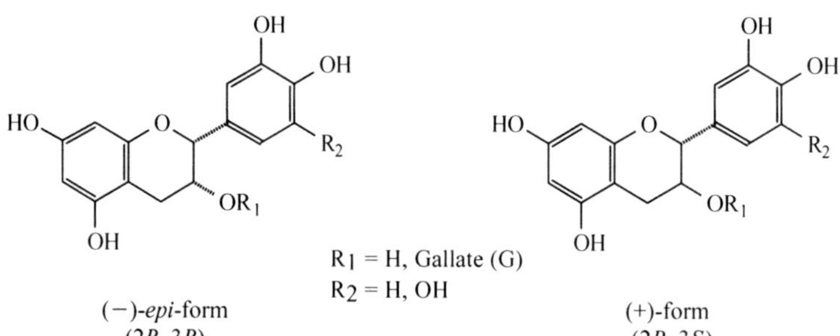

Catechin	Symbol	R1	R2
(−)-Epicatechin	(−)-EC	H	H
(−)-Epicatechin gallate	(−)-ECG	G	H
(−)-Epigallocatechin	(−)-EGC	H	OH
(−)-Epigallocatechin gallate	(−)-EGCG	G	OH
(+)-Catechin	(+)-C	H	H
(+)-Catechin gallate	(+)-CG	G	H
(+)-Gallocatechin	(+)-GC	H	OH
(+)-Gallocatechin gallate	(+)-GCG	G	OH

Figure 1 *Major catechin species in black tea*

of the epimeric counter parts of the major catechin species [(−)EGCG, (−)EGC, (−)ECG, (+)C] is the result of thermal steps in manufacture such as enzyme inactivation and drying in green tea manufacture, drying in black tea manufacture, or in extraction for analysis (typically 70% aqueous methanol at elevated temperatures). A typical procedure for the extraction of tea samples is given below:

Plant materials (1–2 kg, *for source see Physical Properties of Fresh Leaf Flavan-3-ols and Related Compounds* pg 233) were macerated in a high-speed blender (Silverson) with cold methanol (3–4 extractions per batch in 1–2 l of solvent). After removal of the plant debris the methanolic extract was concentrated *in vacuo* to a small volume (*ca.* 200 ml) and an equal volume of water added. The extract was then further reduced in volume to give an aqueous solution of plant polyphenols (*ca.* 200 ml). This solution was filtered (Muslin or Watman No. 1) to remove insoluble materials such as chlorophylls, fats and waxes, and was then extracted with ethyl acetate (300 ml, 6–8 times). The combined ethyl acetate extracts give the crude phenolic extract (usually ~2% of the fresh plant material, but occasionally up to 10%).

Separation Protocols

The preparation of flavan-3ols from their various sources can be achieved by a number of methods. The classical approach has been to apply the extract to a large open column containing an appropriate stationary phase and elute the compounds of interest with a suitable solvent system. This approach is detailed in the next section. Other methods which will be touched upon in the following sections are co-precipitation with caffeine and high speed counter current chromatography.

Column Chromatography

Column chromatography of phenolic extracts can be carried out on a number of solid support phases that include silica gel, alumina and cross-linked dextrose, the latter being the most commonly used support and also known by the trade name of Sephadex or the more robust Toyopearl. Various types of Sephadex are available, the differences being particle size and charge. Sephadex LH-20 is a neutral support that we have found to have the widest application in tea-related flavan-3-ols. Using standard alcoholic and mixed solvent systems this is basically a molecular filtration media.

The extract is dissolved in a small amount of ethanol, applied to a column containing Sephadex LH-20 (Pharmacia) and eluted with ethanol. The size of the column is dependent upon the amount of phenolic extract to be purified. Generally speaking, it is desirable to have the weight ratio of the Sephadex gel to a phenolic extract at 15–20:1. The separation process is monitored by thin-layer chromatography (TLC) conducted in the following two complimentary solvent systems:

(a) 6% aqueous acetic acid (v/v)
(b) butan-2-ol-acetic acid–water [14:1:5 (v/v)].

The cellulose-F plates (Merck) are visualised by spraying a fresh aqueous mixture of $FeCl_3$ and $K_3Fe(CN)_6$ (1% each, 1:1 v/v), and appropriate fractions (based upon tlc R_f values) are then combined. Pure polyphenolic components are then obtained *via* evaporation of the solvent from the combined fractions.

Elution order from Sephadex LH-20 when eluted with ethanol is as follows:

EA – Epiafzelechin

C – Catechin

EC – Epicatechin

GC – Gallocatechin

EGC – Epigallocatechin

CG – Catechin gallate

ECG – Epicatechin gallate

GCG – Gallocatechin gallate

EGCG – Epigallocatechin gallate

EGCmg – Epigallocatechin-3-methyl gallate

EC-3, 5-dg – Epicatechin-3, 5-digallate

TsA – Theasinensin A

TsB – Theasinensin B

Caffeine Precipitation

Catechins and other tea components have a propensity to self-associate as well as associate with other planar molecules (*e.g.* caffeine) through Π–Π interactions. This natural phenomena is termed 'creaming' in black tea infusions and causes the infusion to become turbid or cloudy as the tea infusion cools. This behaviour is especially pronounced with gallate esters[6–8] and has been utilised by Clifford *et al.*[9] to develop a method for isolating, amongst other catechins, EGCG from green tea. Caffeine is added to an infusion of tea until the turbidity reaches a maximum whereupon the precipitate is removed by centrifugation. Solvent partitioning then leaves the desired catechin(s) in enhanced purity. A typical procedure is given below:

To a green tea infusion (2 × 350 ml, 25 g leaf tea) at 70 °C was added caffeine to give a final concentration of 30 mM. This solution was then cooled to 4 °C and left at this temperature for one hour. Centrifugation (31000*g*, 20 min) gave a pellet of material which is then resuspended in distilled water (260 ml). This yielded a gallate-rich fraction. Decaffeination was carried out at 60 °C using solvent partitioning (chloroform, 4 × equal volumes). Sequential solvent partitioning of the aqueous phase yielded in turn a ECG-rich fraction (ethyl hexanoate), and an EGCG-rich fraction (propyl acetate).

High Speed Counter Current Chromatography

High speed counter current chromatography is the term introduced by Ito[10] to describe the most recent advance in counter current chromatography. This is liquid–liquid chromatography system originally developed by Cornish[11] in the 1930s without a solid sorbent, requiring two immiscible solvent phases moving over each other. HSCCC is the application of centrifugal forces to aid the separation process by increasing stationary-phase retention and the range of useable flow rates and solvent systems. HSCCC has a number of advantages over classic CCC including reduced run times, a wider range of available solvent systems and increased sample sizes. These separations allow preparation of materials in shorter times than can be achieved using traditional chromatography and are not prone to material loss through irreversible adsorption. As this chapter is aimed at general extraction/isolation without the use of specialist equipment CCC will not be discussed further. The authors direct interested readers to general reviews on HSCCC[12,13] and specifically to catechin isolation.[14,15]

Once the sample has been isolated and purified it is necessary to use chemical and physical measures to positively identify the compound and assess the purity.

Physical Properties of Fresh Leaf Flavan-3-ols and Related Compounds

(−)-Epiafzelechin

Sources:	Green tea leaves (*Camellia sinensis*) or leaves of leguminosae (*Cassia fistula*)
Re-purification:	On Sephadex LH-20 elution with chloroform and ethanol (7:1 v/v) or with distilled water
Crystallisation:	From ethanol–water to give fine white needles
Melting point:	240 °C (decomp.)
$[\alpha]^{20}$	−58.9° (*c* 1.0 in ethanol)
R_f (a)	0.43
R_f (b)	0.65

(+)-Catechin

Sources:	Green tea leaves (*Camellia sinensis*) or leaves and catkins of Sallow willow (*Salix caprea*)

Re-purification: On Sephadex LH-20 (5 × 45 cm), elution with chloroform and ethanol (7:1 v/v) or distilled water
Crystallisation: From water to give fine needles
Melting point: 177 °C
$[\alpha]^{20}$ +17.1° (*c* 0.1 in acetone)
R_f (a) 0.47
R_f (b) 0.51

(–)-Epicatechin
Source: Green tea leaves (*Camellia sinensis*) or Hawthorn berries (*Crataegus monogyna*)
Crystallisation: From water
Melting point: 218 °C (decomp.)
$[\alpha]^{20}$ –59.0° (*c* 1.0 in acetone)
R_f (a) 0.37
R_f (b) 0.51

(+)-Gallocatechin
Source: Green tea (*Camellia sinensis*)
Re-purification: On Sephadex LH-20 elution with water followed by chloroform and ethanol (7:1 v/v) or eluted with distilled water and then with an aqueous acetone solution in which acetone was gradually increased to 25%
Melting point: 188 °C[16]
$[\alpha]^{20}$ +14.7° (*c* 1.0 in acetone)
R_f (a) 0.50
R_f (b) 0.55

(–)-Epigallocatechin
Source: Green tea (*Camellia sinensis*)
Re-purification: On Sephadex LH-20 elution with water followed by chloroform and ethanol (7:1 v/v) or with distilled water and then with an aqueous acetone solution in which acetone was gradually increased to 25%
Crystallisation: From water
Melting point: 218 °C[16]
$[\alpha]^{20}$ –67.5° (*c* 1.0 in ethanol)
R_f (a) 0.45
R_f (b) 0.40

(+)-Catechin gallate
Sources: Green tea leaves (*Camellia sinensis*) or roots and rhizomes of *Bergenia crassiflora*
Re-purification: On Sephadex LH-20 elution with chloroform and ethanol (5:1 v/v) or eluted with an aqueous acetone solution in which acetone was steadily increased from 10% to 25%

Crystallisation:	From water
Melting point:	N/D
$[\alpha]^{20}$	+56° (*c* 0.9 in acetone)
R_f (a)	0.18
R_f (b)	0.68

(−)-Epicatechin-3-O-gallate

Source:	Green tea (*Camellia sinensis*)
Re-purification:	On Sephadex LH-20 eluted with an aqueous acetone solution in which acetone was steadily increased from 10% to 25%
Crystallisation:	From water
Melting point:	253 °C[16]
$[\alpha]^{20}$	−179° (*c* 0.28 in ethanol)
R_f (a)	0.36
R_f (b)	0.60

(−)-Epigallocatechin-3-O-gallate

Source:	Green tea (*Camellia sinensis*)
Re-purification:	On Sephadex LH-20 eluted with an aqueous acetone solution in which acetone was steadily increased from 10% to 25%
Crystallisation:	From water
Melting point	254 °C[16]
$[\alpha]^{20}$	−179° (*c* 0.28 in ethanol)
R_f (a)	0.36
R_f (b)	0.60

(−)-Epicatechin-3,5-O-digallate

Source:	Green tea (*Camellia sinensis*)
Re-purification:	On Sephadex LH-20 elution with chloroform and ethanol (5:1 v/v) and appropriate fractions re-chromatographed on Sephadex LH-20 elution with water and ethanol (1:1 v/v)
Melting point	NA
$[\alpha]^{20}$	−13.0° (*c* 1.0 in ethanol)
R_f (a)	0.15
R_f (b)	0.65

Theasinensin A

Source:	Green tea (*Camellia sinensis*)
Repurification:	On Diaion HP-20 elution with water/methanol (4:1–2:3 v/v) and appropriate fractions re-chromatographed on Sephadex LH-20 elution with aqueous methanol (80%)
Melting point:	N/D
$[\alpha]^{15}$	−226.8° (*c* 0.73 in acetone)
R_f (a)	N/D
R_f (b)	N/D

Theasinensin B

Source:	Green tea (*Camellia sinensis*)
Repurification:	On Diaion HP-20 elution with water/methanol (4:1–2:3 v/v) and appropriate fractions re-chromatographed on Sephadex LH-20 elution with aqueous methanol (80%)
Melting point:	N/D
$[\alpha]^{21}$	−147.2° (*c* 0.5 in acetone)
R_f (a)	N/D
R_f (b)	N/D

NMR Spectroscopy of Fresh Leaf Flavan-3-ols and Related Compounds

NMR spectra were measured on a Bruker AMX400 spectrometer operating at a probe temperature of 303 K using either a dual $^{1}H/^{13}C$ 5 mm probe or a

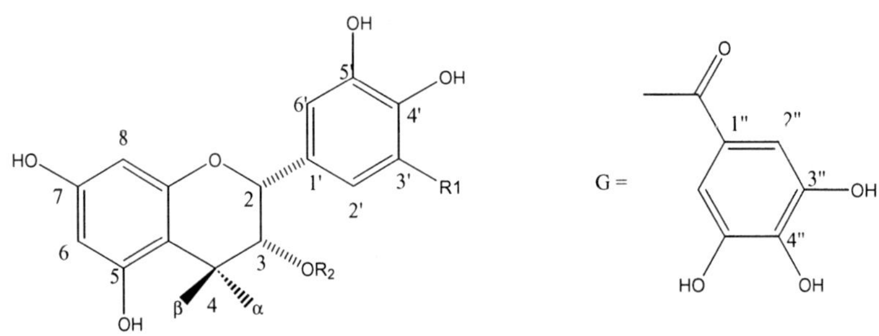

Figure 2 *Flavan-3-ol NMR numbering system*

Table 1 *^{1}H NMR chemical shift data for green tea polyphenols*[a]

	C	*GC*	*CG*	*GCG*	*EC*	*EGC*	*ECG*	*EGCG*	*EA*	*EGCMG*
2	4.563	4.512	5.153	5.135	4.882	4.820	5.134	5.067	4.933	5.100
3	3.994	3.970	5.401	5.402	4.209	4.190	5.555	5.562	4.221	5.512
4	2.912	2.870	2.888	2.824	2.740	2.727	2.925	2.909	2.749	2.948
4	2.531	2.521	2.764	2.765	2.867	2.849	3.052	3.040	2.880	3.051
6	6.024	6.018	6.072	6.064	6.022	6.016	6.063	6.062	6.027	6.049
8	5.881	5.877	5.993	5.988	5.922	5.913	6.037	6.036	5.927	6.033
2′	6.898	6.457	6.924	6.481	7.054	6.574	7.059	6.624	7.362	6.647
3′	–	–	–	–	–	–	–	–	6.816	–
5′	6.798	–	6.794	–	6.787	–	6.764	–	6.816	–
6′	6.758		6.794		6.840		6.891		7.362	
2″	–	–	7.045	7.050	–	–	7.030	7.028	–	7.056
6″	–	–			–	–			–	7.134
OMe	–	–	–	–	–	–	–	–	–	3.816

[a] Spectra measured in acetone-d_6 at 303 K and referenced relative to internal TMS. Magnetically equivalent protons are tabulated only once, a '–' indicates that the given proton is not present in the molecule.

Table 2 *^{1}H NMR coupling constant data for green tea polyphenols*[a]

	C	*GC*	*CG*	*GCG*	*EC*	*EGC*	*ECG*	*EGCG*	*EA*	*EGCMG*
$J(6, 8)$	2.3	2.3	2.3	2.3	2.3	2.3	2.3	2.3	2.3	2.3
$J(4_{\beta}, 4_{a})$	16.1	16.1	16.6	16.7	16.5	16.6	17.4	17.4	16.7	17.4
$J(4_{\beta}, 3)$	8.4	8.2	6.0	5.5	4.6	4.5	4.5	4.6	4.5	4.4
$J(4_{a}, 3)$	5.5	5.5	5.2	5.2	3.2	3.5	2.1	2.2	3.2	2.8
$J(3, 2)$	7.8	7.4	6.0	5.3	1.6	1.7	1.4	1.4	1.5	1.6
$J(4_{\beta}, 2)$	n.r	n.r.	1.0	n.r.	0.8	0.8	0.7	0.8	0.8	n.r.
$J(4_{a}, 2)$	n.r.	n.r.	n.r.	n.r.	0.9	0.8	0.9	1.0	1.1	n.r.
$J(2, 2')$	n.r.	0.5	0.7	n.r.	0.5	0.6	0.5	0.6	0.6	0.6
$J(2, 6')$	n.r.		0.7		0.7		0.7		0.6	
$J(2', 6')$	1.9		n.m.		2.0		2.0		2.5[b]	
$J(5', 6')$	8.1	–	n.m.	–	8.1	–	8.2	–	8.4[b]	–
$J(3', 5')$	–	–	–	–	–	–	–	–	2.5[b]	–
$J(2'', 6'')$	–	–			–	–			–	1.9

[a] Measured in Hz and accurate to 0.15 Hz; n.r. = not resolved, n.m. = not measured. Magnetically equivalent nuclei are listed only once.
[b] Calculated from the $[AX]_2$ spin system of this molecule, $J(2', 5') = J(3', 6') = 0.4$.

Table 3 *^{13}C NMR chemical shift data for green tea polyphenols*[a]

	C	*GC*	*CG*	*GCG*	*EC*	*EGC*	*ECG*	*EGCG*	*EA*	*EGCMG*
2	82.76	82.80	78.80	78.64	79.46	79.46	78.11	78.10	79.53	78.01
3	68.38	68.35	70.46	70.36	66.97	67.02	69.36	69.28	66.90	69.83
4	28.84	28.46	24.20	23.58	29.01	28.82	26.63	26.64	29.16	26.36
4a	100.69	100.63	99.24	99.14	99.85	99.92	99.03	99.09	99.79	98.99
6	96.19	96.15	96.47	96.40	96.22	96.18	96.56	96.54	96.25	96.50
8	95.50	95.49	95.61	95.57	95.75	95.73	95.87	95.87	95.78	95.79
5	157.23	157.24	157.25	157.23	157.61	157.57	157.50	157.46	157.63	157.45
7	157.77	157.74	158.08	158.02	157.60	157.59	157.81	157.78	157.66	157.88
8a	156.95	156.90	156.32	156.24	157.19	157.14	157.14	157.12	157.24	157.06
1′	132.27	131.62	131.47	130.95	132.32	131.56	131.42	130.77	131.55	130.85
2′	115.27	107.25	114.30	106.20	115.31	106.97	114.95	106.81	129.17	106.69
3′	145.66	146.30	145.82[b]	146.57	145.42	146.17	145.60	146.27	115.50	146.35
4′	145.73	133.34	145.89[b]	133.37	145.31	132.97	145.53	133.17	157.69	133.17
5′	115.74		115.99		115.51		115.67			
6′	120.08		119.13		119.41		119.25			
1″	–	–	121.79	121.80	–	–	121.87	121.92	–	121.81
2″	–	–	109.98	109.99	–	–	109.99	110.04	–	106.00
3″	–	–	146.00	146.03	–	–	145.94	145.90	–	148.47
4″	–	–	138.93	138.97	–	–	138.83	138.79	–	139.80
5″	–	–			–	–			–	145.79
6′	–	–			–	–			–	111.67
CO	–	–	166.04	166.15	–	–	166.04	166.07	–	166.16
OMe	–	–	–	–	–	–	–	–	–	56.54

[a] Spectra measured in acetone-d_6 at 303 K and referenced relative to internal TMS. Equivalent carbons are tabulated only once; '–' indicates that the carbon atom is not present in the molecule.
[b] Assignments interchangeable.

Table 4 *Long-range proton–carbon correlations measured for EGCMG using the HMBC experiment*[a]

Proton	*Carbon*
6″	1″, 2″, 4″, 5″, CO
2″	1″, 3″, 4″, 6″, CO
2′	1′, 3′, 4′
6	4a, 5, 7, 8
8	4a, 6, 7, 8a
2	1′, 2′, 3
Ome	3″
4_β	3, 4a, 5, 8a
4_α	2, 3, 4a, 5, 8a

[a] Measured at 303 K in acetone-d_6.

multinuclear 5 mm inverse probe as appropriate. The solvent used was acetone-d_6 and spectra were referenced relative to internal TMS. Sample concentrations were typically in the range 1–10 mg per 0.5 ml.

NOE (Nuclear Overhauser Effect) difference experiments were acquired with typically a 2 s pre-saturation time and with spectra at each pre-saturation position interleaved in groups of 16 scans to minimise artifacts due to instrument instabilities. HMQC experiments were acquired with a 400 ms delay after the BIRD pulse, a de-phasing/refocusing delay of 3.571 ms and a relaxation delay of 0.5 s between scans; GARP de-coupling was applied during acquisition. HMBC experiments were recorded with a 3 ms delay for the low-pass J filter, a 60 ms delay for evolution of long-range couplings and a relaxation delay of 0.6 s. For each spectrum 512 or 1024 increments of 2k data points were collected covering a typical spectral width of 3000 Hz in F2 and 15000 Hz in F1. The HMQC was recorded phase sensitive using TPPI and processed with $\pi/2$ shifted sine-bell windows in both dimensions. The HMBC spectra were processed with a Gaussian window function in both dimensions. The flavan-3-ol NMR numbering system is shown in Figure 2 and NMR data are presented in Tables 1–4.

3 Flavan-3-ol Dimeric Oxidation Products

Theaflavins (TFs) are unique to black tea and are formed through the oxidative condensation of a simple catechin quinone (*i.e.* that formed from EC or ECG) with a gallocatechin quinone (*i.e.* that formed from EGC or EGCG). The name is derived from the latin names for the commodity (*thea* meaning tea) and the appearance (*flavus* meaning orange). Theaflavins constitute 2–4% of the dry weight of black tea. The four major species of TFs can be classified as:

EC + EGC = theaflavin (TF)
EC + EGCG = theaflavin 3-gallate (TF3G)

ECG + EGC = theaflavin 3′-gallate (TF3′G)
ECG + EGCG = theaflavin digallate (TFDG)

Figure 3 shows the structures of these compounds.

Formation and structure elucidation of the theaflavins and related compounds found in black tea have been recently reviewed.[17] All TF species are highly coloured; however, due to its lack of gallate groups, TF transfers more efficiently from leaf to liquor and therefore its relative contribution to liquor colour will be higher. TFs can be further oxidised during fermentation to TRs, a heterogeneous mixture of higher molecular weight polymers (see Section 4).

Extraction

Black tea (3.0 in 1.0 kg batches) was extracted with cold methanol (2 × 1000 ml) with stirring (30 minutes). The spent leaf was discarded and the methanol extracts were combined and concentrated to a known volume (500 ml). An equal volume of distilled water was added and the residual methanol removed *in vacuo*. The aqueous extract was heated (60 °C) and extracted with dichloromethane (2 × 1000 ml) and then with ethyl aceate (8 × 1000 ml). The dichloromethane extracts containing caffeine and chlorophylls were discarded. The ethyl acetate

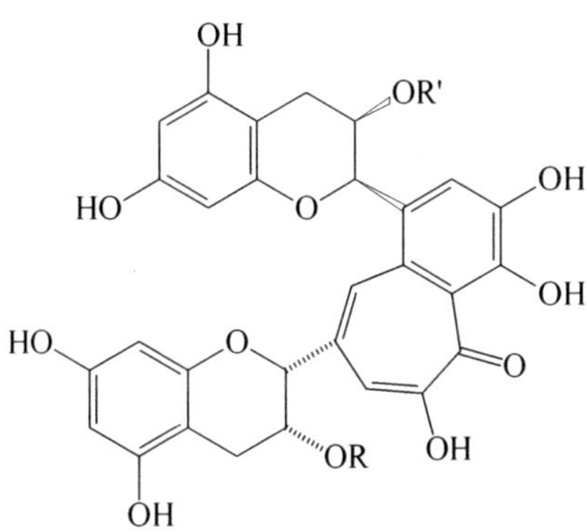

Benztropolone	Symbol	R	R'
Theaflavin	TF	H	H
Theaflavin-3-*O*-monogallate	TF-3-MG	Gallate	H
Theaflavin-3′-*O*-monogallate	TF-3’-MG	H	Gallate
Theaflavin-3,3′-*O*-digallate	TFDG	Gallate	Gallate

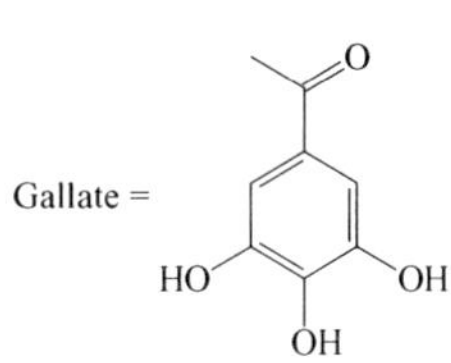

Figure 3 *Major benztropolone species in black tea*

washings were combined and concentrated to dryness and the residual red/purple solid was retained.

Separation Protocols

Column Chromatography

The ethyl acetate soluble solids (100 g) were applied to a column of Sephadex LH-20 pre-swollen in ethanol.[18] Elution with ethanol (6300 ml) followed by aqueous acetone (28% v/v, 10 000 ml) afforded seven fractions (Table 5). The progress of the separation was followed by thin layer chromatography (TLC) on cellulose plates (Merck Aluminium-backed Cellulose-F), which were developed in a mixture of butan-2-ol:acetic acid:water [14:1:5]. The TLC plates were visualised by spraying with freshly prepared aqueous mixture of iron (III) chloride and potassium hexacyanoferrate (III) (1% each, 1:1).

Chromatography of Fraction IV on Sephadex LH-20 (26 cm × 12 cm i.d.) eluted with distilled water containing increasing proportions of acetone (28%–35% v/v) yielded six fractions (Table 6).

Table 5 *Fractions produced from sequential ethanol and aqueous acetone elution of ethyl acetate soluble solids*

Fraction numbers	*Colour*	*TLC identification (yield/g)*
I	Yellow/green	Chlorophyll
II	Purple	Unidentified (3.06)
III	Orange/red	Mixed Catechins (10.78)
IV	Bright red	Theaflavin rich (30.28) R_f 0.35
V	Red	Theaflavin monogallates (12.0) R_f 0.46 and 0.56
VI	Orange	Theaflavin digallate (14.0) R_f 0.62
VII	Yellow/brown	Unidentified (14.87); thearubigins?

Table 6 *Fractions produced from aqueous acetone elution of fraction IV*

Fraction	*Colour*	*HPLC/TLC identification (yield/g)*
A	Colourless	–
B	Yellow	Mixed catechins (3.67)
C	Orange	ECG, theaflavin and (4.57) theaflavic Acid
D	Orange/red	Theaflavin R_f 0.35 (3.14) major component
E	Orange/red	'Pure' theaflavin (1.64)
F	Orange/red	Theaflavin monogallate (4.56) mixture R_f 0.46 (T_f-3-mg) R_f 0.56 (T_f-3′-mg)

Fractions B and E were found to contain mainly a mixture of catechins and the four standard theaflavins (TF, TF-3-MG, TF-3′-MG, TFDG) by HPLC analysis (for analytical conditions see Appendix B). Separation of Fraction C by preparative HPLC (see Appendix A) yielded orange pigments: Theaflavin (100 mg), theaflavic acid (25 mg) and epitheaflavic acid (30 mg). These were further purified on Sephadex LH-20 with ethanol elution to yield pure samples (for structures see Appendix B).

Chromatography of fraction D yielded a mixture of orange/red pigments which were separated on Hypersil ODS and Sephadex LH-20 [water, aqueous acetone (50% v/v)]. This yielded isotheaflavin-3′-monogallate (25 mg), neotheaflavin-3-monogallate (30 mg), epitheaflagallin-3-gallate (20 mg) and theaflavin (200 mg).

Fraction F was separated on Waters NovaPak ODS (for preparative conditions see Appendix B) and Sephadex LH-20 [water, aqueous acetone (50% v/v)]. This yielded pure samples of a further four orange/red pigments: theaflavate B (18 mg), isotheaflavin (25 mg), theaflavin-3-monogallate (120 mg) and theaflavin-3′-monogallate (130 mg).

High Speed Column Chromatography

There are only a few examples reported of HSCCC being applied to Tfs. The interested reader is directed to the work of Donovan *et al.*,[19] Degenhardt *et al.*[20] and Lewis *et al.*[21]

Physical Properties of Black Tea Dimeric Flavonoid Oxidation Products

Theaflavate B

Precursors:	EC + EGCG
Appearance:	Red powder
Melting point:	N/D
Optical rotation:	N/D
UV data:	284 nm
	406 nm
IR data:	None reported.

Isotheaflavin-3′-monogallate

Precursors:	GC + ECG
Appearance:	Red powder
Melting point:	N/D
Optical rotation:	N/D
UV data:	
IR data:	N/D

Neotheaflavin-3-monogallate

Precursors:	C + EGCG
Appearance:	Red powder
Melting point:	N/D
Optical rotation:	N/D
UV data:	N/D
IR data:	N/D

Isotheaflavin

Precursors:	GC + EC
Appearance:	Red powder
Melting point:	N/D
Optical rotation:	N/D
UV data:	207 nm ε 93 500
	231 nm (sh) ε 28 600
	270 nm ε 18 800
	378 nm ε 7500
	462 nm ε 2700
	recorded in ethanol
IR data:	3400 cm^{-1}
	1630 cm^{-1}
	1600–1595 cm^{-1}
	1475 cm^{-1} } benzotrolopone
	1430 cm^{-1} } ring deformations
	1250–1230 cm^{-1}
	recorded in nujol

Epitheaflagallin-3-gallate

Precursors:	EGCG + gallic acid
Appearance:	Red needles (H_2O–MeOH)
Melting point:	218–220 °C
Optical rotation	$[\alpha]^{23}$ – 383.2° (*c* 0.4 in acetone)
UV data:	None reported
IR data:	None reported

Theaflavin

Precursors:	EC + EGC
Appearance:	Red crystalline powder
Melting point	243–246 °C (decomp.)
Optical rotation	$[\alpha]^{25}$ – 246.1°
UV data	461 nm ε 4060
	378 nm ε 10 800
	268 nm ε 18 260
	recorded in methanol
IR data	1630 cm^{-1} tropolone carbonyl

	1600–1595 cm^{-1}
	1475 cm^{-1} } benzotroplone
	1430 cm^{-1} } ring deformations
	1250–1230 cm^{-1}

Theaflavin-3-monogallate

Precursors:	EC + EGCG
Appearance:	Red crystalline powder (H_2O–MeOH)
Melting point:	230–235 °C
Optical rotation	$[\alpha]^{23} = -326.8°$ (*c* 0.4 in acetone)
UV data	455 nm ε 3910
	376 nm ε 10 000
	272 nm ε 26 440
	recorded in methanol
IR data	1630 cm^{-1} tropolone carbonyl
	1600–1595 cm^{-1}
	1475 cm^{-1} } benzotroplone
	1430 cm^{-1} } ring deformations
	1250–1230 cm^{-1}

Theaflavin-3'-monogallate

Precursors:	ECG + ECG
Appearance:	Red crystalline powder; (H_2O–MeOH)
Melting point	236–240 °C (decomp.)
Optical rotation	$[\alpha]^{25} - 370.7°$ (*c* 0.7 in acetone)
UV data	452 nm ε 4080
	376 nm ε 10 040
	278 nm ε 23 670
	recorded in methanol
IR data	1630 cm^{-1} tropolone carbonyl
	1600–1595 cm^{-1}
	1475 cm^{-1} } benzotroplone
	1430 cm^{-1} } ring deformations
	1250–1230 cm^{-1}

Theaflavic acid

Precursors:	C + gallic acid
Appearance:	Red solid
Melting point:	230 °C
Optical rotation:	$[\alpha]^{25} - 42°$ (ethanol)
UV data:	280 nm ε 27 800
	404 nm ε 12 300
	recorded in ethanol
IR data:	3400 cm^{-1}
	1700 cm^{-1} gallic acid or flavanol gallate
	1698 cm^{-1} ester/carboxy carbonyl group

	1630 cm^{-1}
	1600–1595 cm^{-1}
	1475 cm^{-1} } benzotrolopone
	1430 cm^{-1} } ring deformations
	1250–1230 cm^{-1}
	recorded in nujol

Epitheaflavic acid

Precursors:	EC + gallic acid
Appearance:	Red solid (ethyl acetate)
Melting point:	232–236 °C
Optical rotation:	$[\alpha]^{25}{}_D$ – 90° (ethanol)
UV data:	280 nm ε 24 100
	400 nm ε 11 000
	Measured in ethanol
IR data:	3400 cm^{-1}
	1700 cm^{-1} gallic acid or flavanol gallate
	1698 cm^{-1} ester/carboxy carbonyl group
	1630 cm^{-1}
	1600–1595 cm^{-1}
	1600 cm^{-1} } benzotrolopone
	1475 cm^{-1} } ring
	1430 cm^{-1} } deformations
	1250–1230 cm^{-1}
	Measured in nujol

Theaflavin-3,3'-digallate

Precursors:	ECG + EGCG
Appearance:	Red crystalline powder (H_2O–MeOH)
Melting point	225–228 °C (decomp.)
Optical rotation	$[\alpha]^{18}$ – 540.9° (*c* 0.3 in acetone)
UV data	455 nm ε 3540
	378 nm ε 10 230
	278 nm ε 36 350
	recorded in methanol
IR data	1630 cm^{-1} tropolone carbonyl
	1600–1595 cm^{-1}
	1475 cm^{-1} } benzotroplone
	1430 cm^{-1} } ring deformations
	1250–1230 cm^{-1}

The numbering system for theaflavin and related compounds is shown in Figure 4 and NMR data are presented in Tables 7–11.

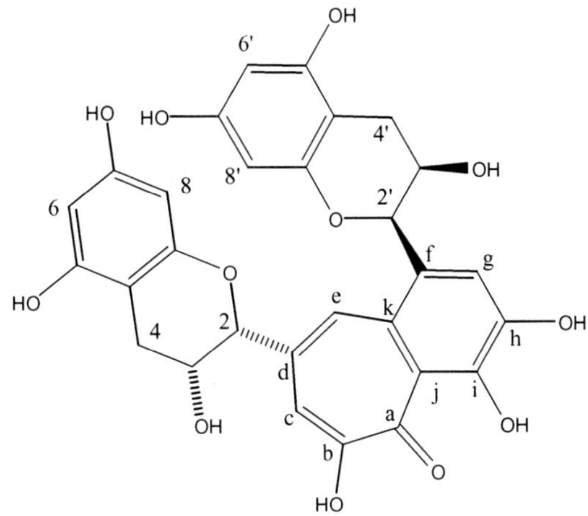

Figure 4 *NMR numbering system for theaflavin and related compounds*

Table 7 *^{13}C–{^{1}H} NMR data for theaflavate B and theaflavin-3,3′-O-digallate*

	Theaflavate B/ppm	*Tfdg/ppm*
α	166.78	–
a	185.99	185.12
b	154.44	155.02
c	115.76	117.81
d	124.21	134.94
e	132.02	126.64
k	126.39	128.44
f	134.77	130.34
g	123.76	123.15
h	148.97	146.51
i	151.52	151.12
j	121.92	121.16
2	77.43	80.53
3	71.65	69.69
4	26.37	26.72
4a	98.81	98.82
5	157.40	157.97[cd]
6	96.90	96.97[c]
7	157.95	157.64[cd]
8	96.27	95.90[c]
8a	156.77	156.44[c]
2′	76.67	75.10
3′	66.02	68.02
4′	29.81	27.06
4′a	100.15	99.36
5′	157.78	157.97[cd]

continued overleaf

Table 7 *(continued)*

	Theaflavate B/ppm	*Tfdg/ppm*
6′	96.90	96.97[c]
7′	157.62	157.54[cd]
8′	96.04	96.04[c]
8′a	157.04	157.07[c]
G0	–	166.04
G1	–	121.37
G2	–	109.99
G3	–	145.98
G4	–	139.01
G′0	–	166.04
G′1	–	121.24
G′2	–	109.99
G′3	–	145.98
G′4	–	139.01
1″	130.95	–
2″	113.97	–
3″	145.82	–
4″	145.48	–
5″	115.86	–
6″	118.72	–

[a] Measured at 300 K in acetone-d_6 unless otherwise stated and referenced relative to internal TMS or the solvent peak = 29.83 ppm.
[c] Assignment follows from HMBC connectivities and unequivocal data from analogous compounds very strongly suggesting that δC-6 > δC-8.
[d] Assignment interchangeable.

Key to Compounds in Tables 7–11
Theaflavate B = Theaflavate B
ITf′mg = Isotheaflavin-3′-monogallate
NTfmg = Neotheaflavin-3-monogallate
ITF = Isotheaflavin
ETFG = Epitheaflagallin-3-gallate
Tf = Theaflavin
Tfmg = Theaflavin-3-monogallate
Tf′mg = Theaflavin-3′-monogallate
Tfa = Theaflavic acid
Etfa = Epitheaflavic acid
PPGA = Pupurogallin carboxylic acid
Tfdg = Theaflavin-3,3′-digallate

4 Flavan-3-ol Polymeric Oxidation Products

The thearubigins (TRs) are a heterogeneous group of orange-brown, weakly acidic compounds formed by oxidative transformations (usually enzmatic or initiated by enzymatic activation) of flavonols during the manufacture of black tea.[22] The name is derived from the latin names for the commodity (*thea* meaning tea) and the appearance (*rubus* meaning brown). TRs are the most abundant polyphenolic fraction in black tea, associated with taste, depth of colour, and body.[23] Due to

Table 8 $^{13}C-\{^{1}H\}$ *NMR data for compounds 3–12, including purpurogallin for reference*

	ITf' mg/ppm	*NTfmg/ppm*	*ITf/ppm*	*ETFGg/ppm*	*Tf/ppm*	*Tfmg/ppm*	*Tf' mg/ppm*	*ETfA/ppm*	*PPGA/ppm*	*Pur/ppm*
α	–	–	–	–	–	–	–	168.18	167.88	–
A	185.39	185.22	182.74	185.31	184.90	185.03	185.00	186.03	183.76	183.26
B	155.44	154.60	154.59	155.21	154.72	154.99	154.90	154.48	154.25	155.70
C	117.88	117.79	116.65	118.04	199.24	117.61	119.44	116.10	114.89	116.98
D	135.19	133.94	134.58	134.63	135.20	133.80	135.88	124.73	125.52	124.73
E	128.08	128.40	133.60	128.57	127.17	126.36	126.81	132.40	139.39	135.61
K	128.43	129.66	134.08	128.62	128.83	128.50	128.73	126.86	132.23	134.96
F	130.28	132.15	111.98	131.77	131.77	131.61	130.32	134.55	114.65	111.37
G	123.31	122.90	152.14	124.11	123.90	124.00	123.16	123.60	151.88	152.06
H	146.44	146.76	135.32	146.40	146.15	146.42	146.22	148.65	137.41	135.37
I	151.13	151.11	152.46	150.82	150.16	150.83	150.92	151.01	153.26	155.56
J	121.68	121.84	115.70	121.72	121.68	121.55	121.69	122.23	116.13	115.94
G0	–	166.43	–	–	–	166.02	–	–	–	–
G1	–	121.08	–	–	–	121.31	–	–	–	–
G2	–	109.98	–	–	–	109.88	–	–	–	–
G3	–	145.99	–	–	–	145.90	–	–	–	–
G4	–	139.18	–	–	–	139.04	–	–	–	–
G′0	165.74	–	–	–	–	–	165.69	–	–	–
G′1	121.38	–	–	–	–	–	121.46	–	–	–
G′2	109.93	–	–	–	–	–	109.90	–	–	–
G′3	145.99	–	–	–	–	–	145.96	–	–	–
G′4	138.96	–	–	–	–	–	138.90	–	–	–

continued overleaf

Table 8 *(continued)*

	ITf' mg/ppm	*NTfmg/ppm*	*ITf/ppm*	*ETFGg/ppm*	*Tf/ppm*	*Tfmg/ppm*	*Tf' mg/ppm*	*ETfA/ppm*	*PPGA/ppm*	*Pur/ppm*
2	85.97	80.29	85.99	79.86	81.37	79.70	81.54	–	–	–
3	68.24	70.00	68.28	68.87	66.42	69.21	66.48	–	–	–
4	29.87	26.70	29.82	26.84	29.49	26.62	29.55	–	–	–
4a	100.79	98.49	100.76	98.82	99.47	98.79	99.55	–	–	–
5	157.23[c]	157.92[cd]	157.21	157.59	157.83[d]	158.02[cd]	157.93[d]	–	–	–
6	96.70[c]	96.89[c]	96.66	96.98	96.60	97.03[c]	96.66	–	–	–
7	158.03[c]	157.84[cd]	157.98	157.99	157.72[d]	157.75[cd]	157.76[d]	–	–	–
8	95.54[c]	95.62[c]	95.53	95.90	95.70	95.80[c]	95.74	–	–	–
8a	156.44[c]	156.23[c]	156.41	156.50	156.66	156.24[c]	156.73	–	–	–
2′	75.30	79.19	76.66	–	76.84	76.58	75.48	76.89	–	–
3′	67.85	68.96	65.46	–	65.25	65.56	67.94	65.86	–	–
4′	26.95	29.17	29.67	–	29.72	29.76	27.01	29.92	–	–
4′a	99.28	100.49	99.95	–	100.04	99.77	99.37	99.62	–	–
5′	157.61[c]	157.53[d]	157.79	–	157.72[d]	157.36[cd]	157.76[d]	157.77	–	–
6′	97.02[c]	96.66	96.66	–	96.60	96.74[c]	96.99	96.60	–	–
7′	157.93[c]	157.34[d]	157.66	–	157.63[d]	157.55[cd]	157.65[d]	157.68	–	–
8′	96.02[c]	95.34	95.94	–	95.90	95.98[c]	96.02	95.94	–	–
8′a	156.97[c]	156.60	157.16	–	157.17	157.09[c]	157.03	157.09	–	–

[a] Measured at 300 K in acetone-d_6 unless otherwise stated and referenced relative to internal TMS or the solvent peak = 29.83.
[b] Measured at 303 K
[c] Assignment follows from HMBC connectivities and unequivocal data from analogous compounds very strongly suggesting that δC-6 > δC-8.
[d] Assignment interchangeable

Table 9 *{^{1}H} NMR data for theaflavate B and theaflavin-3,3′-O-digallate*

	Theaflavate B/ppm	*Tfdg/ppm*
c	7.69	7.76
e	8.46	7.97
g	8.04	8.04
2	5.17	5.46
3	5.69	5.74
4α	3.04	2.96
4β	3.17	3.07
6	6.09	6.11[e]
8	6.21	6.14[e]
2′	5.65	6.10
3′	4.22	5.79
4′α	2.93	2.99
4′β	3.27[d]	3.28
6′	6.17[e]	6.13[e]
8′	6.09[e]	6.15[e]
G2	–	7.02
G′2	–	6.98
2″	7.01	–
5″	6.97	–
6″	6.65	–
OH-i	14.77[f]	14.87[f]

[a] Measured at 300 K in acetone-d_6 plus D_2O and referenced relative to internal TMS or the solvent peak at 2.087 ppm unless otherwise stated.
[b] Measured at 303 K.
[c] The following assignments were obtained from additional spectra measured at 263 K in acetone-d_6.
[d] Distinguish from 4α (or 4′α as appropriate) by coupling chemical shift.
[e] Assignment follows from HMQC *and* unequivocal data from analogous compounds showing δC-6 > δC-8.
[f] Measured in acetone-d_6 and referenced relative to internal TMS or the solvent peak = 2.052 ppm.
[g] Not measured.
[h] Measured at 303 K in methanol-d_4 solvent and quoted in ppm relative to solvent peak at 3.304 ppm.

limitations of analytical techniques, the oxidative processes which lead to these polymers and their detailed chemical structure are poorly understood. Separations and fractionations of TRs are based on bulk characteristics and generally rely on solvent partitioning (Roberts fractionation), column chromatography (theafulvin) or solvent precipitation. These three methods are detailed below.

Roberts Fractionation

After E.A.H. Roberts[24]

Table 10 *{^1H} NMR data for compounds 3–12, including purpurogallin for reference*

	ITf′mg/ppm	*NTfmg[b]/ppm*	*ITf[b]/ppm*	*ETFGg[c]/ppm*	*Tf/ppm*	*Tfmg/ppm*	*Tf′mg[b]/ppm*	*ETfA/ppm*	*PPGA/ppm*	*Pur/ppm*
c	7.42	7.67	7.40	7.52	7.58	7.66	7.59	7.90	7.77	7.14
d	–	–	–	–	–	–	–	–	–	6.81
e	7.83	8.26	7.79	7.69	7.99	8.02	8.02	8.78	8.33	7.40
f	–	–	–	7.03	–	–	–	–	7.23	7.02
g	8.07	7.70	8.05	–	8.05	8.06	8.07	8.09	–	–
2	4.80	5.32	4.71	5.21	5.03	5.36	5.09	–	–	–
3	4.04	5.69	4.03	5.69	4.41	5.76	4.42	–	–	–
4α	3.10	2.96	3.06	3.00	2.85	3.00	2.87	–	–	–
4β	2.57[d]	3.05[d]	2.56	3.13	2.94[d]	3.10[d]	2.95[d]	–	–	–
6	6.08[e]	6.08	6.06	6.11	6.07	6.11[e]	6.07	–	–	–
8	5.96[e]	6.08	5.95	6.12	6.00	6.16[e]	6.00	–	–	–
2′	6.02	5.67	5.72	–	5.74	5.81	6.00	5.73	–	–
3′	5.74	4.25	4.40	–	4.49	4.56	5.77	4.46	–	–
4′α	2.96	2.97	2.84	–	2.81	2.92	2.95	3.06[d]	–	–
4′β	3.17[d]	2.70[d]	3.00	–	3.00[d]	3.11[d]	3.15[d]	2.93	–	–
6′	6.09[e]	6.11	6.07	–	6.09	6.12[e]	6.11	6.11	–	–
8′	6.13[e]	5.98	6.03	-	6.04	6.07[e]	6.14	6.05	–	–

G2	–	6.99	–	7.01	–	6.94	–	–	–	–
G′2	6.97	–	–	–	–	–	6.96	–	–	–
OH-5	–	–	–	8.77[c]	–	–	–	–	–	–
OH-7	–	–	–	8.56[c]	–	–	–	–	–	–
OH-G3	–	–	–	8.58[c]	–	–	–	–	–	–
OH-G4	–	–	–	8.34[c]	–	–	–	–	–	–
OH-b	–	–	–	8.90[c]	–	–	–	–	–	–
OH-g	–	–	–	9.58[c]	–	–	–	–	–	–
OH-h	–	–	–	8.85[c]	–	–	–	–	–	–
OH-l	14.92[f]	n.m.[g]	14.94[f]	15.03[f]	14.98[f]	14.86[f]	14.96[f]	14.82[f]	14.89[f]	15.04[f]

[a] Measured at 300 K in acetone-d_6 plus D_2O and reference relative to internal TMS or the solvent peak at 2.087 ppm unless otherwise stated.
[b] Measured at 303 K.
[c] The following assignments were obtained from additional spectra measured at 263 K in acetone-d_6.
[d] Distinguish from 4α (or $4'\alpha$ as appropriate) by coupling chemical shift.
[e] Assignment follows from HMQC and unequivocal data from analogous compounds showing δC-6 > δC-8.
[f] Measured in acetone-d_6 and referenced relative to internal TMS or the solvent peak = 2.052.
[g] Not measured.
* Measured at 303 K in methanol-d_4 solvent and quoted in ppm relative to solvent peak at 3.304 ppm.

Table 11 *^{1}H NMR coupling constant* data for compounds 1 to 9, 11 and theaflavin-3,3′-O-digallate*[a]

	Theaflavate B/Hz	*Theaflavate C[b]/Hz*	*ITf′mg/Hz*	*NTfmg[b]/Hz*	*ITf[b]/Hz*	*ETFGg[b]/Hz*	*Tf/Hz*	*Tfmg/Hz*	*Tf′mg[b]/Hz*	*Etfa/Hz*	*Tfdg/Hz*
J (4a, 4ß)	17.7	17.8	15.8	17.2	15.9	17.5	16.8	17.5	16.8	–	17.4
J (3, 4ß)	4.6	4.6	9.6	4.5	9.6	4.5	4.3	4.6	4.1	–	4.4
J (3, 4a)	2.2	2.3	5.6	n.m.[d]	5.8	1.9	2.4	2.2	2.7	–	2.2
J (2, 4a)	n.r.[e]	n.r.	n.r.	n.m.	n.r.	0.4	n.r.	0.8	0.6	–	n.r.
J (2, 3)	1.2	1.2	8.6	2	8.6	1.3	1.3	1.3	1.3	–	1.3
J (6, 8)	2.3	2.3	2.3	n.r.[c]	2.3	2.3	2.3	2.3	2.3	–	2.3
J (4′a, 4ß′)	17.1	16.0	17.7	16.1	16.8	–	16.9	17.0	17.5	16.8	17.6
J (3, 4′ß)	4.6	9.3	4.6	8.8	4.4	–	4.6	4.4	4.6	4.5	4.6
J (3′, 4′a)	1.8	5.7	1.7	5.4	2.2	–	2.1	1.9	1.9	2.2	1.8
J (2′, 4′a)	n.r.	n.r.	n.r.	n.r.	n.r.	–	0.6	0.6	n.r.	0.7	n.r.
J (2′, 3′)	n.r.	8.6	0.8	8.2	1.2	–	0.9	0.8	0.8	0.8	0.7
J(6′, 8′)	2.3	2.3	2.3	2.4	2.3	–	2.3	2.3	2.3	2.3	2.3
J (2, 2″)	0.6	n.r.	–	–	–	–	–	–	–	–	–
J (2, 6″)	0.7	0.6	–	–	–	–	–	–	–	–	–
J (2″, 6″)	2.1	2.1	–	–	–	–	–	–	–	–	–
J (5″, 6″)	8.2	8.2	–	–	–	–	–	–	–	–	–
						0.8					
J (2, e)	–	–	n.r.	n.r.	n.r.	0.8	n.r.	n.r.	n.r.	–	n.r.
J (e, f)	–	–	–	–	–	1.6	–	–	–	–	–
J (c, e)	1.3	1.4	1.2	1.1	1.2		1.3	1.3	1.2	1.2	1.2

* All values in Hertz (Hz).
[a] Measured at 300 K in acetone-d_6 plus D_2O and typically accurate to 0.2 Hz.
[b] Measured at 303 K.
[c] H-6 and H-8 are coincident.
[d] n.m. = not measured.
[e] n.r. = not resolved.

Extraction

A black tea infusion was prepared by refluxing the tea with boiling water (375 ml) for five minutes, and filtering hot through a cotton-wool plug. The infusion (250 ml) was continuously extracted for six hours with ethyl acetate and the extract evaporated to dryness under reduced pressure. (Fraction 1A; yield 1 g).

Polyphenolic substances in the residual aqueous layer were precipitated with saturated lead acetate (5 ml). The precipitate was suspended in methanol and decomposed by hydrogen sulfide and the filtrate from the lead sulfide evaporated to dryness under reduced pressure (Fraction 1B; yield 1.0 g).

Separation of SI from Fraction 1A

Fraction 1A (1 g) was dissolved in a small volume of acetone and precipitated with excess chloroform. The precipitate was washed first with chloroform and then with light petroleum (0.65 g). The filtrate consisted largely of caffeine together with a small amount of gallic acid and some flavanols.

The precipitate was dissolved in a small volume of acetone and precipitated with a large excess of ether. The precipitate was washed with light petroleum and air dried. It was then redissolved in acetone and again precitated with excess ether. This procedure was repeated four times until the precipitate was shown to be free from gallic acid flavanols and theaflavins.

The final product (yield 0.3 g) consisted almost entirely of S1 material with trace levels of flavonol glycosides and chlorogenic acids.

Separation of S1a and S11 from Fraction 1B

The aqueous fraction from the initial extraction was extracted with butanol (2 × 250 ml). The organic extracts were combined and washed with water (250 ml in total). The residual aqueous layer was acidified with sulfuric acid (10 ml, 2N) and extracted with butanol (3 × 150 ml). The butanol extracts were combined and washed with water (150 ml in total).

The butanol washings from both extractions were combined (approximate total volume 950 ml) and evaporated to dryness (yield: 1.0 g). The solid was then taken up in methanol (5 ml) and precipitated by ether (50 ml), this process was repeated four times and yielded fraction S11 (0.4 g).

The ether filtrates were combined and evaporated to dryness *in vacuo*. The resultant solid was taken up in acetone (5 ml) and precipitated by ether (50 ml); this process was repeated four times and yielded fraction S1a (0.3 g).

The ether filtrates were again combined and evaporated to dryness *in vacuo* (0.2 g).

Figure 5 shows the overall process.

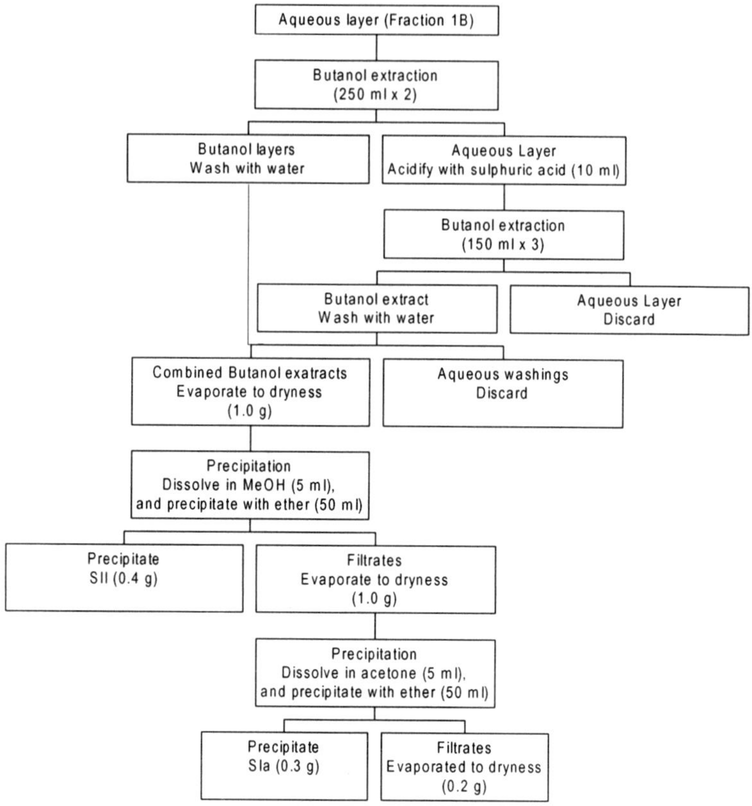

Figure 5 *Roberts fractionation of aqueous tea extract*

Theafulvin Fractionation

Theafulvin is a crude fraction derived from Solka-Floc cellulose chromatography[25] of a tea infusion. The name is derived from the latin names for the commodity (*thea* meaning tea) and the appearance (*fulvus* meaning tawny).

Extraction

A black tea infusion was prepared by refluxing tea (10 g) with boiling water (100 ml) for five minutes, and filtering hot through a cotton-wool plug.

Column Chromatography

Black tea infusion (80 ml) was applied to a column of Solka-Floc previously washed with 1M HCl, distilled water and packed with MeOH. The column was then sequentially eluted with methanol (500 ml), acetone (500 ml) and 50% v/v aqueous acetone (500–1000 ml). The brown pigments eluted with the final solvent were then collected, reduced *in vacuo* to a volume of *ca* 1.0 ml and redissolved in methanol (10 ml). The solution was then filtered through a plug of cotton wool, the plug was washed with methanol (1.0 ml) and diethyl ether (500 ml) was added. The resulting suspension was stored at 4 °C overnight before the buff coloured solid was collected by filtration and dried (0.16 g).

In our experience and that of others[26] the theafulvin protocol is not reproducible, with yields varying ten-fold for the same tea. This said, theafulvin is one of the few sub-fractions to have been isolated and so is of note. Paper chromatograms of the fraction resembled those of classic TR fractions.[22] Theafulvin contained no protein or caffeine, and was free of flavonol glycosides.[25] Reverse-phase HPLC showed the fraction to be a mixture of polymers, the chromatograms showing a convex broad band. IR spectroscopy showed the substance to be phenolic, and ^{13}C NMR showed that it was a flavonol polymer with various linkages to the proanthocyanidin polymers. Theafulvin has also been the subject of other studies wanting a 'model' TR.[27]

Black Tea Fractionation by Solvent Precipitation

Extraction

Black tea leaf (400 g) was extracted in hot water (6.5 l of deionised water, 95 °C, 15 minutes) with constant stirring. The infusion was de-leafed through muslin (four layers) and the leaf pressed (screw tea press) to remove any residual liquor. The liquor was then cooled to ambient temperature in ice and blast frozen into stainless steel trays (1.5 l capacity) and freeze dried. The dried extract was packaged into heat-sealable, aluminium foil pouches for stability.

Preparative Solvent Partition Protocol

Black tea extract (40 g) was dissolved in boiling deionised water (500 ml) and allowed to cool prior to the addition of methanol (100 ml, 16.67%). This liquor was partitioned against ethyl acetate (10 × 500 ml) and the organic extracts were combined and concentrated to dryness *in vacuo* (40 °C). A small volume of deionised water (100–200 ml) was added and the sample carefully transferred to a pre-weighed foil tray and frozen at −18 °C overnight prior to freeze drying, yielding a dark red-brown solid (9.7 g).

The residual aqueous fraction was also concentrated *in vacuo* (to remove residual solvent contamination) and the sample carefully transferred to a pre-weighed foil tray and frozen at −18 °C overnight prior to freeze drying. This yielded a light brown solid (28.5 g).

Preparative Solvent Precipitation Protocol

The aqueous fraction (5 g) was dissolved in boiling deionised water (100 ml) and allowed to cool to room temperature with constant stirring. Absolute ethanol (900 ml) was added and stirring continued for a further 10 minutes. The resultant precipitate was separated from the liquor *via* vacuum filtration through a glass sinter (Pyrex, No 3, 500 ml capacity) and washed with a further aliquot of absolute ethanol (100 ml). The ethanol filtrates were combined and concentrated to dryness *in vacuo* (initially at 40 °C and later at 70 °C). A small volume of deionised water (100–200 ml) was added and the sample carefully transferred to a pre-weighed foil tray and frozen at −18 °C overnight prior to freeze drying. This yielded a buff coloured powder (3.0 g).

The precipitate was re-dissolved in boiling deionised water (100 ml) and allowed to cool to room temperature with constant stirring. Methanol (900 ml) was added and stirring continued for a further 10 minutes. The resultant precipitate was separated as above and washed with methanol (100 ml). The methanol filtrates were combined and concentrated to dryness *in vacuo* (initially at 40 °C and later at 70 °C). A small volume of de-ionised water (100–200 ml) was added, the sample carefully transferred to a pre-weighed foil tray and frozen at −18 °C overnight prior to freeze drying. This yielded a buff coloured powder (0.9 g).

The precipitate was again re-dissolved in boiling deionised water (100 ml) and the solution centrifuged (4000*g*, 45 minutes, 15 °C). The clarified solution was then concentrated *in vacuo* (initially at 40 °C and later at 70 °C), the sample carefully transferred to a pre-weighed foil tray and frozen at −18 °C overnight prior to freeze drying. This yielded a dark brown powder (0.9 g).

Figure 6 shows the overall process.

Preparative Sample List:

1. Whole infusion
2. Ethyl acetate-soluble solids
3. Aqueous soluble solids
4. Ethanol-soluble solids
5. Methanol-soluble solids
6. Water-soluble solids
7. Water-insoluble solids

Notes:
At the end of each solvent precipitation, it may be necessary to clean the glass sinters due to accumulation of insoluble material. A hot solution of bleach can be used to remove this residue. Following washing the sinter requires washing with approximately 1–2 l of warm distilled water to remove bleach residues (continue flushing until on foaming occurs).

Analysis of Solvent precipitation fractions

Solvent precipitation has been found to be very reproducible both for the same tea and for teas of different origins. Though absolute amounts of the fractions do vary the ratio of ethanol to methanol to water-soluble fractions is, in our

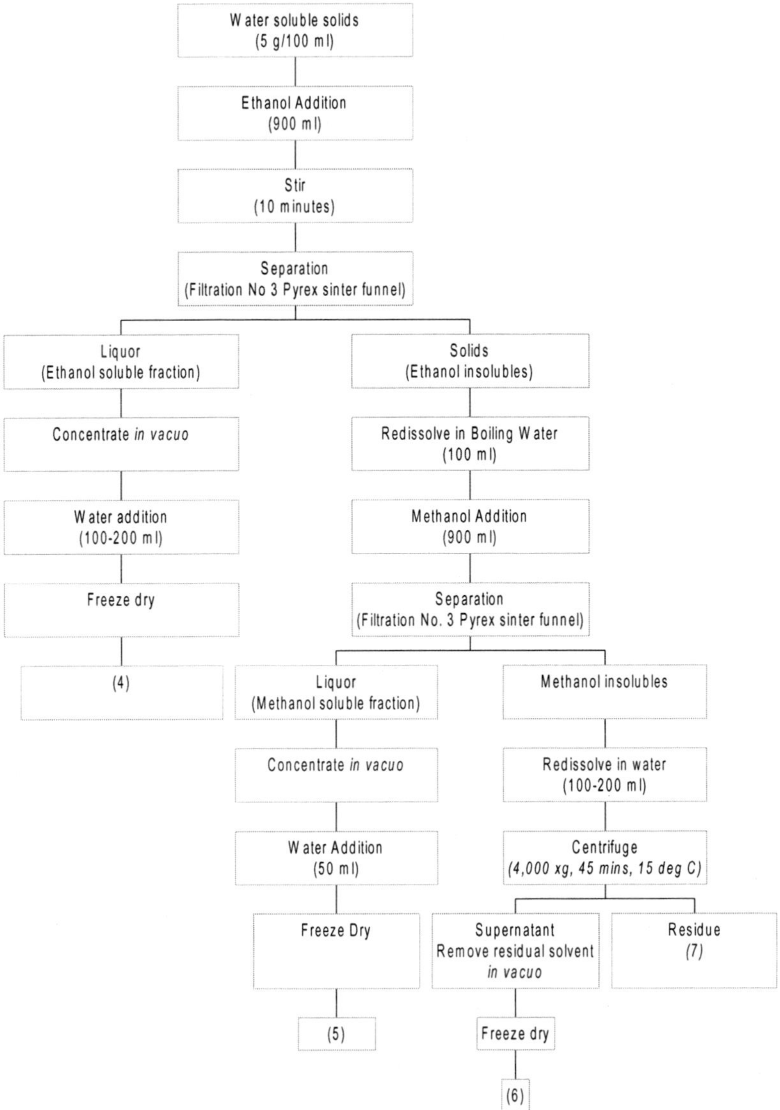

Figure 6 *Partition and solvent precipitation regime*

experience almost constant (±5%). HPLC and NMR analysis of the fractions[28] have shown that fraction 3 and subsequent fractions are free of small molecular weight polyphenols (catechins, flavanol glycosides *etc.*) and caffeine; these have been partitioned into fraction 2. All signal determined by ^{1}H and ^{13}C NMR is only attributed to TR material. Fractionation can also be carried out using ultra-pure solvents and with suitable analysis the fractions can be determined to be food grade, extending the potential uses of these fractions. Further work on solvent-precipitated fractions is currently underway.

5 Determination of Total Polyphenols in Tea Using the Folin–Ciocalteu Reagent

Scope

Total polyphenol determination is carried out using a variety of redox-active reagents including the Folin–Ciocalteu reagent and Prussian Blue assays. Both these reagents are susceptable to interference from other species (*e.g.* corn syrup) in the tea preparation or, and more importantly, in the calibration standards.[29] The following method specifies an improved colorimetric assay using Folin–Ciocalteu reagent to determine the total polyphenol content in leaf and instant teas (applicable to both green and black tea products). This improved method is based on work by the ISO standards committee and numerous international ring tests. Total polyphenol readings resulting from this assay are quoted in gallic acid equivalents due to the unknown molecular weights of the TRs.

Reagents

Use only recognised analytical grade reagents, deionised or distilled water.

1. *Acetonitrile*
2. *Folin–Ciocalteu phenol reagent*, available ready prepared from laboratory suppliers, *i.e.* Merck Ltd (BDH), Sigma-Aldrich. It is recommended to store the reagent in the refrigerator (+4 °C), and to allow the reagent to reach room temperature before use.
3. *Dilute Folin–Ciocalteu working reagent*
Folin–Ciocalteu phenol reagent (20 ml) is diluted to 200 ml with deionised water. Prepare on day of use, and discard any unused reagent.
4. *7.5% Sodium carbonate solution*
Sodium carbonate (anhydrous) (37.5 g) is dissolved in deionised water and made up to volume in deionised water (500 ml). Store solution at room temperature.
5. *Gallic acid stock standard solution*, corresponding to 1.00 mg ml^{-1}. Gallic acid (0.100 g weighed to nearest 0.0001 g) is dissolved and diluted to 100 ml with water. This fresh standard solution is prepared daily.
6. *Gallic acid dilute working standard solutions; 10–50 $\mu g\ ml^{-1}$*

Aliquots of the gallic acid stock standard solution (1, 2, 3, 4 and 5 ml respectively) are diluted to volume with deionised water.

Apparatus

Analytical balance, capable of weighing to an accuracy of ±0.001 g.
Volumetric flasks, 100 ml and 50 ml capacity.
Spectrophotometer, set at λ765 nm and able to accommodate 10 mm path length cells.

Disposable tubes, plastic – 10 ml capacity.
Automatic pipettes, to cover the volume range for standard and sample extract dilutions.

Preparation of test sample

Any instant teas of a granular structure should also be milled. Store samples in well sealed containers, protected from light and kept cool ($T < 10$ °C).

Procedure

Extraction of Instant Teas

Instant tea samples (0.5 g) are weighed to the nearest 0.001 g, in duplicate into 50 ml volumetric flasks. Hot water (25 ml, maximum temperature 60 °C) is added to dissolve the sample (*N.B.* It is important to thoroughly mix samples to ensure complete extraction), and the solution is allowed to cool to room temperature. Acetonitrile (5 ml), is added to the cool sample and diluted to volume with deionised water and mixed well. The sample extract (1.0 ml) is pipetted into separate 100 ml volumetric flasks, and diluted to volume with water and mixed well.

Standard Calibration

Each dilute gallic acid working standard solution (1.0 ml) is pipetted in duplicate into graduated tubes. These correspond to 0, 10, 20, 30, 40 and 50 μg gallic acid respectively. The determination then proceeds as shown below.

Determination

The diluted sample extract (1.0 ml) and dilute gallic acid working standard solutions are pipetted in duplicate (should be between 10–50 μg gallic acid equivalents) into separate disposable tubes. Dilute Folin–Ciocalteu phenol reagent (5 ml) is added into each tube. Within 3–8 min after addition of Folin–Ciocalteu reagent, sodium carbonate solution (4.0 ml) is added into each tube, the tube is stoppered and the solution mixed well.

The samples and standards are allowed to stand at room temperature for 60 minutes, and then their optical densities are measured on a spectrophotometer (4.3) set at λ 765 nm with a 10 mm path length cell, zero on reagent blank.
Note: the sample assay must be repeated with increased dilution if sample optical densities are higher than the 50 μg standard.

Calculation

A linear calibration graph is derived from the gallic acid standard solutions, by plotting standard optical densities (with reagent blank subtracted) against concentration (10–50 μg). It is important that sample optical densities are within calibration range, as deviation from linearity can occur at higher concentrations. The linear is forced through zero and the gallic acid standard slope is obtained from the gradient of this linear.

$$\text{Gallic acid standard slope} = \frac{\text{Standard optical density}}{\text{Standard concentration } (\mu\text{g})}$$

Total polyphenols (μg gallic acid equivalents) in diluted sample extract aliquot (TP_{ext}) is derived using the measured sample optical density and the gallic acid standard slope.

$$TP_{ext} = \frac{\text{Sample optical density}}{\text{Gallic acid standard slope}}$$

Percentage total polyphenols in the undiluted sample (TP_{samp}), calculated on a sample by sample basis:

$$TP_{samp} = \frac{TP_{ext} \times V_1 \times d}{m \times V_2 \times 10\,000}$$

where:

TP_{ext} is the total polyphenols (μg gallic acid equivalents) in the aliquot of diluted sample extract;
m is the mass, in grams of the test sample;
V_1 is the sample extraction volume, in ml (for instant 50 ml and for leaf 10 ml);
d is the dilution factor prior to the colorimetric determination (typically 1–100 ml, thus dilution factor of 100);
V_2 is the aliquot volume, in ml of diluted sample extract taken for the colorimetric determination.

Appendix A: HPLC Conditions for Flavan-3-ol Dimeric Oxidation Products

	Peparative HPLC conditions	*Semi-peparative HPLC conditions*	*Analytical HPLC conditions*
Column	NovaPak C_{18} (30 cm × 4.7 cm) Waters	Hypersil C_{18} (30 cm × 1 cm) Waters	Nucleosil C_{18} (15 cm × 0.47 cm) Waters
Column temperature	293 K	293 K	303 K
Solvent:			
A	CH_3CN	CH_3CN	CH_3CO_2H:CH_3CN
B	H_2O: CH_3CO_2H:CH_3CN [79.5:0.5:20]	H_2O:CH_3CO_2H:CH_3CN [81.5:0.5:18]	CH_3CO_2H:H_2O [2:98]
Elution profile	(1) 100% B: 50 min (isocrat) (2) 50% B (step) (3) 50% B: 5 min (isocrat) (4) 100% B (step) (5) 100% B 10 min (isocrat)	(1) 100% B: 50 min (isocrat) (2) 50% B (step) (3) 50% B: 5 min (isocrat) (4) 100% B (step) (5) 100% B 10 mi (isocrat)	(1) 8–40% A: 55 min (lin) (2) 40% A 5 min (isocrat) (3) 40–8% A: 1 min (lin) (4) 8% A 15 min (isocrat)
Detection	280 nm (UV)	280 nm (UV)	260–500 nm (UV/Vis)
Flow rate	75 ml min^{-1}	4.75 ml min^{-1}	1.0 ml min^{-1}
Sample mass per injection	100 mg	20 mg	–
Injection volume	4 ml	200 ml	20 ml

Appendix B: Structures of Flavan-3-ol Dimeric Oxidation Products

Theaflavate B (1)

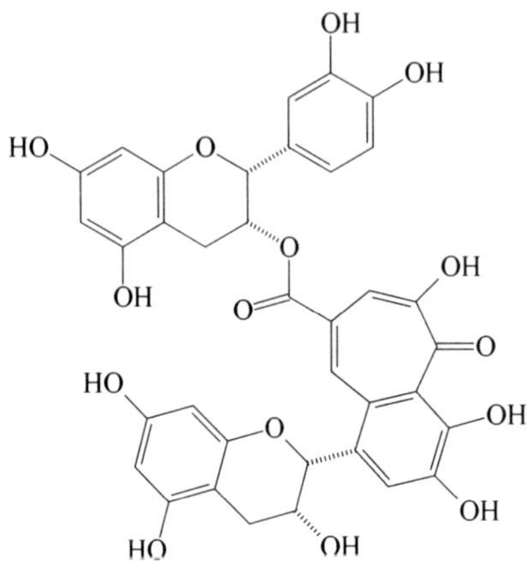

Isotheaflavin-3′-monogallate (3)

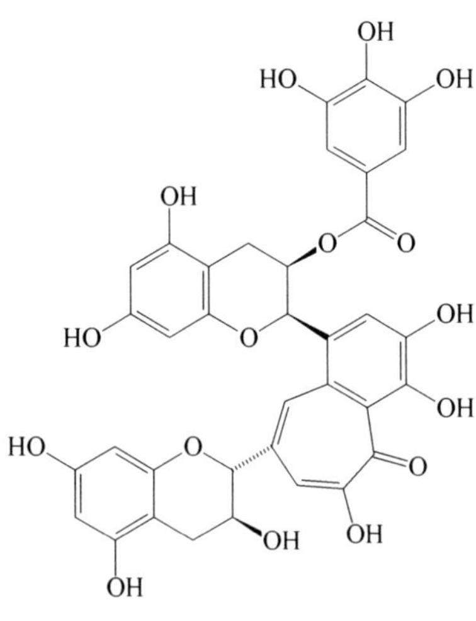

Neotheaflavin-3-monogallate (4)

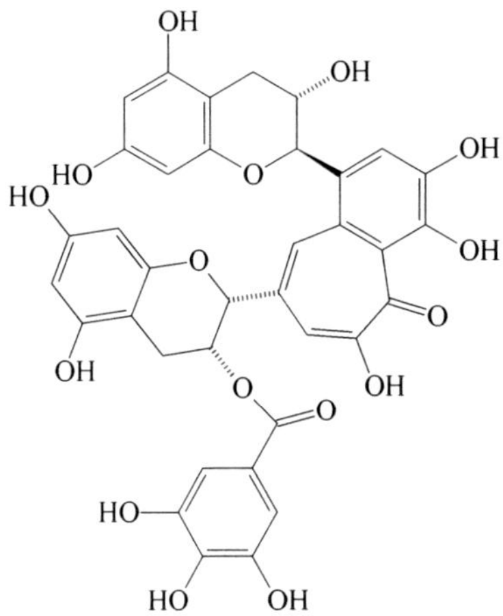

Isotheaflavin (5)

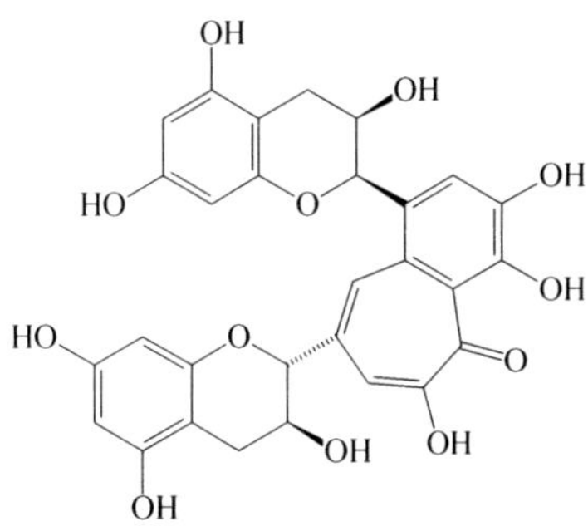

Epitheaflagallin-3-*O*-gallate (6)

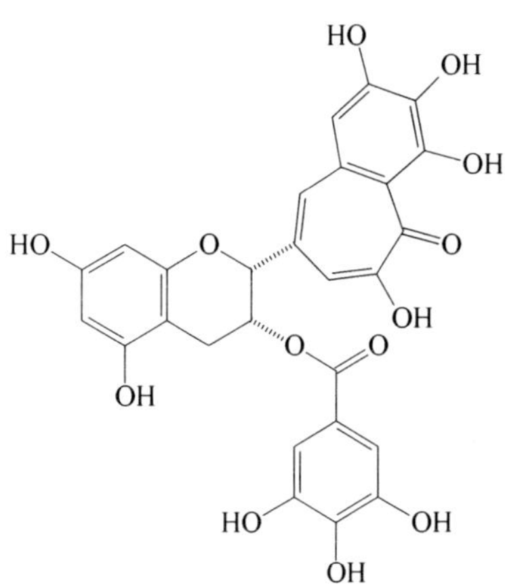

R = R′ = H;
Theaflavin (7)
R = gallate, R′ = H; Theaflavin-3-*O*-monogallate (8)
R = H, R′ = gallate; Theaflavin-3′-*O*-monogallate (9)
R = R′ = gallate; Theaflavin-3,3′-*O*-digallate (13)

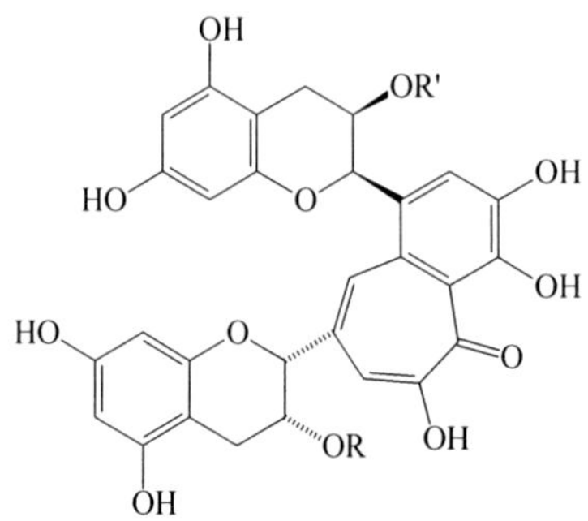

Theaflavic acid (10)

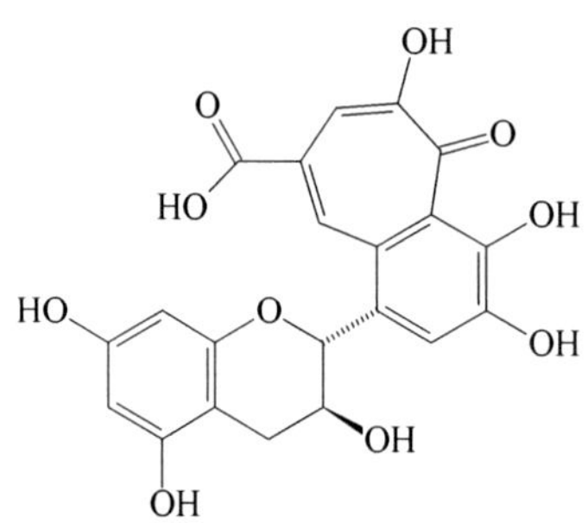

Epitheaflavic acid (11)

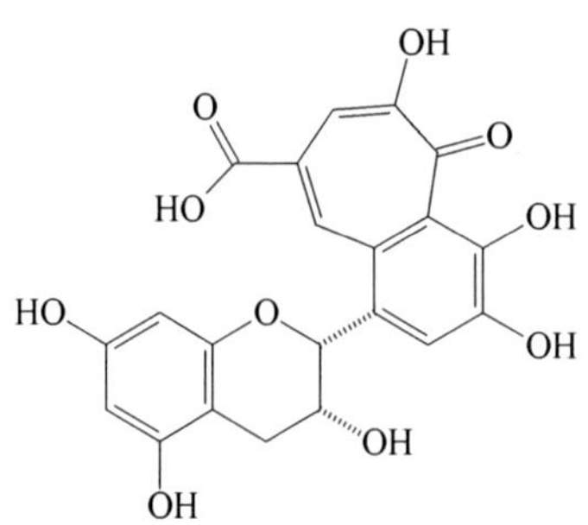

References

1. K. C. Willson, M. N. Clifford, *Tea – Cultivation to Consumption*, Chapman & Hall, 1992.
2. D. A. Balentine, S. A. Wiseman and L. C. M. Bouwens, *Crit. Rev. Food Sci. Nutr.*, 1997, **37**, 693.
3. Epimerisation – see entry in *Comprehensive Organic Chemistry* - Jerry March
4. Y. Komatsu, S. Suematsu, Y. Hisanobu, H. Saigo, R. Matsuda and K. Hara, *Biochem. Biotechnol. Biochem.*, 1993, **57**, 907.
5. R. Seto, Y. Nakamura, F. Nanjo and Y. Hara, *Biochem. Biotechnol. Biochem.*, 1997, **61**, 1434.
6. P. D. Collier, R. Mallows and P. E. Thomas, *Phytochemistry*, 1972, **11**, 867
7. R. Martin, T. H. Lilley, N. A. Bailey, C. P. Falshaw, E. Haslam, D. Magnolato and M. J. Begley, *J. Chem. Soc. Chem. Commun.*, 1986, 105.
8. C. Powell, M. N. Clifford, S. C. Opie, M. A. Ford, A. Robertson and C. L. Gibson, *J. Sci. Food Agric.*, 1992, **63**, 77.
9. E. L. Copeland, M. N. Clifford and C. M. Williams, *Food Chem.,* 1998, **61**, 81.
10. Y. Ito, *CRC Crit. Rev. Anal. Chem.*, 1986, **17**, 65.
11. R. E. Cornish, R. C. Archibald, E. A. Murphy and H. M. Evans, *Ind. Eng. Chem.*, 1934, **26**, 397.
12. A. P. Foucalt and L. Chevolot, *J. Chromotogr. A*, 1998, **808**, 3.
13. I. A. Sutherland, L. Brown, S. Forbes, G. Games, D. Hawes, K. Hostettmann, E. H. McKerrell, A. Marston, D. Wheatley and P. Wood, *J. Liq. Chromatogr. Rel. Technol.*, 1998, **21**, 279.
14. T. Okuda, Y. Yoshi, H. Hatano, K. Mori and T. Fukuda, *J. Liq. Chromatogr.*, 1990, **13**, 3637.
15. Q. Du, M. Li, Q. Cheng, T. Zhang and Y. Ito, *J. Chromatogr. A*, 1994, **687**, 174.
16. T. Yamamoto, L. R. Juneja, D.-C. Chu and M. Kim, *Chemistry and Applications of Green Tea*, CRC Press, 1997.
17. A. P. Davies, C. Goodsall, Y. Cai, A. L. Davis, J. R. Lewis, J. Wilkins, X. Wan, M. N. Clifford, C. Powell, A. Parry, A. Thiru, R. Safford and H. E. Nursten, *The 3rd Tannin Conference, Plant Polyphenols: Chemistry and Biology*, Bend, OR, USA, 1998.
18. J. R. Lewis, 'The Minor Theaflavins of Black Tea', M Phil Thesis, University of Leicester, 1997.
19. B. L. Wedzicha, M. F. Lo, T. J. Donovan, *J. Chromatography*, 1990, **505**(2), 357.
20. A. Degenhardt, U. H. Engelhardt, A. S. Wendt and P. Winterhalter, *J. Agric. Food Chem.*, 2000, **48**, 5200.
21. J. R. Lewis, X. L. Cao, T. J. Bond and A. P. Davies, CCC 2000 *The 1st International Conference on Countercurrent Chromatography*, Brunel University, UK, 2000, abstracts and programme, 2000, 14.
22. E. A. H. Roberts., *J. Sci. Food Agric.*, 1958, **9**, 381.

23. E. A. H. Roberts, *J. Sci. Food Agric.*, 1963, **14**, 700.
24. E. A. H. Roberts, R. A. Cartwright and M. Oldschool, *J. Sci. Food Agric.*, 1957, **8**, 72.
25. R. G. Bailey, H. E. Nursten and I. McDowell, *J. Sci. Food Agric.*, 1992, **59**, 365
26. Prof. M. Clifford, personal communication.
27. C. Powell, M. N. Clifford, S. C. Opie and C. L. Gibson, *J. Sci. Food Agric.*, 1995, **68**, 33.
28. T. J. Bond, J. R. Lewis and A. L. Davis, *in preparation.*
29. P. D. Collier, C. Dacombe, *in preparation.*

CHAPTER 12

Analysis and Purification of Proanthocyanidin Oligomers

SHERYL A. LAZARUS, MARK A. KELM, GERALD A. WÄCHTER, JOHN F. HAMMERSTONE AND HAROLD H. SCHMITZ

1 Introduction

Flavonoids are polyphenols which have a characteristic fifteen-carbon backbone ring structure (C_6–C_3–C_6) consisting of two aromatic rings generally each containing a number of hydroxyl groups (Figure 1). The class of flavonoids called the flavan-3-ols contain varying degrees of hydroxylation on the 5 and 7 positions of the A-ring and 3′, 4′ and 5′ positions on the B-ring (Figure 2). The 3 position on the C-ring is commonly a hydroxyl group or is esterified with gallic acid. Isomers which vary in the spatial orientation at the 3 position are distinguished by using the prefix *epi-* in the cases where $R_2 = H$ (Table 1), and the stereochemistry at the 2 position is noted with a (+) or (−). Proanthocyanidins are oligomers of flavan-3-ol monomer units most frequently linked by C–C bonds either at $4 \rightarrow 6$ or $4 \rightarrow 8$ (Figure 3a). The most common classes of proanthocyanidins are the procyanidins which are chains of catechin, epicatechin, and their gallic acid esters and the prodelphinidins which consist of gallocatechin, epigallocatechin, and their galloylated derivatives as the monomeric units[1] (Figure 2; Table 1). Structural variations to proanthocyanidin oligomers may also occur with the formation of a second interflavanoid bond by C–O oxidative coupling to form A-type oligomers (Figure 3b). Due to the complexity of this conversion, A-type proanthocyanidins are not as frequently encountered in nature in comparison to the B-type (*i.e.* singly linked) oligomers.[2]

Proanthocyanidins are found in commonly consumed fruits such as blueberries, cranberries, grapes, apples, kiwis and pears,[3–7] but have yet to be identified in vegetables. Furthermore, certain grain products such as barley[7,8] and legumes such as lentils[7] have also been reported to contain proanthocyanidins. Additionally, proanthocyanidins are found in the skins of peanuts, the seedcoat of almonds[4] and

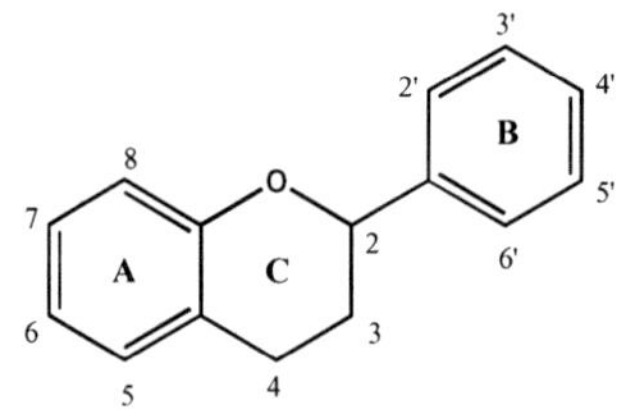

Figure 1 *Basic structure of a flavonoid*

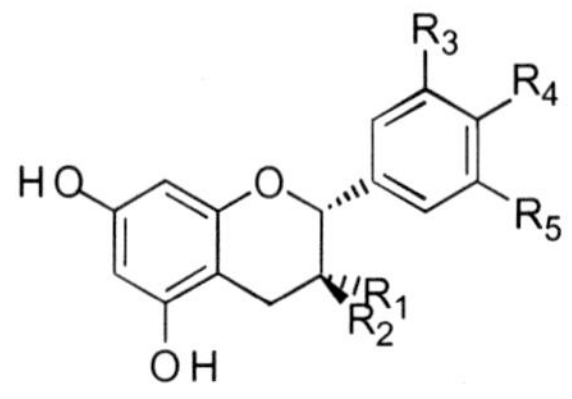

Figure 2 *Structure of flavan-3-ol monomers*

Table 1 *Substitution patterns for common flavan-3-ol monomers*

Flavan-3-ol	R_1	R_2	R_3	R_4	R_5
Catechin	H	OH	OH	OH	H
Epicatechin	OH	H	OH	OH	H
Catechin gallate	H	O-gallate	OH	OH	H
Epicatechin gallate	O-gallate	H	OH	OH	H
Gallocatechin	H	OH	OH	OH	OH
Epigallocatechin	OH	H	OH	OH	OH
Gallocatechin gallate	H	O-gallate	OH	OH	OH
Epigallocatechin gallate	O-gallate	H	OH	OH	OH

in certain cocoas.[9,10] Processed foods made from these raw materials, such as red wine, grape juice, beer and chocolate, can also contain proanthocyanidins although the levels found in the final products are affected by processing techniques.[11,12]

Over the last few years, proanthocyanidins have been shown to have potent antioxidant activity and the ability to scavenge reactive oxygen and nitrogen species.[10,13,14] In addition, recent research suggests that proanthocyanidins may modulate immune function, act as antimicrobials, inhibit cancer cell growth and inhibit platelet activation.[15–23] Given that these compounds have exhibited the ability to influence health in *in vitro* and *in vivo* studies, detailed quantitative information in foods is needed in order to define intake levels in populations and to determine whether epidemiological associations with health and disease endpoints may be present. Food composition data for proanthocyanidins have been limited due to the lack of availability of commercial standards and chromatographic difficulties separating the oligomers. This chapter reviews

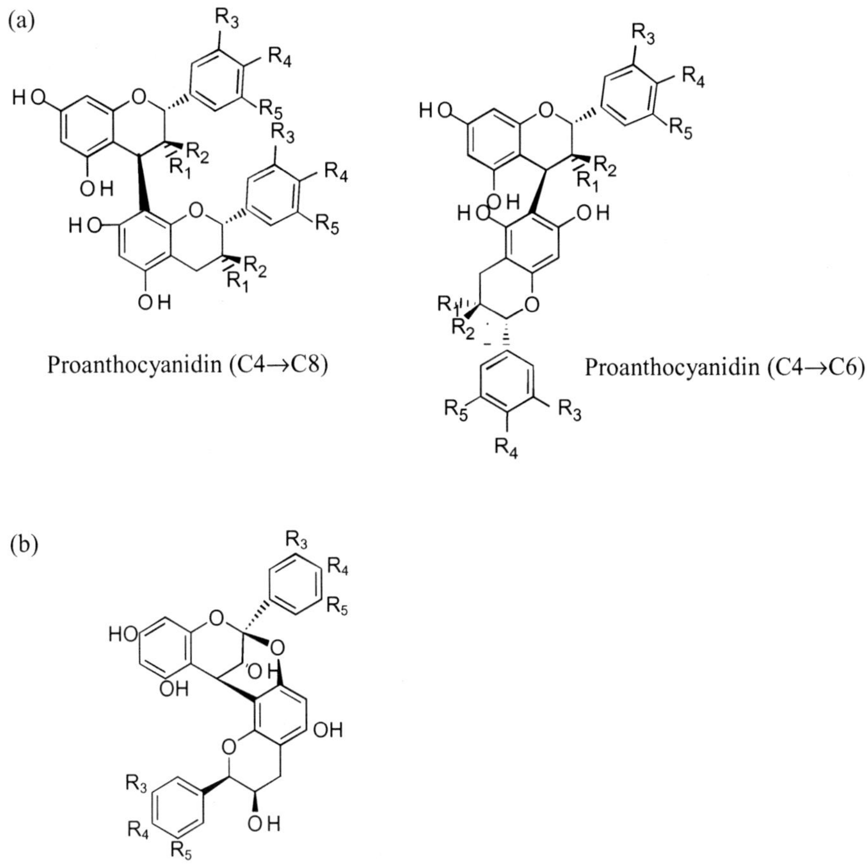

Proanthocyanidin (C4→C8; C2→O→C7)

Figure 3 *Structure of Proanthocyanidin oligomers*

analytical techniques currently available for the extraction and purification of proanthocyanidin oligomers, their subsequent structural characterization and their quantification, particularly in complex matrices such as food products.

2 Extraction and Purification of Proanthocyanidin Oligomers

Extraction and purification of proanthocyanidins have been conducted on a wide array of plants and plant foods. A broader overview of the subject can be obtained through previously published reviews[24,25] as well as the articles cited below.

Extraction

Solid samples have typically been freeze-dried followed by high-speed dry milling to reduce particle size (90–500 μm) to enhance solvent extraction efficiency. Liquid nitrogen or dry ice can be added to prevent thermal degradation and to encourage uniform milling of dry samples. Homogenizations of freeze-dried samples have also been conducted 'wet' with organic solvent. Efficient extractions of proanthocyanidins with acetone and water (*e.g.* 70:30, v/v) have typically been chosen and were reported by Foo and Porter.[26] Solvent extraction mixtures of alcohol and water have also been used.[27,28] In cases where lipid rich samples have been analyzed, such as cocoa (*Theobroma cacao*) and chocolate, a de-fatting step was carried out first.[9]

Extraction conducted under inert atmospheres (*i.e.* CO_2, N_2) or the addition of adjuvants to extraction solvents has been a common practice in order to minimize oxidative degradation of proanthocyanidins. Addition of sulfite (*i.e.* potassium pyrosulfite) in the preparation of fresh apple (*Malus domestica*) juices has been done to minimize oxidation of apple proanthocyanidins.[29] Acetic acid (0.5%) added to acetone/water has been used in the extraction of proanthocyanidins from cinnamon (*Cinnamomum* spp.), peanut (*Arachis hypogaea*) skins, peanut nutmeal, apple and almond (*Prunus amygdalus*) seedcoat.[4] Similarly, Nitao and co-workers[30] added ascorbic acid (0.1%) to increase proanthocyanidin stability during extraction of sugar maple (*Acer saccharum*) leaves. Furthermore, the authors reported that controlling extraction time and post-extraction sample handling does minimize proanthocyanidin degradation. Finally, with many natural products, to prevent degradation, exposure to high temperatures (>35–45 °C), light and air should be avoided.

Purification

A wide range and combination of techniques have been used for the purification of crude proanthocyanidin extracts. The examples given below illustrate the different kinds of current techniques used for the purification of proanthocyanidins and are generally used as a part of a more complex purification scheme.

The simplest purification of crude extracts has involved removal of sugars and organic acids prior to analytical analyses. For example, extracts of green tea (*Camellia sinensis*), red wine and fruit juices were conveniently purified by elution through Supelcosil Envi-18 solid-phase extraction prior to high performance liquid chromatography/mass spectrometry (HPLC/MS) analysis.[4] However, often the removal of extraneous material has been the first step in a larger purification protocol. Enrichment of proanthocyanidin extracts by elution through polyamide followed by fractionation over Sephadex LH-20 afforded the novel compounds, epiafzelechingallate-(4$\beta \rightarrow$ 8)-epicatechingallate and epiafzelechingallate-(4$\beta \rightarrow$ 6)-epicatechingallate.[31] Sephadex LH-20 has also been used successfully in isolation of proanthocyanidin polymers.[27,32,33] Other stationary phases such as MCI-gel CHP 20 in conjunction with Sephadex LH-20[34,35] have been utilized for preparative chromatographic isolation of proanthocyanidins. The size exclusion media TSK HW-50F[36] and TSK HW 40-F[37] were used in preparation of samples for proanthocyanidin composition analysis of grape (*Vitis vinifera*) stems and skins. McMurrough *et al.*[38] demonstrated the semi-preparative isolation of dimeric and trimeric proanthocyanidins from barley on Superdex 75 HR.

Preparative normal-phase separations in party with other chromatographic methods have been conducted as well. Cranberry (*Vaccinium macrocarpon*) extracts fractionated over TSK HW-40C+F, silica gel, and LH-20 led to the isolation of chemopreventative proanthocyanidin dimers and oligomers.[39] A preparative normal-phase method was used to fractionate cocoa procyanidin oligomers according to their degree of polymerization over a Supelcosil LC-Si column.[40] Similarly, Labarbe[41] used an inert glass powder column to fractionate grape proanthocyanidins. Semi-preparative separations with PrepNova-Pak HR silica were used to fractionate cider apple extract prior to fractionation on PrepNova-Pak HR C_{18}.[42]

Recently, alternative approaches have been advanced in the purification of crude proanthocyanidin mixtures.[24] High-speed counter current chromatography (HSCCC) was used to isolate proanthocyanidins on a preparative scale from green and black tea.[43] HSCCC was also incorporated as a step in the isolation of procyanidin dimers A2, B1, B2, B3, B5, B7, and trimer C1 from St. John's wort (*Hypericum perforatum*).[44] Degenhardt *et al.*[45] reported the use of a related method, centrifugal precipitation chromatography (CPC) to fractionate polymeric pigments in black tea and red wine and this method may prove adaptable in the purification of proanthocyanidins.

Due to the vast structural complexity of proanthocyanidins, the purification and isolation of discreet oligomeric sets (*i.e.* dimers, trimers, *etc.*) or individual compounds has been very labour intensive. Further experimentation, invention of new solid-phase supports and advances in novel instrumentation have the potential to drive forward the ability of researchers to pursue more efficient purification and isolation methods.

3 Characterization of Proanthocyanidin Oligomers

Thiolysis

Depolymerization by thiolysis has been demonstrated as an efficient method for determining the nature of the flavan-3-ol units within proanthocyanidins and for determining average degree of polymerization. Thiolysis is performed in the presence of acid and a nucleophilic reagent such as toluene-α-thiol or benzyl mercaptan. The extension subunits of proanthocyanidins are attacked by the nucleophilic reagent to form the corresponding benzyl thioether and the terminal unit is released as the free flavan-3-ol. The resulting flavan-3-ols and their benzyl thioethers are generally quantified by reversed-phase HPLC and the average degree of polymerization determined by calculating the ratio between total units (terminal plus extension) and terminal units. Interestingly, A-type proanthocyanidins are resistant to degradation by thiolysis, thus rendering this technique of little value in identifying their monomeric units.[46,47]

Inconsistencies in thiolysis may be attributed to epimerization of the flavan-3-ol units and poor yields.[25,48] The epimerization rates of pure solutions of (−)-epicatechin and (+)-catechin have been reported as 15% and 4%, respectively.[49] Given that free flavan-3-ols exhibit different epimerization rates, it is difficult to estimate the composition of terminal units. The yield of thioether derivatives measured on purified proanthocyanidins has been reported to be 34–63%.[48] Thiolysis yields can be influenced by instability of the reaction products, the presence of non-proanthocyanidin compounds, and secondary side reactions during thiolysis.[48] For example, the gallic acid ester commonly found on the 3-position can be cleaved under the acidic conditions of the thiolysis reaction.

Mass Spectrometry

The use of mass spectrometry (MS) in the analysis of proanthocyanidins has a relatively short history compared with other spectroscopic techniques. Early drawbacks in the use of MS arose from the limited volatility of proanthocyanidins and thus the inability to hyphenate the technique with other more common isolation techniques such as gas chromatography or HPLC. Initial efforts to overcome this limitation were focused on creating volalite derivatives. For example, Dadic and Belleau[50] purified proanthocyanidins from beer, acetylated with acetic anhydride then analysed using Electron Impact (EI) MS. Likewise, Nonaka *et al.*[51] used dimethyl sulfate or sodium methoxide to form methyl ethers of flavan-3-ols, proanthocyanidin dimers and oxidative derivatives extracted and purified from tea. Field Desorption (FD) or EI MS were then employed to obtain suitable mass spectra. Methylation has also been successfully applied with diazomethane used as the derivatizing reagent to give EI spectra up to trimers.[52] An alternative derivatization method was employed by Nonaka *et al.*,[53] who exploited pyridinium hydrobromide perbromide to brominate A-typed procyanidin dimers and trimers purified from cinnamon, and obtain mass spectra employing EI. In most cases, EI-MS and FD-MS could yield molecular ions for the monomeric flavan-3-ols or dimers only, while fragment ions were used to infer structures of higher oligomers. The limited mass range of these techniques has led to their abandonment for methods that could accommodate the high molecular weights of larger oligomeric proanthocyanidins.

The next leap in technology, which extended the usefulness of MS for the structural elucidation of proanthocyanidins, was the development of Fast Atom Bombardment (FAB) and Liquid Secondary Ion (LSI) MS. Morimoto *et al.*[54,55] used positive ion FAB-MS to obtain molecular ions $(M+H)^+$ and potassium adducts $(M+K)^+$ of A-type and B-type procyanidins in cinnamon up to hexamers. Subsequently, these techniques were applied to the analysis of proanthocyanidins in numerous plant materials, including apples, chestnut (*Castanea sativa*), cocoa, grapes, oak (*Quercus spp.*), peanuts and tea[46,56–59] with an increase in detection of degree of polymerization to 8. Moreover, the work of Karchesy *et al.*[60] suggested that high-resolution analysis may be used to elucidate some structural differences. For example, they observed distinct difference in fragmentation patterns between a procyanidin trimer that was linear compared with a procyanidin trimer that was branched.

A complementary technique to FAB and LSIMS is Matrix Assisted Laser Desorption-Ionization (MALDI) Time-of-Flight (TOF) MS. Whereas FAB and LSIMS analyses were primarily used on purified or semi-purified proanthocyanidin fractions, MALDI analysis was successfully applied to crude extracts with a broad range of polymers. For example, pentadecameric procyanidins were detected in unripe apple using positive ion MALDI with the aid of lithium, potassium and sodium ions.[61] Likewise polygalloyl proanthocyanidins were detected in extracts from grape and grape seed[62,63] in the positive ion mode without the addition of cations. Nevertheless, both molecular ion $(M+\mathrm{H})^+$ and sodium adducts $(M+\mathrm{Na})^+$ were observed for monomers through undecamers. The technique clearly has the promise of rapidly profiling the proanthocyanidin composition of plant materials; however, quantification with MALDI still remains a challenge.

Recently, a number of hyphenated mass spectral techniques have come into use. These include, Thermospray, Particle Beam and Atmospheric Pressure-Electrospray Ionization (ESI) techniques that are commonly used in combination with HPLC. Thermospray-MS was used to analyze proanthocyanidins in tea and beer.[64,65] Although effective with relatively small masses, the technique lacks the sensitivity for analyzing the wide range of masses observed in plants.

The introduction of ESI/MS for the analyses of proanthocyanidins was first used as a stand alone technique where the purified extracts were infused into the spray chamber. Results from these experiments generated results similar to those observed for FAB/MS and MALDI experiments. More recently, ESI/MS has been used in combination with HPLC. The earliest reports of HPLC–ESI/MS for the analyses of proanthocyanidins in foodstuffs were by Cheynier *et al.*[66] and Hammerstone *et al.*,[9] where cider apples and cocoa were directly analyzed. Although FAB has proven greater utility in the positive ion mode, increased sensitivity was demonstrated in the negative ion mode for ESI/MS. Furthermore, multiple charges were likely to be observed when using ESI/MS as compared with FAB/MS. For example, procyanidin pentamers exhibited a molecular ion at m/z 1441 $(M-\mathrm{H})^{-1}$ and a doubly charged species at m/z 720 $(M-2\mathrm{H})^{-2}$. Similarly, procyanidin octamers gave doubly and triply charges species at m/z 1152 for $(M-2\mathrm{H})^{-2}$ and m/z 768 for $(M-3\mathrm{H})^{-3}$ but no molecular ion.

Most recently, HPLC in combination with ESI has been applied to the analysis of proanthocyanidins in a host of foods, including almonds, blueberry (*Vaccinium* spp.), cranberry (*Vaccinium macrocarpon*), cinnamon (*Cinnamomum* spp.), grape seed, peanuts, tea and wine.[3,4,67,68]

NMR

Much of the progress in the structural elucidation of natural products in the last two decades has been due to the development of new nuclear magnetic resonance (NMR) techniques and it is not surprising that much of what we know today about proanthocyanidins comes from the application of these methods.[69–71] As 500–800 MHz proton NMR spectra on samples of 1 mg or less have become routine, less time has to be spent on the isolation of large quantities of compounds or acquisition of NMR data. With the development of inverse detected methods such as HMBC and HMQC heteronuclear 2D data can now be obtained as quickly as homonuclear 2D NMR (COSY, NOESY, TOCSY, *etc.*). It is therefore not surprising that there are an increasing number of reports on structural elucidation of procyanidin dimers and trimers using these techniques.[31,35,72–76] NMR has been able to answer most of the fundamental problems of procyanidin structural analysis: the stereochemistry of the 3-hydroxy group (2,3-*cis* *vs.* 2,3-*trans*) (Figure 2), the location ($4 \rightarrow 6$ *vs.* $4 \rightarrow 8$) and stereochemistry (α, β) of the interflavan linkage (Figure 3), the conformation of the heterocyclic C-ring and the degree of polymerization, branching and composition of proanthocyanidin polymers.

While NMR can, in principle, distinguish between 2,3-*cis* (epicatechin) and 2,3-*trans* (catechin) geometry in the C-ring, it cannot provide information on the absolute stereochemistry of the 2-position. Degradation to monomeric units remains necessary to answer the question as to whether the building blocks are (+)-catechin or (−)-catechin, (+)-epicatechin or (−)-epicatechin.[77] Typically, relative orientation on a six-membered ring can be determined by measurement of ^3J coupling constants. This approach has also been applied to procyanidins where doublets for H-4 and H-2 with ^3J coupling constants of 8–10 Hz indicate quasi *trans*-diaxial configuration whereas broad singlets can be due to either quasi *trans* or *cis* configurations. Analysis of Overhauser enhancements obtained from ROESY spectra has been used to elucidate the *trans* 2,3-configuration of the terminal unit of a catechin trimer where H-2 appears as a broad singlet.[75]

The first unequivocal proof of the position of an interflavan linkage (IFL) was provided for procyanidin B3 (catechin-($4\alpha \rightarrow 8$)-catechin) by synthesis from a bromo derivative whose structure was proven several years later by X-ray analysis.[78,79] Haslam *et al.*[80] concluded from molecular models that dimers showed equal amounts of the two conformers when the IFL is $4 \rightarrow 6$. Otherwise they considered it to be $4 \rightarrow 8$; in the latter case the preponderance of conformers is different for native forms and derivatives. The observation that H-8 of 6-substituted catechin derivatives resonates at higher field than H-6 of 8-substituted catechins[81] has been used extensively for the assignment of the IFL as have other shift rules for carbon resonances. This is reviewed in some detail in the introduction of Balas and Vercauteren,[74] who also describe the first use of heteronuclear 2D NMR to determine the position of the IFL in two peracetylated catechin dimers. Several studies using this methodology on dimers and trimers with and without peracetylation followed.[35,73–76] The addition of $Cd(NO_3)_2$ to acetone-d_6 solutions has been shown to be an alternative to peracetylation as it leads to sharp OH signals which provide useful structural information in 2D experiments.[73] The determination of the stereochemistry of the IFL is less straightforward especially when no 8–10 Hz coupling for H-4 can be observed. This particular problem hasn't received much attention from NMR spectroscopists since 3,4-*cis* stereochemistry is quite rare and has prompted synthetic studies to clarify the situation.[82–84]

The characterization of oligomeric procyanidins by NMR includes most of the aspects that are relevant for dimers and trimers such as regio- and stereo-chemical aspects but in addition requires some measure of the degree of polymerization and branching. Both ^{13}C and ^{1}H NMR have been used to obtain this information. From ^{13}C spectra such detailed information as the ratio of bottom terminal units linked through their 6 or 8 positions, the relative number of top starter units linking to the 6 or 8 position of extension units, the proportion of extension units linking to other extension units through their 6 or 8 position and the proportion of branching units substituted at 6 and 8 positions, have been deduced based on comparison with spectra of monomers and oligomers of known structures.[85] This information is useful for the determination of the overall three-dimensional structure and allows the calculation of the degree of polymerization. Although this methodology has been developed for epicatechin oligomers it can be extended to mixed catechin-epicatechin oligomers and prodelphinidin oligomers.[35] The analysis of carbon spectra has also been extended to solid-state ^{13}C NMR where it has been used to show the presence of proanthocyanidin oligomers in biomass without isolating and purifying them.[86] In ^{1}H spectra the ratio of protons 6 or 8 over proton 4 of the terminal units has been shown to correlate well with data obtained from thiolysis.[87]

4 Quantification of Proanthocyanidin Oligomers

Quantification of proanthocyanidins is complicated by difficulties in separating the oligomers and the lack of commercially available reference materials. Generally, compounds which are commercially available such as gallic acid or catechin have been used as the reference materials and quantitative results expressed as gallic acid or catechin equivalents. However, these compounds have different response factors than the proanthocyanidins, and therefore the results obtained are only estimates. Alternatively, standards have been rigorously purified from plant sources to use as standards for quantification. Methods used for quantification of proanthocyanidins in foods and plants can be divided into three groups: colorimetric methods, reversed-phase and normal-phase HPLC.

Colorimetric Methods

Historically, the analytical method used to estimate the amount of proanthocyanidins involved their colorimetric measurement of their total content after reaction with various reagents. Proanthocyanidins contain numerous phenolic groups which reduce the Folin reagent or iron in the Prussian Blue reaction[88–90] and have been used to estimate the quantity of proanthocyanidins in food and plant materials. These methods are non-specific for proanthocyanidins since other phenolic compounds and non-phenolic reducing compounds also react with these reagents. Hydrolysis of proanthocyanidins in a solution of hydrochloric acid in butanol yields anthocyanidins which can be colorimetrically measured to estimate the proanthocyanidins. However, Porter and co-workers[91] reported that the reaction is greatly influenced by the proanthocyanidin structure, the presence of transition metals and side reactions, which resulted in a maximal yield of anthocyanidins from proanthocyanidins of less than 58%. Other colorimetric methods employed measure the reaction of flavanols and proanthocyanidins with aromatic aldehydes such as dimethylaminocinnamaldehyde (DMACA) or vanillin in strongly acidic media.[92–96] However, these methods give estimations on the total flavanol and proanthocyanidin content instead of the quantitative contribution of each compound within these classes. Furthermore, both methods are severely influenced by the presence of co-existing anthocyanins and also depend on the proanthocyanidin structure and degree of polymerization which makes it difficult to compare results across samples which differ in their proanthocyanidin content.

Reversed-phase HPLC

Reversed-phase HPLC has been the primary method of analysis for proanthocyanidins in food samples. For example, reversed-phase C_{18} columns have been used to analyze the oligomers to trimer in the Spanish diet[97] and to tetramer in French wines.[12] Reversed-phase columns have the ability to separate oligomers of the same degree of polymerization. However, there is difficulty in determining the degree of polymerization using reversed-phase methods since the oligomers elute in non-sequential order (Figure 4). Furthermore, analysis of higher oligomeric proanthocyanidins (*i.e.* $\geqslant$ tetramers) is not feasible since the number of isomers increases with increasing degrees of polymerization.[68,98] This effect results in a retention time overlap of isomers containing differing degrees of polymerization, causing the higher oligomers (>trimer) to co-elute as a large unresolved peak.[99] Thus, quantification of oligomeric proanthocyanidins is underestimated because only the resolved peaks corresponding to dimers and trimers are considered.

Normal-phase HPLC

Lea[100] was the first to describe the use of normal-phase HPLC for the analysis of proanthocyanidins. His work attempted to separate the procyanidins in cider using a mixture of methanol and acetic acid under isocratic conditions. He observed that the use of normal-phase packings resulted in an elution order in which the larger oligomers had longer retention times and that the retention times could be increased with the addition of water to the mobile phase. A few years later, Wilson[98] used a cyano column for normal-phase separation of apple juice procyanidins under gradient conditions with a mixture of THF–hexane–acetic acid–formic acid–IPA. He noted that the addition of acid helped reduce peak tailing and the presence of alcohol helped maintain column efficiency. This method was able to partially separate the oligomers based on degree of polymerization up to the heptamers. Since the separation was incomplete, the procyanidin peaks overlapped resulting in a substantial rise in the baseline beginning when the trimers eluted.

Normal-phase methodology improved substantially when Rigaud *et al.*[101]

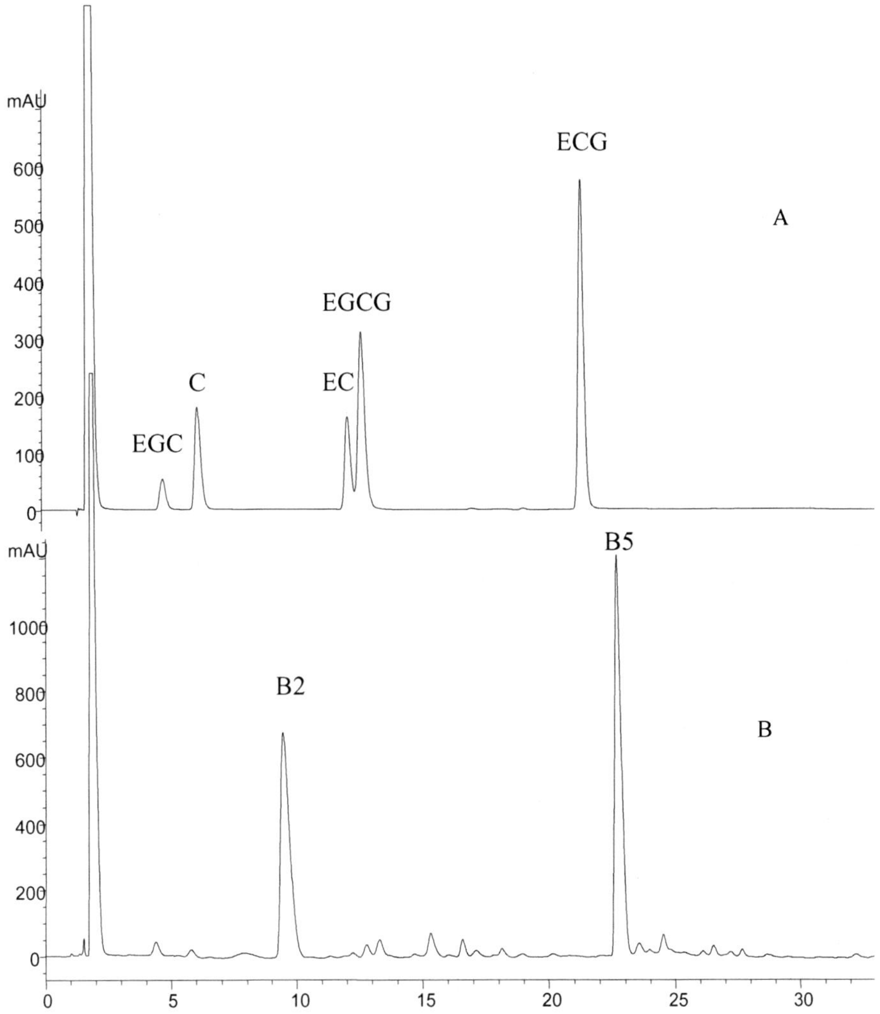

Figure 4 *Reversed-phase HPLC separation with UV detection at 280 nm of flavanol monomers* (A) *and procyanidin dimers* (B). *C, EC, EGC, ECG, and EGCG refers to (+)-catechin, (−)-epicatechin, (−)-epigallocatechin, (−)-epicatechin gallate and (−)-epigallocatechin gallate, respectively. B2 and B5 refer to dimers which consist of (−)-epicatechin as its monomeric units linked 4 → 8 and 4 → 6, respectively*

developed an HPLC method based on a silica thin layer chromatography method. The normal-phase HPLC method used a silica column and a gradient of dichloromethane–methanol–formic acid–water mixtures to achieve better separation of procyanidin oligomers through pentamer in cocoa beans. Several years later, the same group reported that slowing the gradient of methanol in the mobile phase allowed better resolution of larger oligomers.[102] Furthermore, Rigaud *et al.*[101] applied the normal-phase methodology to galloylated proanthocyanidins for the first time and were able to separate through tetramer in grape seeds. It was noted that the galloylated forms eluted later than their non-galloylated counterparts and this observation was later confirmed by Lazarus *et al.*[4]

Hammerstone and colleagues[9] modified the normal-phase HPLC method reported by Rigaud *et al.*[101] for improved separation based on degree of polymerization. This new method reduced the gradient time from 70 minutes to 50 minutes and utilized a series of linear gradients of methanol into dichloromethane with constant amount of acetic acid and water. Hammerstone *et al.*[9] were the first to achieve separation of oligomers up to decamers in cocoa and chocolate (Figure 5). Since cocoa and chocolate contains procyanidins primarily consisting of epicatechin as the monomeric units, this same method was later applied to foods with more complex mixtures of proanthocyanidins[4] such as peanut skins, which contain both singly and doubly linked procyanidins, and grape seeds, which contain galloylated oligomers. These structural variations resulted in retention time shifts making it difficult to completely separate the higher oligomers into discrete peaks based on degree of polymerization. Nevertheless, this method has been successfully used to quantify proanthocyanidins in various food samples.[10,103]

Detection

HPLC with UV detection at 280 nm has been the primary method for quantitative analysis of proanthocyanidins. However, UV detection at 280 nm is not selective for proanthocyanidins relative to other polyphenols. In foods, proanthocyanidins are usually accompanied by other UV-absorbing polyphenols, sometimes in larger amounts or with higher UV extinctions than the proanthocyanidins. Thus, a prepurification step has typically been required to remove interfering compounds

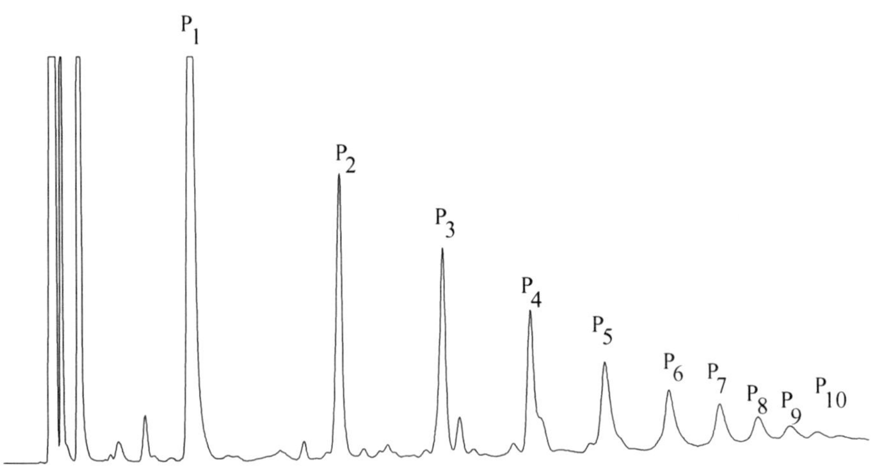

Figure 5 *Normal-phase HPLC separation with FLD detection of flavanols and procyanidins in cocoa (*Theobroma cacao*). P_1 refers to monomeric flavanols and P_2 through P_{10} refer to procyanidin dimers through decamers*

before the analysis of proanthocyanidins.[104,105] Alternatively, DMACA has been used as a post-column reagent which forms products unique to proanthocyanidins that can be detected by UV at 640 nm, thus eliminating the need for pre-purification.[93,97] Electrochemical detection has also become more widespread due to the smaller number of electroactive substances compared to UV-absorbing compounds.[106–108] Additionally, fluorescence detection (excitation 276 nm and emission 316 nm) has been reported to be more sensitive and selective than UV detection at 280 nm.[4] However, it should be noted that fluorescence detection is insensitive to proanthocyanidins containing a gallic acid ester and/or the gallocatechins as a monomeric unit, and hence, should only be used to quantify procyanidins consisting of (+)-catechin or (−)-epicatechin as the monomeric units.[68]

5 Conclusion

Recent developments in analytical science have provided methodologies to allow for the purification of proanthocyanidin standards and the separation and subsequent quantification in food matrices. Proanthocyanidins are currently under investigation for their potential benefits to health, particularly cardiovascular health. However, food composition data for the content of proanthocyanidins in the diet are lacking. Future research should focus on developing food composition databases for proanthocyanidins so that intake levels can be estimated and relationships with health and disease endpoints determined.

References

1. L. J. Porter, in *The Flavonoids: Advances in Research since 1986*, ed. J. B. Harborne, Chapman and Hall, London, 1994, p. 23.
2. S. Morimoto, G.-I. Nonaka and I. Nishioka, *Chem. Pharm. Bull.*, 1985, **33**, 4338.
3. R. L. Prior, S. A. Lazarus, G. Cao, H. Muccitelli and J. F. Hammerstone, *J. Agric. Food Chem.*, 2001, **49**, 1270.
4. S. A. Lazarus, G. E. Adamson, J. F. Hammerstone and H. H. Schmitz, *J. Agric. Food Chem.*, 1999, **47**, 3693.
5. H. M. Dawes and J. B. Keene, *J. Agric. Food Chem.*, 1999, **47**, 2398.
6. M. J. Amiot, M. Tacchini, S. Aubert and W. Oleszek, *J. Agric. Food Chem.*, 1995, **43**, 1132.
7. C. Santos-Buelga and A. Scalbert, *J. Sci. Food Agric.*, 2000, **80**, 1094.
8. D. Madigan, I. McMurrough and M. R. Smyth, *Analyst.*, 1994, **119**, 863.
9. J. F. Hammerstone, S. A. Lazarus, A. E. Mitchell, R. Rucker and H. H. Schmitz, *J. Agric. Food Chem.*, 1999, **47**, 490.
10. G. E. Adamson, S. A. Lazarus, A. E. Mitchell, R. L. Prior, G. Cao, P. H. Jacobs, B. G. Kremers, J. F. Hammerstone, R. B. Rucker, K. A. Ritter and H. H. Schmitz, *J. Agric. Food Chem.*, 1999, **47**, 4184.
11. A. L. Waterhouse and R. L. Walzem, in *Flavonoids in Health and Disease*,

eds C. A Rice-Evans and L. Packer, Marcel Dekker Inc., New York, 1998, p. 359.
12. S. Carando, P.-L. Teissedre, L. Pascual-Martinez and J.-C. Cabanis, *J. Agric. Food Chem.*, 1999, **47**, 4161.
13. A. E. Hagerman, K. M. Riedl, G. A. Jones, K. N. Sovik, N. T. Ritchard, P. W. Hartzfeld and T. L. Riechel, *J. Agric. Food Chem.*, 1998, **46**, 1887.
14. G. E. Arteel and H. Sies, *FEBS Lett.*, 1999, **462**, 167.
15. C. Sanbongi, N. Suzuki and T. Sakane, *Cell Immunol.*, 1997, **177**, 129.
16. T. Mao, J. Van de Water, C. L. Keen, H. H. Schmitz and M. E. Gershwin, *J. Nutr.*, 2000, **130**, 2093S.
17. I. Ofek, J. Goldhar, D. Zafriri, H. Lis, R. Adar and N. Sharon, *N. Engl. J. Med.*, 1991, **324**, 1599.
18. A. B. Howell, N. Vorsa, A. Der Marderosian and L. Y. Foo, *N. Engl. J. Med.*, 1998, **339**, 1085.
19. D. Rein, T. G. Paglieroni, D. A. Pearson, T. Wun, H. H. Schmitz, R. Gosselin and C. L. Keen, *J. Nutr.*, 2000, **130**, 2120S.
20. D. Rein, T. G. Paglieroni, T. Wun, D. A. Pearson, H. H. Schmitz, R. Gosselin and C. L. Keen, *Am. J. Clin. Nutr.*, 2000, **72**, 30.
21. S. Carnesecchi, Y. Schneider, S. A. Lazarus, D. Coehlo, F. Gosse and F. Raul, *Cancer Lett.*, 2002, **175**, 147.
22. H. U. Gali, E. M. Perchellet, X. M. Gao, J. J. Karchesy and J. P. Perchellet, *Planta Med.*, 1994, **60**, 235.
23. J. Zhao, J. Wang, Y. Chen and R. Agarwal, *Carcinogenesis*, 1999, **20**, 1737.
24. L. J. Porter, in *The Flavonoids: Advances in Research since 1986*, ed. J. B. Harborne, Chapman and Hall, London, 1994, p. 46.
25. L. J. Porter, in *The Flavonoids: Advances in Research since 1980*, ed. J. B. Harborne, Chapman and Hall, London, 1988, p. 21.
26. L. Y. Foo and L. J. Porter, *Phytochemistry,* 1980, **19**, 1747.
27. Z. Peng, Y. Hayasaka, P. G. Iland, M. Sefton, P. Høj and E. J. Waters, *J. Ag. Food Chem.*, 2001, **49**, 26.
28. B. Bartolomé, T. Hernàndez, M. L. Bengoecha, C. Gómez-Cordovés and L. Estrella, *J. Chromatogr. A*, 1996, **723**, 19.
29. A. Yanagida, T. Kanda, T. Takahashi, A. Kamimura, T. Hamazono and S. Honda, *J. Chromatogr. A.*, 2000, **890**, 251.
30. J. K. Nitao, B. A. Birr, M. G. Nair, D. A. Herms and W. J. Mattson, *J. Agric. Food Chem.*, 2001, **49**, 2207.
31. C. Lakenbrink, U. H. Engelhardt and V. Wray, *J. Agric. Food Chem.*, 1999, **47**, 4621.
32. Z. Czochanska, L. Y. Foo, R. H. Newman and L. J. Porter, *J. Chem. Sci. Perkin Trans. 1*, 1980, 2278.
33. T. L. Eberhardt and R. A. Young, *J. Agric. Food Chem.*, 1994, **42**, 1704.
34. M. Hör, M. Heinrich and H. Rimpler, *Phytochemistry,* 1996, **42**, 109.
35. L. Y. Foo, Y. Lu, A. L. Molan, D. R. Woodfield and W. C. McNabb, *Phytochemistry,* 2000, **54**, 539.
36. J.-M. Souquet, B. Larbarbe, C. L. Guenevé, V. Cheynier and M. Moutounet, *J. Agric. Food Chem.*, 2000, **48**, 1076.

37. J. A. Kennedy, Y. Hayasaka, S. Vidal, E. J. Waters and G. P. Jones, *J. Agric. Food Chem.*, 2000, **49**, 5348.
38. I. McMurrough, D. Madigan, and M. R. Smyth, *J. Agric. Food Chem.*, 1996, **44**, 1731.
39. F. E. Kandil, M. A. L. Smith, R. B. Rogers, M.-F. Pepin, L. L. Song, J. M. Pezzuto and D. S. Seigler, *J. Agric. Food Chem.*, 2002, **50**, 1063.
40. S. A. Lazarus, J. F. Hammerstone, G. E. Adamson and H. H. Schmitz, *Methods Enzymol.*, 2001, **355**, 53.
41. B. Labarbe, V. Cheynier, F. Brossaud, J. M. Souquet and M. Moutounet, *J. Agric. Food Chem.*, 1999, **47**, 2719.
42. S. Guyot, N. Marnet and J.-F. Drilleau, *J. Agric. Food Chem.*, 2001, **49**, 14.
43. A. Degenhardt, *J. Agric. Food Chem.*, 2000, **48**, 3425.
44. O. Ploss, F. Petereit and A. Nahrstedt, *Pharmazie,* 2001, **56**, 509.
45. A. Degenhardt, U. H. Engelhardt, P. Winterhalter and Y. Ito, *J. Agric. Food Chem.*, 2001, **49**, 1730.
46. J. J. Karchesy and R. W. Hemingway, *J. Agric. Food Chem.*, 1986, **34**, 966.
47. D. Jacques, E. Haslam, G. R. Bedford and D. Greatbanks, *J. Chem. Soc., Perkin Trans. 1*, 1974, 2663.
48. S. Matthews, I. Mila, A. Scalbert, B. Pollet, C. Lapierre, C. L. M. Hervé du Penhoat, C. Rolando and D. M. X. Donnelly, *J. Agric. Food Chem,* 1997, **45**, 1195.
49. C. Prieur, J. Rigaud, R. Cheynier and M. Moutounet, *Phytochemistry*, 1994, **36**, 781.
50. M. Dadic and G. Belleau *J. ASBC*, 1976, **34**, 158.
51. G.-I. Nonaka, O. Kawahara and I. Nishioka, *Chem. Pharm. Bull.*, 1983, **31**, 3906.
52. J. A. Delcour, D. Ferreira and D. G. Roux, *J. Chem. Soc., Perkin Trans. 1*, 1983, 1711.
53. G.-I. Nonaka, S. Morimoto and I. Nishioka, *J. Chem. Soc., Perkin Trans. 1*, 1983, 2139.
54. S. Morimoto, G.-I. Nonaka and I. Nishioka, *Chem. Pharm. Bull.*, 1986, **33**, 4338.
55. S. Morimoto, G.-I. Nonaka and I. Nishioka, *Chem. Pharm. Bull.*, 1986, **34**, 633.
56. R. Self, J. Eagles, G. C. Galletti, I. Mueller-Harvey, R. D. Hartley, A. G. H. Lea, D. Magnolato, U. Richli, R. Gujer and E. Haslam, *Biomed. Environ. Mass Spect.*, 1986, **13**, 449.
57. G. Galletti and R. Self, *Ann. Chim. (Rome)*, 1986, **76**, 195.
58. J. M. Ricardo da Silva, J. Rigaud, V. Cheynier, A. Cheminant and M. Moutounet, *Phytochemistry*, 1991, **30**, 1259.
59. J. Clapperton, J. F. Hammerstone, L. J. Romanczyk, S. Yow, D. Lim and R. Lockwood, *Proceedings of the 16th International Conference of Groupe Polyphenols, Lisbon, Portugal*, 1992, Tome II, p. 112.
60. J. J. Karchesy, R. W. Hemingway, Y. L. Foo, E. Barofsky and D. F. Barofsky, *Anal. Chem.,* 1986, **58**, 2543.

61. M. Ohnishi-Kameyama, A. Yanagida, T. Kanda and T. Nagata, *Rapid Comm. Mass Spectrum.*, 1997, **II**, 31.
62. C. G. Krueger, N. C. Dopke, P. M. Treichel, J. Folts and J. D. Reed, *J. Agric. Food Chem.*, 2000, **48**, 1663.
63. Y. Yang and M. Chein, *J. Agric. Food Chem.*, 2000, **48**, 3990.
64. A. Roeder, T. M. L. Lam and R. Galensa, *Monatsschr. Brauwissenschat*, 1995, **11/12**, 390.
65. A. Kiehne and C. Lakenbrink, *Lebensm. Unters Forch. A*, 1997, **205**, 153.
66. V. Cheynier, T. Doco, H. Fulcrand, S. Guyot, E. Le Roux, J. M. Souquet, J. Rigaud and M. Moutounet, *Analysis Magazine*, 1997, **25**.
67. H. Fulcrand, S. Remy, J.-M. Souquet, V. Cheynier and M. Moutounet, *J. Agric. Food Chem.*, 1999, **47**, 1023.
68. J. F. Hammerstone and S. A. Lazarus in *Caffeinated Beverage Health Benefits, Physiological Effects, and Chemistry*, eds T. H. Parliment, C. T. Ho and P. Schieberle, American Chemical Society, Washington, DC, 2000.
69. D. Ferreira and E. V. Brandt, in *Chemistry and Significance of Condensed Tannins*, ed. R. W. Hemingway and J. J. Karchesy, Plenum Press, New York, 1989, p. 153.
70. R. W. Hemingway, S. Ohara, J. P. Steynberg, E. V. Brandt, and D. Ferreira, in *Plant Polyphenols*, ed. R. W. Hemingway and P. E. Laks, Plenum Press, New York, 1992, p. 321.
71. J. Vercauteren, in *Polyphenols 1994*, ed. INRA, Paris, 1995, p. 29.
72. C.-B. Cui, Y. Tezuka, T. Kikuchi, H. Nakano, T. Tamaoki and J.-H. Park, *Chem. Pharm. Bull.*, 1991, **39**, 2179.
73. T. De Bruyne, Luc A. C. Pieters, R. A. Dommisse, H. Kolodziej, V. Wray, T. Domke and A. J. Vlietinick, *Phytochemistry*, 1996, **43**, 265.
74. L. Balas and J. Vercauteren, *Magn. Res. Chem.*, 1994, **32**, 386.
75. L. Balas, J. Vercauteren and M. Laguerre, *Magn. Res. Chem.*, 1995, **33**, 85.
76. M. L. Khan, E. Haslam and M. P. Williamson, *Magn. Res. Chem.*, 1997, **35**, 854.
77. F. Geiss, M. Heinrich, D. Hunkler and H. Rimpler, *Phytochemistry*, 1995, **39**, 635.
78. K. Weinges, J. Perner and H.-D. Marx, *Chem. Ber.*, 1970, **103**, 2344.
79. D. W. Engel, M. Hattingh, H. K. L. Hundt and D. G. Roux, *J. Chem. Soc. Chem. Commun.*, 1978, 695.
80. A. C. Fletcher, L. J. Porter, E. Haslam and R. K. Gupta, *J. Chem. Soc., Perkin Trans. 1*, 1977, 1628.
81. H. K. L. Hundt and D. G. Roux, *J. Chem. Soc., Perkin Trans. 1*, 1981, 1227.
82. J. A. Delcour, E. J. Serneels, D. Ferreira and D. Roux, *J. Chem. Soc., Perkin Trans. 1*, 1985, 669.
83. A. P. Kozikowski, W. Tückmantel and C. George, *J. Org. Chem.*, 2000, **65**, 5371.
84. A. P. Kozikowski, W. Tückmantel and Y. Hu, *J. Org. Chem.*, 2001, **66**, 1287.
85. R. H. Newman, L. J. Porter, L. Y. Foo, S. R. Johns and R. I. Willing, *Magn. Res. Chem.*, 1987, **25**, 118.

86. G. R. Gamble, D. E. Akin, H. P. S. Makkar and K. Becker, *Appl. Envir. Microbiol.,* 1996, **62**, 3600.
87. S. Guyot, C. Le Guerneve, N. Marnet and J.-F. Drilleau, *Basic Life Sci.*, 1999, **66** (Plant Polyphenols 2), 211.
88. T. Swain and W. E. Hillis, *J. Sci. Food Agric.*, 1959, **10**, 63.
89. M. L. Price and L. G. Butler, *J. Agric. Food Chem.,* 1977, **25**, 1268.
90. H. D. Graham, *J. Agric. Food Chem.,* 1992, **40**, 801.
91. L. J. Porter, L. N. Hrstich and B. G. Chan, *Phytochemistry*, 1986, **25**, 223.
92. L. J. Porter, in *Methods in Plant Biochemistry, Volume I Plant Phenolics*, eds P. M. Dey and J. B. Harborne, Academic Press Inc., San Diego, 1989.
93. D. Treutter, *J. Chromatogr.,* 1989, **467**, 185.
94. J. A. Delcour and D. J. Varebeke, *J. Inst. Brew.,* 1985, **91**, 37.
95. R. B. Broadhurst and W. T. Jones, *J. Sci. Food Agric.,* 1978, **29**, 788.
96. B. Sun, J. M. Ricardo-da-Silva and I. Spranger, *J. Agric. Food Chem.,* 1998, **46**, 4267.
97. S. de Pascual-Teresa, C. Santos-Buelga and J. C. Rivas-Gonzalo, *J. Agric. Food Chem.*, 2000, **48**, 5331.
98. E. L. Wilson, *J. Sci. Food Agric.*, 1981, **32**, 257.
99. S. Guyot, T. Doco, J.-M. Souquet, M. Moutounet and J.-F. Drilleau, *Phytochemistry*, 1997, **44**, 351.
100. A. G. H. Lea, *J. Sci. Food Agric.,* 1979, **30**, 833.
101. J. Rigaud, M. T. Escribano-Bailon, C. Prieur, J. M. Souquet and V. Cheynier, *J. Chromatogr.*, 1993, **654**, 255.
102. V. Cheynier, J. Souquet, E. L. Roux, S. Guyot and J. Rigaud, *Methods Enzymol.*, 1999, **299**, 179.
103. J. F. Hammerstone, S. A. Lazarus and H. H. Schmitz, *J. Nutr.*, 2000, **130**, 2086S.
104. G. A. Spanos and R. E. Wrolstad, *J. Agric. Food Chem.*, 1990, **38**, 1565.
105. J. M. Ricardo da Silva, J.-P. Rosec, M. Bourzeix and N. Heredia, *J. Agric. Food Chem.,* 1990, 53, 85.
106. G. Chiavari, P. Vitali and G. C. Galletti, *J. Chromatogr.*, 1987, **392**, 426.
107. G. Achilli, G. P. Cellerino and P. H. Gamache, *J. Chromatogr.*, 1993, **632**, 111.
108. I. McMurrough and T. Baert, *J. Inst. Brew.*, 1994, **100**, 409.

CHAPTER 13

Analysis of Polymeric Proanthocyanidins and Complex Polyphenols

VÉRONIQUE CHEYNIER AND HÉLÈNE FULCRAND

1 Introduction

Polymeric phenolics comprise a substantial portion of the phytochemicals in the human diet.[1] They are often designated by the term tannin, as defined by Bate-Smith and Swain, which refers to water-soluble phenolic compounds having the ability to precipitate alkaloids, gelatin and other proteins.[2] This property has been exploited when using plant extracts for converting raw animal hides and skins to durable leather for over 2000 years.[3] It is also responsible for numerous characteristic features of tannins, including in particular astringency, which results from interactions with salivary proteins,[4,5] formation of haze in protein-rich beverages[6] and antinutritional properties of tannin-rich animal feeds.[7]

Tannins are classically divided in two groups: hydrolysable tannins, deriving from gallic and ellagic acids, and condensed tannins which are also called proanthocyanidins because they release red anthocyanidin pigments when heated in acidic media. In addition, transformation products of plant polyphenolic compounds formed during food processing (especially in wine or tea) may bind to proteins, and thus be regarded as tannins. These derived phenolic compounds are called here tannin-like compounds as recently proposed.[8] Although some foods may also contain hydrolysable tannins, polymeric dietary polyphenols belong mostly to the last two categories, which will be treated in this chapter.

There have been extensive reviews covering the nomenclature, structure and distribution of proanthocyanidins in plants[9,10] as well as the dietary significance of proanthocyanidins and tannin-like compounds.[1,8] Therefore, this chapter will address mostly analytical aspects, with special emphasis on the respective advantages and limits of techniques currently available to extract, purify, characterize and assay these polydisperse molecules in complex plant and food matrices.

2 Structure of Proanthocyanidin Polymers and Tannin-like Compounds

Proanthocyanidins

As described in Chapter 12, proanthocyanidins are oligomers and polymers of flavans (2-phenylchromanes). They are subdivided into a number of classes, depending on the substitution pattern of their constitutive flavan units and named after the corresponding red anthocyanidin (2-phenylchromenium) pigments released from them under acidic conditions (Figure 1; see also the more detailed list in ref. 10). The main proanthocyanidins encountered in foods are procyanidins, based on (epi)catechin units [2-(3,4-dihydroxyphenyl)chromane-3,5,7-triol] and prodelphinidins, based on (epi)gallocatechin [2-(3,4,5-trihydroxyphenyl)chromane-3,5,7-triol] (*cf.* Chapter 12, Table 1 and Figure 2) whereas the major industrial proanthocyanidins, isolated from mimosa and quebracho, are predominantly composed of their 5-deoxy analogues, namely profisetinidins based on fisetinidol [2-(3,4-dihydroxyphenyl)chromane-3,7-diol] and prorobinetinidins, based on robinetinidol [2-(3,4,5-trihydroxyphenyl)chromane-3,7-diol]. Although some plants synthesize only proanthocyanidins based on a single constitutive unit (*e.g.* epicatechin-based procyanidins in the case of apple or cacao), several types of units often co-occur within a single proanthocyanidin chain, in which they can

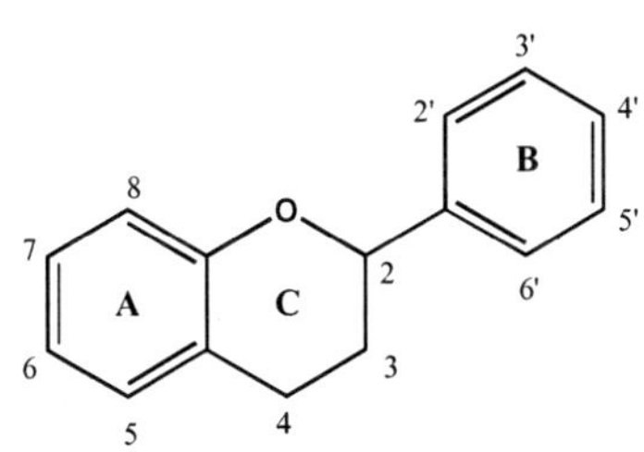

		Substitution pattern					
Proanthocyanidin class		*3*	*5*	*7*	*3'*	*4'*	*5'*
Propelargonidin	Flavan-3-ols	OH	OH	OH	H	OH	H
Procyanidin		OH	OH	OH	OH	OH	H
Prodelphinidin		OH	OH	OH	OH	OH	OH
Profisetinidin	5-Deoxyflavan-3-ols	OH	H	OH	OH	OH	H
Prorobinetinidin		OH	H	OH	OH	OH	OH

Figure 1 *Structure of most common proanthocyanidin classes*

be linked by C-4–C-6 and/or C-4–C-8 bonds (B-type) or doubly linked, with an additional C-2–O–C-7 or C-2–O–C-5 linkage (A-type) (*cf.* Chapter 12, Figure 3) and eventually substituted (*e.g.* glycosylated, galloylated).

A large number of oligomers (up to the pentamer) have been isolated from various plant sources and fully characterized, as listed in regularly up-dated reviews.[9,10] However, these lower molecular weight compounds make up only a relatively small proportion of proanthocyanidins found in plant tissues which consist mostly of higher oligomers and polymers.[11,12] The original definition proposed by Bate-Smith and Swain[2] described tannins as water-soluble phenolic compounds having molecular weights between 500 and 3000 but proanthocyanidins have been described up to 20 000 (around 70 units) in molecular weight[5] and more recently polymeric fractions containing 190 units in average (molecular weight around 55 000) have been isolated from cider apple varieties.[13]

Tannin-like Compounds

Biochemical (*e.g.* enzymatic oxidation) and chemical reactions of phenolic compounds taking place during handling, storage and processing of plants lead to various types of tannin-like phenolic compounds. The occurrence of tannin-like compounds is well documented in black tea[14] and in wine.[15–17]

Tannin-like Compounds of Tea

Enzymatic oxidation of native green tea flavanols, catalysed by polyphenoloxidase, is involved in the so-called 'fermentation' of oolong and black teas.[18] The first step of the reaction process is the enzymically-catalysed transformation of *o*-dihydroxyphenyl (and hydroxyphenyl) substrates to *o*-quinones. The resulting oxidized species are highly unstable, being both strong electrophiles and powerful oxidants,[19] and rapidly proceed to various reaction products. In black tea fermentation, about 85% of tea leave flavanols are thus converted into reaction products. Among them, two major groups of hydrosoluble pigments, namely thearubigins and theaflavins, are responsible for the characteristic dark brown colour of black tea.[20,21] Other oxidation products, such as theaflavic acids[22] and bisflavanols, also called theasinensins,[23,24] are also present in lower quantities. Similar bisflavanols and pigments resulting from further oxidation of these products have also been obtained by enzymatic oxidation of catechin catalysed by polyphenoloxidase and characterized by NMR[25]. Examples of tea tannin-like compounds are given in Figure 2.

Several theaflavins have been identified.[26–28] These red-orange compounds show a benzotropolone structure (Figure 2, **1–4**) resulting from oxidative coupling of flavanol units containing pyrogallol (*i.e.* 3′,4′,5′-trihydroxyphenyl) moieties (epigallocatechin and epigallocatechin gallate).[26] Further oxidation products of theaflavin have been recently obtained in model solution and identified.[29] Enzymatic oxidation of a mixture of (−)-epicatechin and (−)-epigallocatechin yielded theaflavin, theanaphtoquinone (Figure 2, **5**) and bistheaflavin A (Figure 2,

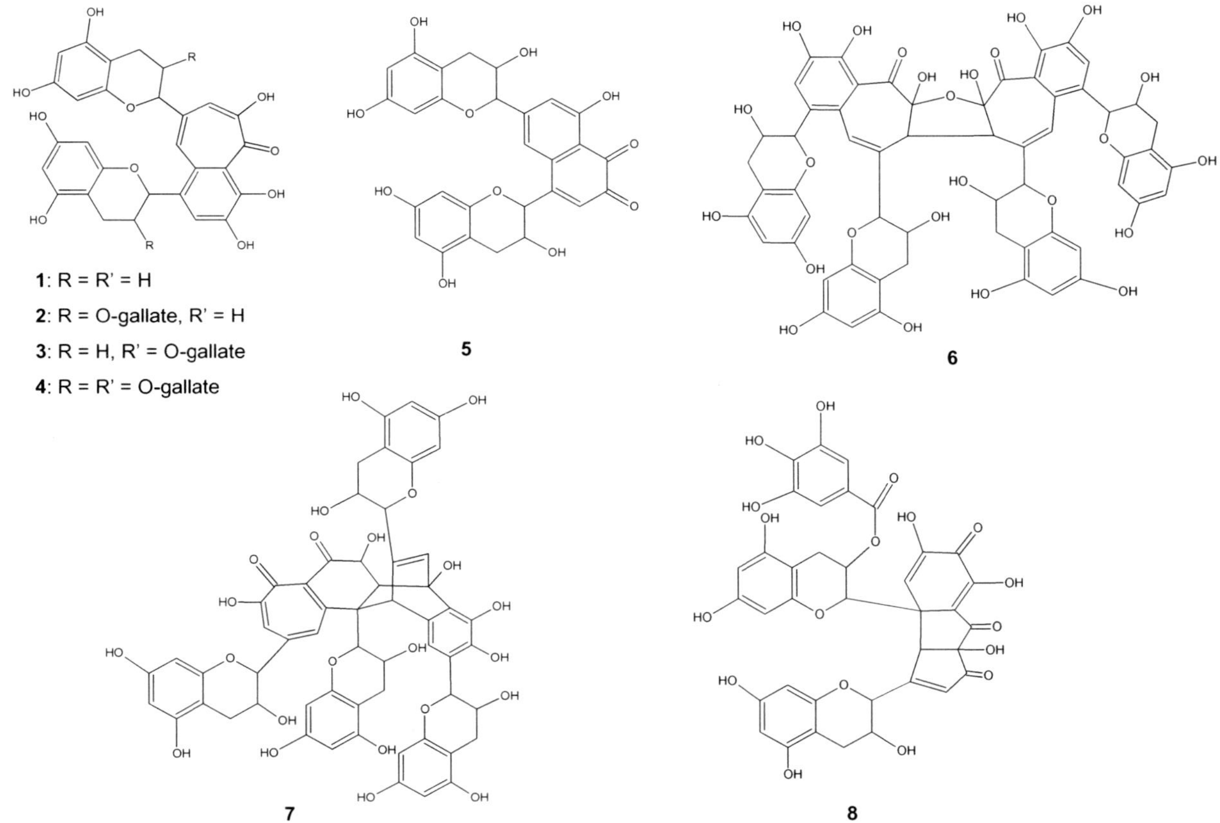

Figure 2 *Examples of structures of tannin-like compounds from tea*

6) whereas autoxidation of theaflavin gave bistheaflavin B (Figure 2, **7**) and theanaphtoquinone. Identification of bistheaflavins suggested that bistheaflavin A resulted from oxidative C–C coupling of two theaflavin molecules whereas bistheaflavin B was probably formed by intramolecular cyclization between dehydrotheaflavin and dihydronaphtoquinone.

Thearubigins are poorly defined complex polymeric brown pigments. They probably arise from addition reactions between theaflavins and flavan-3-ols by random polymerization and oxidative reactions[30,31] involving polyphenol-oxidases[32,33] and possibly also peroxidases.[34,35] The only thearubigin pigment identified so far is theacitrin A[28] (Figure 2, **8**).

Tannin-like Compounds in Red Wine

Colour and taste changes occurring during wine ageing are attributed to conversion of grape phenolic compounds, including in particular red anthocyanin pigments and proanthocyanidins, into tannin-like products.[15,16] Two major groups of reactions, namely enzymatic oxidation processes and chemical reactions, are involved in the formation of tannin-like species in wine. Both take place simultaneously but the former prevails in the early stages of wine-making whereas the latter becomes predominant in later stages, when enzymatic activities decrease. However, enzymatic oxidation appears rather limited in red wine, as most of the available oxygen is consumed by the fermenting yeast.[36]

Major polyphenolic reactions involved in red wine aging are anthocyanin–tannin additions, which can either be direct, generating orange xanthylium salts, or involve acetaldehyde, leading to purple pigments.[15,37]

Studies of these reactions in wine-like model systems have enabled to obtain and characterize various products.[38,56] Besides, the occurrence of some other anthocyanin-derived pigments in wines has been demonstrated.[45,48–61]

Major products obtained by incubating tannins and anthocyanins in the presence of acetaldehyde in wine-like model systems include ethyl-linked anthocyanin-flavanol adducts (Figure 3, **1**),[37,39,40,41,47,54,62,63] ethyl-linked flavanol oligomers (Figure 3, **2**)[46] and pyranoanthocyanin-flavanol adducts (Figure 3, **3**).[44] All three groups have also been demonstrated to occur in red wines.[59,61] Acetaldehyde arises from yeast metabolism or from oxidation of ethanol, involving autoxidized polyphenols.[64] Consequently, ethyl-linked species and acetaldehyde derived pyranoanthocyanins were shown to accumulate in wines submitted to oxygen exposure.[65] The latter were recently formally identified after isolation from Port wine, where they are especially abundant.[61]

Glyoxalic acid (COOH–CHO), which is a product of tartaric acid oxidation, reacts with flavanols in the same way as acetaldehyde.[48] The resulting carboxymethine-linked catechin oligomers proceeds to yellow pigments. Structural elucidation of these pigments by NMR allowed to demonstrate that they were xanthylium salts[42,66] and their ethyl esters.[43] Similar reactions of catechin with furfural and hydroxymethylfurfural also yield xanthylium pigments and may play a part in non enzymatic oxidative browning.[67]

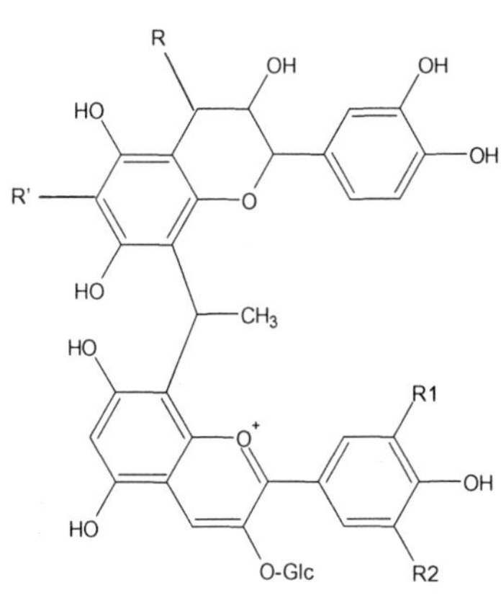

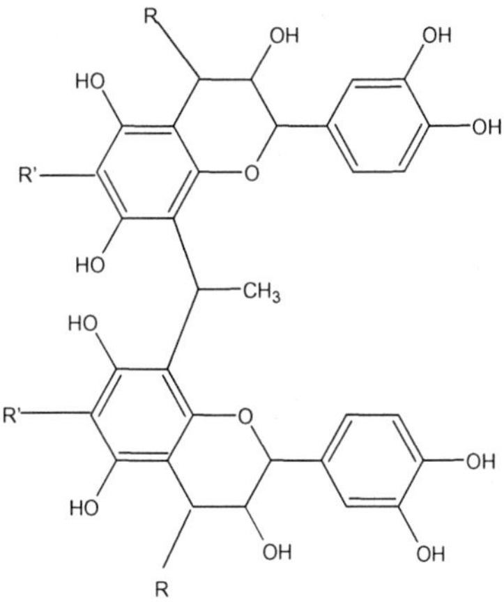

1 **2**

R = H, flavanol-6-yl, flavanol-8-yl

R' = H, flavanol-4-yl, 8 (or 6)-ethylflavanol, 8 (or 6)-ethylanthocyanin

R1, R2 = H, OH, OCH_3

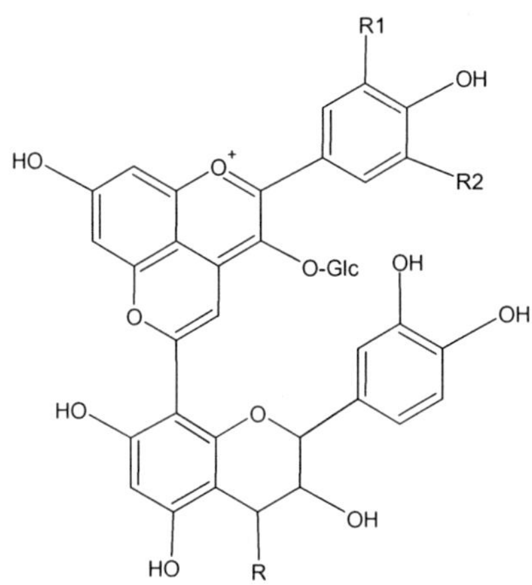

3

Figure 3 *Examples of structures of tannin-like compounds arising from acetaldehyde-mediated reactions in wine*

Two direct reactions between anthocyanins and proanthocyanidins, respectively leading to anthocyanin–tannin (A-T) and to tannin–anthocyanin (T-A) adducts, have been postulated. The former consists in nucleophilic addition of a carbon C-8 (or C-6) of the tannin molecule to the positively charged C-4 of the flavylium cation, yielding a colourless flavene (Figure 4, **1**) which proceeds to the corresponding flavylium (A^+-T)[15] (Figure 4, **2**) and xanthylium[37,51,68] (Figure 4, **3**). Alternatively, the anthocyanin under its hemiketal form may add to the electro-

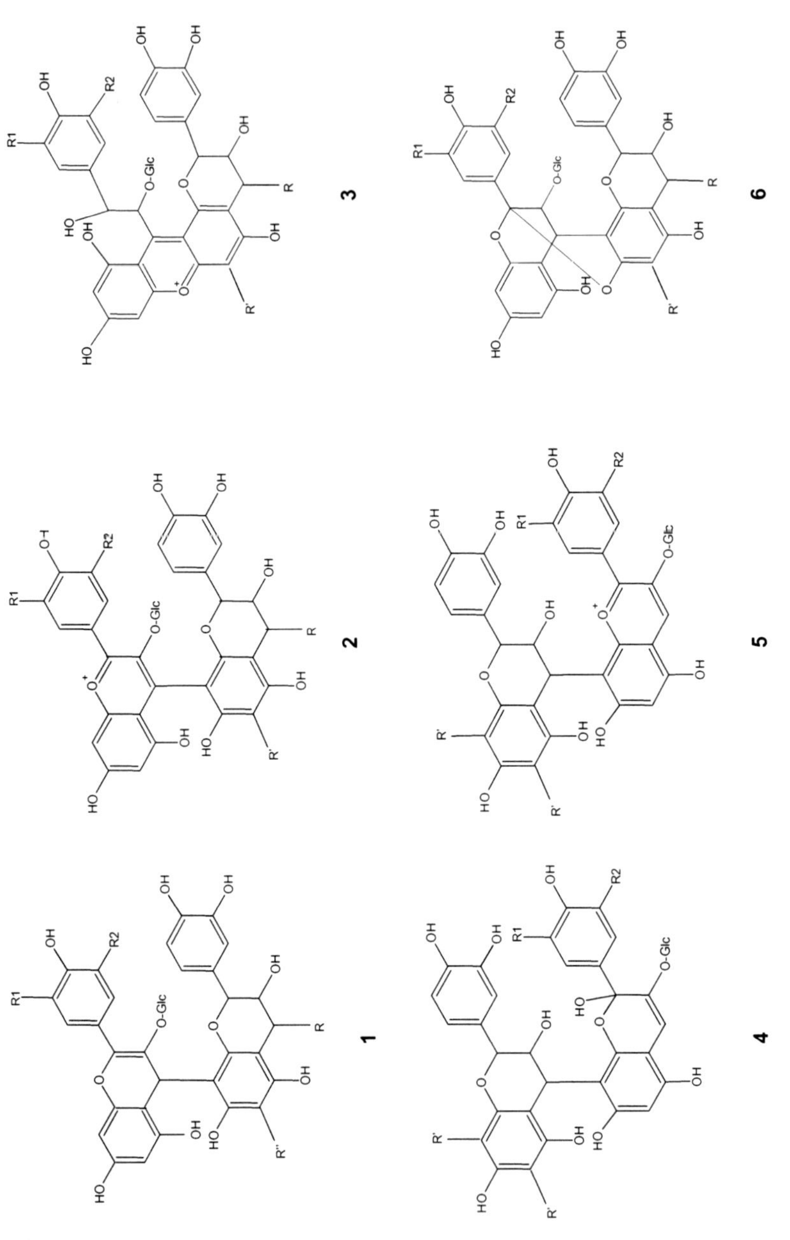

R = H, flavanol-6-yl, flavanol-8-yl ; R' = H, flavanol-4-yl ; R1 = OH, CH_3, R2 = H, OH, OCH_3

Figure 4 *Structures of tannin-like compounds arising from direct reactions of tannins and anthocyanins in wine*

philic species arising from breakage of the tannin interflavanic bond to yield a flavanol-hemiketal adduct (T-AOH) (Figure 4, **4**), which can dehydrate to the corresponding flavylium pigment (T-A$^+$) (Figure 4, **5**).

Pigments of the T-A$^+$ type were detected in wine,[60] in agreement with the latter mechanism. Analysis of the wine extract also showed the presence of another series of colourless compounds, which were tentatively identified to A-T adducts in the flavene form[55] or bicyclic A-T adducts[60] (Figure 4, **6**) analogous to that obtained by reaction of malvidin 3,5-diglucoside with catechin.[69]

3 Analysis and Characterisation

Tannin properties, including radical scavenging effects and protein-binding ability, largely depend on their structure and particularly on the nature and number of constitutive units. Formal identification of proanthocyanidins can be achieved by bidimensional NMR techniques.[70] The same approach enables to identify tannin-like derivatives.[45,49,61,71] However, this is restricted to pure compounds, which become increasingly difficult to isolate as their degree of polymerization (DP) increases, owing to the larger number of possible isomers, smaller amounts of each individual compound, and poorer resolution of the chromatographic profiles. Therefore, specific methods have been developed to characterize proanthocyanidins polymers in rather crude plant extracts and fractions.

Extraction from Plant Tissues

Immediate processing of the fresh material is preferable but storage in the frozen or freeze-dried form is possible. Comparison of conventional freeze drying of frozen material, freeze drying without prefreezing and vacuum drying procedures on leaves concluded that only heat drying methods resulted in significant proanthocyanidin losses.[72] Lyophilisation of fresh leaves has also led to slight enhancement of condensed tannin extraction (usually less than 10%) in other studies.[73] Care must be taken to avoid enzymatic oxidation (*e.g.* immediate sample stabilization in liquid nitrogen or organic solvent, use of antioxidants such as sulfur dioxide or ascorbic acid). Spraying of the fresh material with aqueous formic acid (3% v/v) during peeling and grinding enabled the prevention of enzymatic oxidation of apple tissues.[74]

Extraction of proanthocyanidins is performed by grinding or homogenizing the ground material in methanol, ethanol or acetone solutions.[75–78] Relative efficiencies of methanol, acetone and acetone–water (3:1) for extraction of total flavanols from barley were 6.5:1:15, meaning that 100% acetone is a very poor solvent for flavanol extraction.[79] 80% methanol and 70–75% acetone are the best extraction solvents for catechin monomers and proanthocyanidins from grape seeds. Ethyl acetate extracted only catechins and small amounts of lower molecular weight oligomers whereas diethylether extracted only monomers.[80] Species-dependent differences in solvent extraction efficiencies are commonly observed and, although usually 70% acetone is found to be superior over aqueous methanol[73,81] on plant leaf samples, the reverse is also encountered.[82]

Higher molecular weight proanthocyanidins are better extracted with aqueous acetone than with methanol.[74,78,83,84] Successive extraction first with methanol and then with aqueous acetone thus enable the recovery of lower oligomers and higher molecular proanthocyanidins in different fractions. The average degrees of polymerization in extracts from cider apple were ~3 in the methanol extract and ~11 in the 70% acetone extract. Additional diethyl ether extraction steps may be added to remove monomeric flavanols.[80] Preliminary extraction with hexane is often performed to remove lipids, carotenoids and chlorophyll.[85]

Ascorbic acid (5 mM) is sometimes added to prevent proanthocyanidin oxidation during the extraction procedure.[82,86,87] However, ascorbic acid has recently been shown to induce browning of catechin and act as a pro-oxidant rather than as an anti-oxidant in solutions containing flavanols[88] and should thus be used with caution. Slight acidification of the extraction solvent (for instance with 0.05% TFA or 1% acetic acid) is recommended to reduce autoxidation which is driven by the proportion of phenolate ions and thus increases as a function of pH.

Part of the proanthocyanidins present in a plant tissue is not extractible as the insoluble residues obtained after extensive solvent extraction still release anthocyanidins when heated in acidic medium.[89] This material is often described as cell-wall bound proanthocyanidins.

Sample Clean-up by Liquid–Liquid or Solid-phase Extraction

Sample clean-up can be achieved by hexane[87] or petroleum ether[90] partition to remove lipids and lipophilic pigments such as chlorophylls and carotenoids. Solid-phase extraction on C_{18} Sep-Pack cartridges allows the removal of sugars and other highly polar compounds (*e.g.* organic acids, amino acids, proteins, caffeine) which are eluted with water, eventually acidified with 2.5% acetic acid. Phenolic compounds can be recovered by elution with methanol or aqueous acetonitrile (1:1, v/v), evaporation of the organic solvent and freeze drying.[13] Removal of polar compounds can also be achieved by adsorption on other chromatographic phases including Sephadex LH-20, Fractogel TSK[91] or Amberlite XAD-7,[92] followed by rinsing with water and elution of phenolic compounds with methanol or aqueous acetone.

Enrichment in higher molecular weight proanthocyanidins can be achieved by liquid–liquid extraction using ether or ethyl acetate. Flavanol monomers, flavonol aglycones and phenolic acids (if the aqueous phase is acidic) are partitioned into the ether phase, whereas proanthocyanidins (but also anthocyanins and flavonol glycosides) remain in the aqueous phase. Ethyl acetate extracts oligomeric proanthocyanidins (DP ~4) but also acylated anthocyanins and flavonols, in addition to diethyl ether soluble compounds.

Replacement of liquid–liquid extraction by solid-phase extraction on C_{18} Sep-Pak cartridges permits similar separations with significant reduction of solvent volumes and of fractionation time.[93] Successive elutions with ethyl acetate and methanol allow to separate monomers and oligomers from polymers. Diethyl ether elution allows the removal of flavanol monomers[94] and the recovery of all proanthocyanidins in methanol. Other chromatographic materials, including polyvinylpyrolidone–divinylbenzene (Oasis HLB), cellulose, silica, polyamide and polyvinylpyrolidone do not allow satisfactory recovery of proanthocyanidins, due to irreversible adsorption.

A major limitation of all SPE and liquid partition processes developed so far is that they do not enable the separation of polymeric proanthocyanidins from anthocyanins and tannin-like pigments. As well, oligomeric proanthocyanidins recovered in ethyl acetate may be contaminated by flavonols, acylated anthocyanins and various anthocyanin-tannin adducts if present in the original sample.

Fractionation According to Molecular Weight

Several methods have been developed to separate proanthocyanidins from other phenolic compounds and fractionate them on a molecular weight basis.

Adsorption Chromatography on Sephadex LH-20 or Fractogel (Toyopearl)

Sephadex LH-20[95] and Toyopearl (*syn.* Fractogel) TSK HW-40 (F)[96,97] or TSK HW-50(F)[91] columns are classically used to fractionate proanthocyanidins by low-pressure chromatography. Lower molecular weight compounds (*e.g.* sugars, organic acids, phenolic acids, anthocyanins and flavanol monomers) are eluted with EtOH–H_2O–TFA (55:45:0.02, v/v/v). The tannin fraction is then recovered by elution with 60–70% acetone. However, oligomeric proanthocyanidins and wine tannin-like compounds are distributed between the two fractions.[60] Chromatography of a grape seed extract on Fractogel or Sephadex using methanol as the eluent allows to fractionate monomers and oligomers, up to the tetramers, in increasing molecular weight order.[97] Larger molecular weight procyanidins are then recovered with aqueous acetone (70:30, v/v), with no further separation.

Gel Permeation Chromatography

Although Sephadex and Toyopearl phases are often used to perform exclusion chromatography, elution of procyanidins in increasing MW order indicates that separation of these molecules actually relies upon adsorption processes. Gel permeation chromatography of epicatechin-based oligomers has been performed by using a dimethylformamide–3M aqueous ammonium formate (95.5:0.5, v/v) solvent and polystyrene–divinylbenzene column[98] and by chromatography on a Toyopearl HW-40F column with a mobile phase of acetone–8M urea (60:40, v/v adjusted to pH 2)[99] but the resolution was very poor beyond the tetramer. Resolution was even poorer when tested on procyanidins from grape seeds which also contain catechin and epicatechin gallate units. Application of this method to rosé cider and red wine allowed the recovery of early eluting pigments which were believed to be tannin-like pigments.[100] Size exclusion chromatography was also applied to tannin-like pigments from black tea[101] using a Phenomex (Macclesfield, Cheshire, UK) Biosep-SEC-S2000 column eluted isocratically with water containing sodium azide (0.05%).[102] The method was calibrated with epicatechin monomer, trimer and tetramer and ellagitannins but no further characterization of the tea component fractions was provided.

Chromatography on Normal Phase Silica

As described in the previous chapter, flavanol monomers and oligomers have been separated on a molecular weight basis, without derivatization, using TLC on silica plates.[97,103] Adaptations of this procedure to HPLC enabled the separation of monomers and procyanidin oligomers from cacao,[104,105] cider apple[106] and grape seed[104] extracts, in increasing molecular weight order up to the pentamer series. However, some overlapping takes place as the molecules become increasingly substituted and the separation of oligomeric procyanidins is rather poor compared with that obtained by reversed-phase HPLC.

Further adaptation of the elution conditions allowed to improve separation of the polymeric procyanidins from grape seeds.[84] This method was successfully applied to grape skin tannins, containing both (epi)catechin and (epi)gallocatechin units, along with small amounts (3–6%) of galloylated epicatechin units[91] and to epicatechin-based procyanidins of A- and B-types from litchi pericarp.[107] Although the chain length increased with the retention time in each proanthocyanidin extract, no relationship between these two parameters could be established when comparing the three extracts studied, due to differences in proanthocyanidin structures.[105] Besides, the presence of other UV absorbing material (*e.g.* anthocyanins, which elute much later than procyanidin dimers under these chromatographic conditions) may lead to misinterpretation of the molecular weight distribution. Therefore, the method cannot be readily transposed to other plant extracts of unknown composition and the correspondence between retention times and average DP should be established in each particular case, especially when heterogeneous polymers are to be encountered.

It is worth noting that fractionation of the 60% acetone soluble procyanidin polymers from apples was also achieved by reversed-phase HPLC[108] and a procyanidin polymer peak was obtained by reversed-phase chromatography of a grape seed extract on a C_{18} column using a gradient of acetonitrile in 0.04% phosphoric acid.[109]

Counter Current Chromatography

Counter current chromatography (CCC) methods, including droplet counter current chromatography (DCCC) and centrifugal partition chromatography (CPC) rely upon partition between two liquid phases, eliminating the need for a solid support and hence the risk or irreversible adsorption and clogging. Early application of DCCC using distribution between ethyl acetate and water allowed the separation of flavanol monomers, dimers, oligomers (*i.e.* trimers and tetramers) and polymers from white wines.[110] More recent developments using high-speed countercurrent chromatography (HSCCC) have enabled purification of flavanol monomers, proanthocyanidin dimers and dimeric tannin-like compounds from tea leaves using hexane–ethyl acetate–methanol–water (1:5:1:5).[111] Theaflavins and thearubigins from black tea have been recovered by using hexane–ethyl acetate–methanol–water (1.5:5:1.5:5) and ethyl acetate–*n*-butanol–water (2:3:5),[92] after preliminary sample clean-up. Hydrophilic tannin-like pigments from red wine were similarly separated from the more hydrophobic anthocyanins and pyranoanthocyanins by HSCCC using *tert*-butyl methyl ether–*n*-butanol–acetonitrile–water (2:2:1:5) (less dense layer as the stationary phase) containing 0.1% TFA.[112]

Precipitation Methods

Methods involving selective precipitation can be used to isolate proanthocyanidins and tannin-like compounds from complex mixtures. For instance, thearubigins have been fractionated by precipitation with caffeine.[101] Methods involving precipitation/dissolution with chloroform–methanol mixtures have enabled workers to obtain proanthocyanidin fractions with different degrees of polymerization but showing important overlapping, from grape seed and skin extracts.[113,114]

Quantitation and Characterization of Polymeric Tannins

The heterogeneity of proanthocyanidins increases with their chain length, resulting in poorer resolution of all separation methods. This makes it virtually impossible to obtain proanthocyanidins in the pure state beyond the pentamer. Diverse methods have thus been developed to characterize proanthocyanidins in rather crude plant extracts. Major structural features to be determined include the nature and proportions of constitutive units, number average molecular weight (MWn) and mean degree of polymerisation (mDP), and types of interflavanic linkages.

Various chemical methods and NMR spectrometry enable to determine the mean composition of an extract whereas gel permeation chromatography and mass spectrometry provide information on the distribution of molecular weights. Chemical methods can also be used for quantitative purposes. Each of these methods will be discussed separately in the following sections.

Depolymerization in Butanol–HCl (Porter's Reagent)

This method, initially referred to as the leucoanthocyanidin or Bate–Smith method,[4] relies on the acid-catalysed oxidative cleavage of interflavanic linkages in alcohol–water mixtures to produce anthocyanidin chromophores from the chain extension units (C-4-substituted in the original polymer). It is specific for proanthocyanidins but the reaction yield is low, due to the formation of side-products, and highly dependent on the conditions used and polymer structure. In particular, the yield is significantly higher for 4–8 linked compared with 4–6 linked derivatives.[115] After cooling, released coloured anthocyanidins can be assayed spectrophotometrically, usually at around 550 nm.[4] The maximum absorbance wavelength and extinction coefficient at a given wavelength depends on the anthocyanidin structure. The released anthocyanidins can be separated and tentatively identified by comparison with standard compounds by paper chromatography, TLC or HPLC. HPLC can also be used to determine the respective amounts of the different proanthocyanidin types.

The most widespread method uses heating of the proanthocyanidin extract (around 10 mg l^{-1}) in butanol–12N HCl (95:5, v/v) containing 100 μl of $(NH_4)Fe(SO_4)_2.12H_2O$ (2% in 2N HCl, final concentration of Fe^{3+} ions 0.6 mM) (Porter's reagent).[115] Addition of iron salts as catalyst of the final oxidation step increases the sensitivity and reproducibility of the assay.[115] Heating is usually performed at 95 °C for 30 min but optimal colour development has been obtained after 50 min in the case of birch leaves tannins.[89] The presence of water affects the yield of anthocyanidins released, maximum yield being obtained with 6%[115] to 10%[89] of water. The specific chemical structures of proanthocyanidins in various plants demand optimization of the assay conditions.

The molar concentration of upper and extension procyanidin units theoretically equals that of released cyanidin (noted $[cya]_M$) which is measured spectrophotometrically, using the molar extinction coefficient of cyanidin (35 000) at 550 nm.

Calibration with known solutions of procyanidin trimer C1 showed that the Beer–Lambert law was obeyed up to absorbance values of 0.6–0.7 ($Abs_{550} = 16825\ [cya]_M + 0.0095$, $r^2 = 0.99$).[99,116] The reaction yield obtained with pure procyanidin C1 was only 48%,[116] which is close to those calculated for a series of oligomers and polymers (40–44%).[115] These rather poor yields should be kept in mind when using Porter's reagent for proanthocyanidin quantitation. Besides, since all anthocyanidins have different extinction coefficients at a given wavelength, the proportions of the various tannins units within a given polymer should be taken into account when converting the absorbance values into molar concentrations of upper and extension units.

A similar method employing sulfuric acid instead of HCl has been adapted for an automated assay.[82] It enables rapid and stable colour development, with yields similar to those obtained with Porter's reagent and is unaffected by the presence of water or methanol. It can be used at ambient temperature and thus enables to analyse 3-deoxyproanthocyanidins for which Porter's reagent is not suitable. Maximum absorbance after reaction with 3-hydroxyproanthocyanidins is at 580 nm rather than at 550 nm.

These methods allow detection of the presence of proanthocyanidins in a plant extract or tissue and give access to the type and proportions of constitutive proanthocyanidin if combined with analysis of the released anthocyanidins.

Estimation of the mDP using Porter's reagent relies upon the following equation:

$$\text{mDP} = [\text{tannin units}]_M/([\text{tannin units}]_M - [\text{upper \& extension units}]_M)$$

where all concentrations are expressed on a molar basis, $[\text{tannin units}]_M$ is calculated by gravimetry from $[\text{tannin units}]_M = [\text{tannin units}]_W/\text{mUMW}$, in which mUMW is the mean unit molecular weight, and $[\text{upper \& extension units}]_M$ is estimated from $[\text{anthocyanidin}]_M$.

Values of mDP calculated for a series of procyanidin fractions isolated from grape seeds using this equation and the calibration curve established for procyanidin trimer C1 increased with the chain length but were lower than those determined by using other methods (Table 1).

Reactions with Aromatic Aldehydes

Flavanol monomers and polymers react with aldehydes such as vanillin (VA)[117] in acidic medium to yield chromophores absorbing at 500 nm. Replacement of vanillin with dimethylaminocinnamaldehyde (DMCA)[118–120] gives a green chromophore (λ_{max} 640 nm) which can be easier to measure in coloured extracts absorbing around 500 nm. Reactions with vanillin or DMCA are specific of *m*-dihydroxyphenyl moieties and thus of the free A-rings of the upper flavanol unit in a proanthocyanidin polymer.

The rate and extent of colour development after vanillin reaction is greatly solvent dependent.[121,122] The colour intensity increases with HCl concentration but concentrations above 30% in methanol should be avoided.[121] Exposure to light reduces the reaction yield. No linear response is obtained with catechin when the vanillin–HCl assay is performed in methanol[123] and the reaction kinetics vary with the flavanol chain length.[122] The reaction is more sensitive and more specific of end units when performed in a solution containing 4% HCl and 0.5% vanillin in glacial acetic acid, prepared immediately before use.[122] The absorbance is measured at 510 nm after 5 minutes of incubation. Under these conditions, the

Table 1 *mDP calculated for grape seed procyanidin fractions obtained as described by Prieur et al.*[87] *using different methods*

Fraction	*I*	*II*	*III*	*IV*	*V*
Bate-Smith[a]	1.8	2.4	2.7	2.8	3.4
VA (calibrated with catechin)	0.9	1.4	2.6	3.5	3.4
VA (calibrated with B2)	1.2	1.4	2.6	3.5	6.9
DMCA (calibrated with catechin)	3.0	3.9	4.1	5.1	5.2
Thiolysis	2.3	3.6	5.4	7.8	15.1
Phloroglucinol	2.5	4.8	7.7	12.2	19.1
Gel permeation[b]	2.5	4.6	7.2	12.9	18.9

[a] Using the yield of cyanidin released determined with the procyanidin trimer C1.
[b] Using 500 as the Mw of acylated unit.

concentrations of chromophore produced are proportional to the concentration of end-groups (E_{510}1% = 257).[122] Thus the mDP of a sample can be obtained as the ratio of absorbance of monomer to polymer, both determinations being made on equal weights of material. However, the molar extinction coefficients after vanillin reaction published for flavanol monomers are lower than those obtained for procyanidin oligomers (9800 and 8500, for epicatechin and catechin, 11 200 and 12 300 for B2 and B3, 12 200 for C1;[122] 18 600 for catechin and 25 200 for procyanidin B2[116]) meaning that the use of catechin for the standard curve results in overestimation of tannin end units and underestimation of mDP values. Values of mDP determined using the calibration curve established with B2 were higher than those calculated from the vanillin response of catechin but lower than those obtained using other methods.[116]

Reaction with DMCA is reported to be more sensitive[120] and more specific[124] than that with vanillin but, to our knowledge, it has not been tested with dihydrochalcones, flavanones and flavanonols and other *m*-diphenols which react positively with vanillin.[125]

The reaction is performed by dissolving the tannin sample (around 4 mg l^{-1}) in a 0.1% solution of DMCA in MeOH–12N HCl (3:1, v/v). The absorbance is measured at 640 nm after 1 minute of incubation. The molar extinction coefficient at 640 nm determined for catechin after DMCA reaction is 74 141, and thus much higher than that of the vanillin reaction product.[116] Comparison of the mDP values calculated after vanillin and after DMCA reactions for a series of grape seed procyanidin fractions suggested that DMCA was less specific of end units[116] (Table 1).

Methods relying upon the coupling of VA or DMCA reaction with depolymerization using Porter's reagent are also commonly applied. Although each of these methods has been calibrated to determine proanthocyanidin concentrations,[88] the resulting conversion factors are seldom used in the calculation of DPn values. The results are often given as arbitrary index values calculated as the ratio of absorbance after vanillin[117,126] or DMCA[120] reaction to absorbance after Bate-Smith's reaction. Nevertheless, calculation of mDP as mDP = ([extension units]$_M$ + [end units]$_M$)/[end units]$_M$), using proper calibration factors is unlikely to provide accurate values, given the imprecision in the determination of both end units (given by the VA or DMCA reaction) and extension units (given by the Bate-Smith reaction).

Depolymerization in the Presence of Nucleophilic Agents

Concentrations of both extension and terminal units and consequently mDP can also be determined in a single reaction, based on acid-catalysed cleavage in the presence of a nucleophilic agent.[127] Breakage of the interflavanic C–C bond under mild acidic conditions releases the terminal units as the corresponding flavanols and the upper and intermediate units as carbocations which react with the nucleophilic reagent to form stable adducts.

The most commonly used nucleophile is benzyl hydrosulfide (phenylmethanethiol, toluene-α-thiol) (the method being referred to as thiolysis)[127] but 1,3,5-trihydroxybenzene (phloroglucinol)[128–130] is sometimes preferred because it is odourless. Some investigators have also used ethanethiol especially for ethyl-linked tanin-like compounds.[131,132] The reaction products (*i.e.* flavanol monomers and their derivatives, *e.g.* benzyl thioethers in the presence of benzyl hydrosulfide) can be analysed by gas chromatography after derivatization to their trimethylsilyl ethers[133,134] or, directly, by HPLC.[84,91,104,107,130,134–137] A ^{1}H NMR method relying upon the integration of the H-2 signals characteristic of each unit released by thiolysis of procyanidin–prodelphinidin heteropolymers has also been proposed[138] in spectra recorded in acetone-d_6. The latter method is quantitative because the intensity of each proton signal is proportional to the concentration of the corresponding flavanol monomer or benzyl thioether in the solution whereas methods based on gas or liquid chromatography require identification and calibration of each released unit. Provided each unit can be individually quantified, depolymerization in the presence of nucleophilic agent gives access to the nature and average proportions of monomeric constituents of a given tannin or fraction and allows calculation of its mDP (mDP = ([upper & extension units]$_M$ + [end units]$_M$)/[end units]$_M$). It also allows the determination of tannin content by summing the concentrations of all released units. Another advantage of these methods is that they preserve substituents such as gallates[139] and glucosides[140] as well as A-type linkages.[107] mDP calculation must be adapted for proanthocyanidins containing A-type linkages which are resistant to cleavage: the released A-type dimers must then be counted as two units in the numerator and as only one in the denominator.[107]

Low yields due to important side-reactions have been reported for thiolysis[141] but this is probably due to the rather drastic conditions (24 h at 105 °C) used. Various parameters have been reported to alter the reaction yield. When carried out in 1% acetic acid in ethanol, increasing the temperature from 70 to 105 °C reduced the reaction time from 70 to 4.5 h and increased the yield from 56 to 63%.[34] The presence of 20% water in the reaction mixture resulted in 24–36% yield reduction. Replacement of acetic acid, the usual hydrolysing agent[87,127,134]) by hydrochloric acid (0.1 N) reduces the reaction time, minimizes degradation of the products and improves the yield.

The following experimental conditions are recommended: 50 μl of a proanthocyanidin solution (1 g l^{-1} in MeOH) are put together with an equal volume of a 5% solution of benzyl hydrosulfide in MeOH containing 0.2 N HCl into a glass ampoule. After sealing, the mixture is shaken and heated. Optimum yields for thiolysis of galloylated grape seed procyanidins are obtained by heating for 2 minutes at 90 °C. These conditions have been adapted for procyanidin oligomers (30 min at 40 °C) and polymers (10 min at 50 °C) from apple[74] and for mixed prodelphinidin–procyanidin polymers from grape skins (10 min at 60 °C).[91] Yields determined on a procyanidin trimer and fractions from grape seeds and apple comprised between 65% and 83%.[108,116] Rather important epimerization of the terminal units takes place under these conditions. It can be avoided by performing the reaction at room temperature; increased reaction times (16 h) are then required.[84,116]

Thiolysis is also useful for the analysis of derived tannins. Ethyl-linked oligomers resulting from acetaldehyde condensation of flavanols yield several adducts when submitted to acid-catalysed cleavage in the presence of ethanethiol, providing information on the position and stability of linkages in the original ethyl-linked compounds.[131,132] Thiolysis of a polymeric fraction from red wine released anthocyanins in addition to the expected tannin cleavage products, indicating that anthocyanins were originally linked through their C-6 or C-8 position in $T\text{-}A^+$ structures.[60] Benzyl thioether derivatives of anthocyanin–flavanol adducts[60] and of pyranoanthocyanin–flavanols[65] were also detected after thiolysis, indicating that proanthocyanidin oligomers and polymers were also involved in formation of such structures. However, some of the tannin-like structures present in wine (*e.g.* flavanol–ethyl–anthocyanins) are rather resistant to thiolysis, and were only partly cleaved when the reaction time was increased to 5 h (at 60 °C).

Thiolysis thus appears like a rather simple, sensitive and powerful tool for quantitation and characterization of tannins. However, its use is somewhat limited by the unpleasant odour and toxicity of the reagent, especially since preparation of rather large quantities of derivatives are needed for identification and calibration of the method. Differences in sensitivity to acid-catalysed cleavage often make it difficult to find optimum conditions, enabling cleavage of rather resistant molecules without degradation of the monomers released from the more labile species. Besides, proanthocyanidins based on 5-deoxyflavanol moieties are remarkably resistant to thiolysis[142] although in molecules containing both 5-hydroxy and 5-deoxy flavanols, linkages adjacent to the 5-hydroxy groups are cleaved.

Phloroglucinol is also commonly used as a trapping reagent. The reaction is usually performed in ethanol or methanol and its yield is much reduced in the presence of water.[134] Acetic or formic acid 0.5 M,[128] hydrochloric acid 0.1 N[134,137] and sulfur dioxide 0.5%[116] have been used as the acid catalyst. Yields obtained with phloroglucinol have been reported to be much lower than with benzylmercaptans,[134,143] due to the high instability of the phloroglucinol derivatives. Addition of ascorbic acid enabled these side reactions to be limited.[137] The following optimized procedure has been proposed: the proanthocyanidin extract ($5\ g\,l^{-1}$) is dissolved in a solution of 0.1 N HCl in methanol containing $50\ g\,l^{-1}$ of phloroglucinol. After 20 minutes of incubation at 50 °C, the reaction is stopped by dilution with five volumes of 40 mM aqueous sodium acetate. The yields obtained under these conditions on proanthocyanidin fractions from grape seeds and skins were equivalent to those obtained using thiolysis.[137] DPn values calculated for grape seed fractions after depolymerization in the presence of phloroglucinol using sulfur dioxide as the acid catalyst[116] were similar to those determined after thiolysis (Table 1).

NMR

Characteristic structural features and mDP values of a tannin extract can also be determined by NMR. The spectra are generally recorded in acetone-d_6. The first method proposed for mDP estimation was based on ratios of ^{13}C signal areas corresponding to the C-3 of the lower units (located between 65 and 68 ppm) and those of the extension units (at 72–73 ppm).[144] More recently, a ^{1}H NMR procedure based on the integration of the signals corresponding on one hand to all A-ring protons and, on the other hand, to H-4 protons of the terminal units has been developed and applied to epicatechin-based polymers extracted from cider apple.[108] However, the accuracy of both methods becomes lower as the DP increases due to imprecision in the integration of the terminal unit signal. The application of the ^{13}C technique is complicated by the presence of A-type linkages, due to downfield shift of the upper C-3 signal of the A-type subunits (68.5 ppm) which thus merges with the C-3 of the terminal flavanol. This results in overestimation of terminal units and thus underestimation of the mDP value.[90]

Additional information on the nature and proportions of constitutive units can be obtained from specific ^{13}C signals. For instance, the C-2 signals of *trans*-units (77 ppm) are well separated from those of *cis*-units (84 ppm) and have been integrated directly to calculate *cis/trans* ratios of extension units in a polymer.[11] The ^{13}C NMR spectra also yield information on the hydroxylation pattern of A and B rings and the presence of substituents. Glycosides are indicated by the presence of extra signals in the 60–80 ppm region and esters are indicated by resonance in the carbonyl region (δ 160–170).[9] The presence of a signal at 166 ppm and of a strong resonance at 110 ppm due to the two unsubstituted carbon signals of the pyrogallol ring can thus be attributed to galloyl groups.[145] Substitution of the C-3 hydroxyl group of extension units induces an upfield shift of the adjacent C-4 carbon (δ 34) compared to the C-4 of non galloylated units (δ 36) and slight modifications in the B-ring signals. Unsubstituted carbons of pyrogallol B rings (C-2′, C-6′) give a characteristic signal at 107 ppm which can be distinguished from that of the unsubstituted galloyl carbons (110 ppm) and those of C-2′, C-5′ and C-6′ of the catechol rings (115, 116 and 119 ppm, respectively). The proportions of prodelphinidins/procyanidins and extent of galloylation can thus be determined by integrating the relevant signals.[145] Another method is based on integration of the B-ring C-OH signals near 145 ppm (C-3′ and C-4′ at 144.5 ppm for procyanidins, C-3′ and C-5′ at 145.5 for prodelphinidins).[11,89] Nevertheless, interpretation of the above listed chemical shifts may prove difficult, and is unreliable in the case of complex polymers and especially tannin-like structures. Formal attribution of the signal requires the use of bidimension ^{3}H-^{1}H and ^{13}C-^{1}H NMR techniques. ^{13}C NMR spectrometry of mature grape skin extracts containing both proanthocyanidins and tannin-like pigments has enabled the detection of characteristic glucose and methoxy carbon signals, indicating the presence of anthocyanin moieties associated to proanthocyanidin polymers.[146]

Mass Spectrometry

Analysis of tannins by mass spectrometry (MS) gives access to the molecular weights of the different proanthocyanidin species present in a fraction or extract and provides important information on their constituent moieties.

The first MS techniques, such as electronic impact coupled to gas chromatography, are poorly suited to the analysis of tannins, because of their thermo-sensitivity and non-volatility in the native (non derivatized) form. Also, electronic impact yields only fragment ions which provide no information on the molecular weight of the original polymer. New alternative techniques allowing direct mass analysis directly or after liquid chromatography (LC) elution are a considerable improvement.

Coupling of MS with normal-phase or reversed-phase HPLC is used to identify oligomeric proanthocyanidins and tannin-like compounds, as discussed in Chapter 12. However, the increasing number of isomers and the poorer separation on the chromatographic systems as the degree of polymerisation increases impedes application of this technique to higher molecular weight species.

Direct mass spectrometry methods, involving soft ionization techniques, namely electrospray ionization (ESI),[59,90,107,147,148] matrix-assisted laser desorption ionization/time-of-flight (MALDI-TOF),[90,149,150] and fast atom bombardment (FAB)[149] have been successfully applied for the analysis of proanthocyanidins from various plant sources without prior separation by HPLC.

Analysis of apple extracts by direct ESI-MS[59,147] enabled the detection of signals corresponding to $[M-H]^-$ ions of procyanidin oligomers up to the octamer but also to doubly charged ($[M-2H]^{2-}$) and triply charged ($[M-3H]^{3-}$) ions for higher molecular weight species. Monocharged signals corresponding to a given DP (n) overlap with doubly charged signals of DP2n species and triply charged signals of DP3n species. Arguments to distinguish between mono-charged DPn, doubly charged DP2n and triply charged DP3n signals are provided by examination of the distance between the isotope peaks mainly due to the natural occurrence of carbon 13, as the spacing between the isotopic peaks is equal to one mass unit for mono-charged species, 0.5 mass units for doubly charged ions and 0.3 mass units for triply charged ions. Signals corresponding to doubly charged ions of uneven DP species or to triply charged species of DP $3x+1$ or $3x+2$ can thus be unambiguously attributed. Interpretation with higher charge states or several charge states in a single mass peak is limited by the resolution of the mass spectrometer.

ESI-MS analysis gave evidence of a complete series of B-type polymeric procyanidins from DP2 to DP17 in cider apple skin and pulp extracts. The results are consistent with the average degrees of polymerization determined by thiolysis degradation of the cider apple extracts (12.5 and 11.5, respectively in skin and pulp). ESI-MS analysis of a grape seed extract also allowed to detect a complete series of oligomers, from DP2 to DP9, as well as the corresponding monogalloylated and digalloylated species.[59] Analysis of grape seed tannins in the positive ion mode enabled to detect $[M+H]^+$ ions up to the galloylated pentamer.[151] A list of the procyanidin ion peaks detected in the negative ion mode in apple and grape seed extracts is given in Table 2. Analysis of a litchi tannin extract[107] showed numerous peaks which can be interpreted as the oligomeric procyanidins from DP2 to DP22, containing at least one A-type interflavanoid linkage. Procyanidins above DP15 can only be detected as triply charged ions, given the detection limit (2400 amu) of the ESI-MS apparatus. The average number of A-type bonds increased with the chain length, the predominant DP11[152] and DP17[107] species

Table 2 *Mass signals detected by ESI-MS in the negative ion mode in apple and grape seed procyanidin extracts*

DP	*c + ec*	*Galloyl*	$[M-H]^-$	$[M-2H]^{2-}$	$[M-3H]^{3-}$
1	1		289.0		
1	1	1	441.0		
2	2		577.0		
2	2	1	729.1		
2	2	2			
3	3		865.2		
3	3	1	1017.1		
4	4		1153.8		
4	4	1	1305.2		
5	5		1441.0	720.2	
5	5	1		796.3	
5	5	2		872.7	
6	6		1729.5	864.9	
6	6	1		940.2	
6	6	2		1017.1	
6	6	3		1092.9	
7	7		2017.5	1008.3	
7	7	1		1084.5	
7	7	2		1160.8	
7	7	3		1237.3	
8	8			1153.1	
8	8	1		1228.8	
8	8	2		1304.5	
8	8	3		1380.6	
9	9			1297.0	
9	9	1		1372.8	
9	9	2		1449.4	
10	10			1441.0	960.3
11	11			1585.0	1056.3
12	12			1729.5	
13	13			1873.2	1248.4
14	14			2017.5	1344.8
15	15			2161.2	1441.0
16	16			2305.5	1536.2
17	17				1633.2

DP: degree of polymerization; c + ec: number of catechin and epicatechin units.

presenting respectively three and four A-type linkages. Similar results were obtained with A-type procyanidin oligomers in cranberry extracts.[90]

Analysis of grape skin and wine polyphenols showed the presence of additional series of species. The 16 mass unit differences to signals given by grape seed procyanidins can be attributed to the presence of (epi)gallocatechin moieties in the mixed procyanidin–prodelphinidin oligomers.[152,153]

ESI-MS has also been applied to investigate acetaldehyde-induced condensation reactions in wine-like model solutions,[46,152,154] enabling to detect ethyl-linked catechin oligomers up to the hexamer as the mono-charged $[M-H]^-$ ions and signals corresponding to dimeric, trimeric and tetrameric species containing one anthocyanin and one, two or three flavanol units as the M^+ flavylium ions.[59,152,154] A tetrameric species containing two malvidin 3-glucoside units and two flavanol units was also found as the doubly charged M^{2+} ion at $m/z = 822$.[59,152]

Thus, ESI-MS analysis enables the detection of high molecular weight proanthocyanidins in plant extracts and the determination, for each mass signal, of the nature and proportion of constitutive units, degree of polymerization, type of interflavanoid bonds, and substituents (gallates). However, MS does not distinguish between isomers (2–3 *cis/trans* units, C-4–C-6/C-4–C-8 linkages, different sequences of constitutive units). Besides, isomers of A- and B-type procyanidins such as oxidation products[25] may be encountered in plant extracts, leading to misinterpretation of the mass spectral data. Further information on the sequence in oligomers containing units with different molecular weights can be obtained by fragmentation using increased orifice voltage in the classical ESI-Q mass spectrometers or more easily with the more recent ion trap (ESI-IT) technique.[59,153] However, application of this method is limited in the infusion mode as each peak signal usually corresponds to a mixture of different species showing different fragmentation patterns.

Although, for a given compound, the *m/z* signal intensity is proportional to the concentration, it depends on the capacity of the molecule to become ionized, which is greatly influenced both by the structure and by the environment. In particular, proanthocyanidins should be more easily negatively charged as the degree of polymerization and/or number of galloyl substituents increases. The detection of increasingly charged ions species as the molecular weight increases confirms this hypothesis but it is not known whether the charge density increases linearly with the DP. Therefore, the use of ESI-MS for quantitative purposes would require a calibration curve to be established for each compound. The presence of several ions formed from a single compound and variations in the response induced by the presence of other constituents further complicate quantitative interpretation of the signal intensity. The mass distribution observed when analysing successive oligomeric procyanidin fractions isolated from litchi pericarp perfectly matched the results of DPn determination by thiolysis.[107] However, due to the overlapping of signals arising from different chain lengths, the multiplicity of charge states for a given polymer and the increasing polydispersity of signals and species as the molecular weight increases, ESI-MS does not give access to the distribution of molecular weights in a polymeric fraction.

MALDI-TOF MS has been selected for its soft ionization energy, high ion transmission yield and tolerance towards contaminants[150]. It also has higher mass range than other mass techniques and usually produces only single ionization, unlike ESI-MS. The sample dissolved in organic solvents such as anhydrous tetrahydrofuran, acetone or methanol is mixed with the matrix and rapidly evaporated under a stream of nitrogen to ensure a homogeneous incorporation of the sample in the matrix. The presence of water should be avoided as it makes the sample difficult to evaporate, resulting in weak mass ion signals and poorer repeatability and reproducibility.[150] Among the numerous matrices tested, only *trans*-3-indolacetic acid (IAA) and 2,5 dihydroxybenzoic acid (DHB) are able to produce molecular ion peaks but the latter requires higher laser power.[149] IAA gave $[M - H]^-$ signals in the negative ion mode and mixtures of $[M + Na]^+$ and $[M + K]^+$ signals in the positive-ion mode. When silver trifluoracetic acid solution is added to the IAA matrix, the two series of alkali adduct ion peaks are replaced by silver ion adducts. This resulted in improved sensitivity and enabled detection of species up to the pentadecamer. Other authors in similar experiments concluded that DHB was the best matrix to analyse partly galloylated grape seed procyanidins by the reflectron mode as IAA tends to generate a high background noise.[150] The use of IAA in the linear mode allowed expansion of the range of detected masses, in spite of the poorer quality of the spectrum.[150] The ratio of 1.8 mg of the procyanidin sample dissolved in 100 μl of acetone–water (80/20, v/v) to 5 mg of the matrix t-IAA was found to give the most reproducible spectra with best signal to noise ratio.[155]

Both linear mode and reflectron mode MALDI-TOF experiments have been conducted. The latter gave superior MALDI spectra, with higher resolution both in the negative and in the positive ion modes, but the intensity was poor in the high-mass region .[149,150,155,156] This is attributed to the breakdown of the larger ions due to the post-acceleration process. Larger ions were detected in the linear mode because the flight path is shorter and the post-acceleration process is not involved.[149,150,156] Measurement in the linear mode at high acceleration voltage allowed intact detection of polymers up to DP 30 as $[M + K]^+$ ions in soybean seed coat extracts, using IAA as the matrix. Although MALDI-TOF cannot be used for quantitation, it has been postulated that the intensity of each peak signal in the mass spectrum can be used as a measure of relative abundance. DPn calculation based on this distribution for a proanthocyanidin oligomer fraction isolated from cranberry (DPn 4.7) was consistent with the DPn values obtained by ^{13}C spectrometry.[90] However, the charge density decreases as the molecular weight increases so that larger molecular weight tannins are more difficult to detect. Peak intensities from heptamer and up are too small to be accurately quantified.[150]

Fast atom bombardment mass spectrometry (FAB-MS) gives access to the molecular weight and some specific fragments.[157,158] It is mostly used for identification of isolated molecules but can also be applied directly on complex mixtures.[149] The 'magic bullet' matrix, consisting of dithiothreitol–dithioerythritol (5:1) is superior to glycerol for FAB-MS analysis of procyanidin oligomers.[158] Its application to apple extracts allowed the detection of a series of procyanidin oligomers up to the undecamer, with similar intensities in both ion modes.[149] Addition of alkali ions (Na^+, K^+) yielded the corresponding addition ions but with low intensity. Silver ions reduced the intensity of protonated ions but did not form any silver adduct. Fragmentation of proanthocyanidins, generating ions with signals identical to these of lower molecular weight species, renders the method unsuitable for accurate measurement of oligomer distribution.[149]

Gel Permeation Chromatography

GPC can supply the number average molecular weight (MWn) but since proanthocyanidins adsorb on all GPC supports, the most reliable techniques require derivatization of all hydroxy groups. This can be achieved by methylation[95] or, more easily, by acetylation.[84,159]

Acetylation is performed by incubating the tannin fractions in pyridin–acetic anhydride (1:1, v/v) at ambient temperature during 1 hour.[159] The reagent in excess is destroyed by addition of water and the precipitate containing the acetylated molecules recovered by centrifugation. After drying, the acetylated procyanidins are dissolved in tetrahydrofuran (THF) and analysed directly by GPC on two columns TSK G2500 Hxl and G 3000 Hxl (particle size 6 μm, 7.8 $\times$ 300 mm each, Tosohaas, Philadelphia, MA) connected in series. Acetylated proanthocyanidins are eluted isocratically with THF at 1 ml min^{-1} flow-rate and detected at 254 nm.

One of the limitations encountered when using GPC is the lack of high molecular weight proanthocyanidin standards to calibrate the separation. However, the molecular weights calculated for a series of acetylated flavanols, from monomers to tetramers, using the calibration curve established with a range of eleven polystyrene standards (MWn from 162 to 50 000) was very close to the actual molecular weights (500 for a fully acetylated monomer, 236 for an acetylated galloyl group), meaning that the regression equation established with polystyrene standard was valid for acetylated flavanols.[87,116] DPn values determined for a series of procyanidin fractions from grape seeds were similar to those obtained by depolymerization in the presence of nucleophilic agents.

Gel permeation chromatography thus provides satisfactory estimations of molecular weight and should give access to polymer size distribution when applied to a proanthocyanidin mixture. However, accurate conversion of these values to mDP requires prior determination of the tannin composition (B-ring hydroxylation pattern, substituents). For instance, the MWn of the acetylated derivatives is equal to mDP $\times$ 500, if the polymer is based on (epi)catechin units, or to mDP $\times$ (500 + (gall% $\times$ 236)), if it contains gall% of galloylated moieties.

4 Conclusion

Important progress has been achieved in the development of methods to analyse proanthocyanidins and tannin-like compounds in complex mixtures. Although classical proanthocyanidin reagents such as Porter's reagent, vanillin or DMCA can still be of use for the detection of proanthocyanidins in plant samples and/or for comparison of samples of identical composition, they appear rather inaccurate for characterization and quantification purposes. Determination of average composition should be based on NMR spectrometry and/or depolymerization in the presence of nucleophilic agents followed by HPLC analysis. Both types of method also give access to calculations of mDP values. However, application of the former to higher polymers is limited by poor integration of signals corresponding to end units, due to higher noise to signal ratios as the molecular weight increases. The latter methods appear more sensitive but require time-consuming purification of each released unit, for identification and calibration. Information on mass distribution can theoretically be obtained by gel permeation and mass spectrometry. However, the lack of appropriate standards to calibrate or validate these methods, due to poor resolution of separation techniques for higher molecular weight and/or highly polydisperse compounds, remains a major problem. Besides, difficulties in ionizing higher molecular weight proanthocyanidins (in MALDI-TOF) or in the interpretation of complex and overlapping signals contributed by multiply charged ions (in ESI-MS) has so far impeded accurate analysis of polymeric fractions.

References

1. A. Hagerman, *Polyphenols Actualités*, 2001, **21**, 19.
2. E. Bate-Smith and T. Swain, in *Comparative Biochemistry*, eds H. Mason and A. Florkin, Academic Press: New York, 1962, Vol. 3, p. 764.
3. E. Haslam, *Practical Polyphenolics. From Structure to Molecular Recognition and Physiological Action*, Cambridge University Press, Cambridge, 1998.
4. E. Bate-Smith, *Food*, 1954, **23**, 124.
5. E. Haslam and T. H. Lilley, *Crit. Rev. Food Sci. Nutr.*, 1988, **27**, 1.
6. J. Gramshaw, *J. Inst. Brew.*, 1967, **73**, 455.
7. L. G. Butler, in *Plant Polyphenols*, eds R. W. Hemingway and P. E. Laks, Plenum Press, New York, 1992, p. 693.
8. C. Santos-Buelga and A. Scalbert, *J. Sci. Food Agric.*, 2000, **80**, 1094.
9. L. J. Porter, in *The Flavonoids: Advances in Research since 1980*, ed. J. B. Harborne, Chapman and Hall, New York, 1988, p. 21.
10. L. J. Porter, in *The Flavonoids: Advances in Research since 1986*, ed. J. B. Harborne, Chapman and Hall, New York, 1994, p. 23.
11. Z. Czochanska, L. Y. Foo, R. H. Newman, and J. L. Porter, *J. Chem. Soc., Perkin Trans. 1*, 1980, 2278.
12. L. J. Porter, R. H. Newman, L. Y. Foo, H. Wong and R. W. Hemingway, *J. Chem. Soc., Perkin Trans. 1*, 1982, 1217.
13. S. Guyot, N. Marnet and J. F. Drilleau, *J. Agric. Food Chem.*, 2001, **49**, 14.
14. M. E. Harbowy and D. A. Balentine, *Crit. Rev. Plant Sci.*, 1997, **16**, 415.
15. T. C. Somers, *Phytochemistry*, 1971, **10**, 2175.
16. E. Haslam, *Phytochemistry*, 1980, **19**, 2577.
17. P. Ribéreau-Gayon, *The Anthocyanins of Grapes and Wines*, Academic Press, New York, London, 1982.
18. D. Balentine, S. A. Wiserman and L. C. M. Bouwens, *Crit. Rev. Food Sci. Nutr.*, 1997, **37**, 693.
19. W. S. Pierpoint *Rothamsted Report for 1970*. 1970, Part 2, p. 199.
20. E. A. H. Roberts, *J. Sci. Food Agric.*, 1958, **9**, 212.
21. E. Roberts and R. Smith, *Analyst*, 1961, **86**, 94.
22. D. Coxon, A. Holmes and W. Ollis, *Tetrahedron Lett.*, 1970, **60**, 5247.
23. G. Nonaka, O. Kawahara and I. Nishioka, *Chem. Pharm. Bull.*, 1983, **31**, 3906.

24. F. Hashimoto, G. Nonaka and I. Nishioka, *Chem. Pharm. Bull.*, 1988, **36**, 1676.
25. S. Guyot, J. Vercauteren and V. Cheynier, *Phytochemistry*, 1996, **42**, 1279.
26. Y. Takino, H. Imagawa, H. Horikawa and A. Tanaka, *Agric. Biol Chem.*, 1964, **28**, 64.
27. A. Davis, Y. Cai and A. Davies. *Mag. Reson. Chem.*, 1995, **33**, 549.
28. A. Davis, J. Lewis, Y. Cai, C. Powell, A. Davies, J. Wilkins, P. Pudney and M. Clifford, *Phytochemistry*, 1997, **46**, 1397.
29. T. Tanaka, K. Inoue, Y. Betsumiya, C. Mine and I. Kouno, *J. Agric. Food Chem.*, 2001, **49**, 5785.
30. R. G. Bailey, H. E. Nursten and I. McDowel, *J. Sci. Food Agric.*, 1992, **59**, 365.
31. T. Ozawa, M. Kataoka, K. Morikawa and O. Negishi, *Biosci. Biotechnol. Biochem.*, 1996, **60**, 2023.
32. G. W. Sanderson, J. E. Berkowitz, H. Co and H. N. Graham, *J. Food Sci.*, 1972, **37**, 399.
33. S. C. Opie, A. Robertson and M. N. Clifford, *J. Sci. Food Agric.*, 1990, **50**, 547.
34. A. Finger, *J. Sci. Food Agric.*, 1994, **66**, 293.
35. N. Subramanian, P. Venkatesh, S. Ganguli and V. P. Sinkar, *J. Agric. Food Chem.*, 1999, **47**, 2571.
36. V. Cheynier, I. Hidalgo-Arellano, J. M. Souquet and M. Moutounet, *Am. J. Enol. Vitic.*, 1997, **48**, 225.
37. C. F. Timberlake and P. Bridle, *Am. J. Enol. Vitic.*, 1976, **27**, 97.
38. J. Bakker, A. Pinicelli and P. Bridle, *Vitis,* 1993, **32,** 111.
39. J. D. Baranowski and C. W. Nagel, *J. Food Sci.*, 1983, **48**, 419.
40. C. Dallas, J. M. Ricardo-da-Silva and O. Laureano, *J. Sci. Food Agric.*, 1996, **70**, 493.
41. M. Escribano-Bailon, O. Dangles and R. Brouillard, *Phytochemistry,* 1996, **41**, 1583.
42. N. E. Es-Safi, C. Le Guerneve, B. Labarbe, H. Fulcrand, V. Cheynier and M. Moutounet, *Tetrahedron Lett.,* 1999, **40**, 5869.
43. N. E. Es-Safi, C. Le Guerneve, V. Cheynier and M. Moutounet *J. Agric. Food Chem.*, 2000, **48**, 4233.
44. E. M. Francia-Aricha, M. T. Guerra, J. C. Rivas-Gonzalo and C. Santos-Buelga, *J. Agric. Food Chem.*, 1997, **45**, 2262.
45. H. Fulcrand, P. J. Cameira dos Santos, P. Sarni-Manchado, V. Cheynier and J. Fabre-Bonvin *J. Chem. Soc., Perkin Trans. 1*, 1996, 735.
46. H. Fulcrand, T. Doco, N. E. Es-Safi, V. Cheynier and M. Moutounet, *J. Chromatogr. A.*, 1996, **752**, 85.
47. N. E. Es-Safi, H. Fulcrand, V. Cheynier and M. Moutounet, *J. Agric. Food Chem.*, 1999, **47**, 2096.
48. H. Fulcrand, V. Cheynier, J. Oszmianski and M. Moutounet, *Phytochemistry*, 1997, **46**, 223.
49. H. Fulcrand, C. Benabdeljalil, J. Rigaud, V. Cheynier and M. Moutounet, *Phytochemistry*, 1998, **47**, 1401.

50. C. Benabdeljalil, V. Cheynier, H. Fulcrand, A. Hakiki, M. Mosaddak and M. Moutounet, *Sci. Aliments,* 2000, **20**, 203.
51. H. Liao, Y. Cai and E. Haslam, *J. Sci. Food Agric.,* 1992, **59**, 299.
52. M. Mirabel C. Saucier, C. Guerra and Y. Glories, *Am. J. Enol. Vitic.*, 1999, **50**, 211.
53. J. Oszmianski, V. Cheynier and M. Moutounet, *J. Agric. Food Chem.*, 1996, **44**, 1712.
54. J. C. Rivas-Gonzalo, S. Bravo-Haro and C. Santos-Buelga, *J. Agric. Food Chem.*, 1995, **43,** 1444.
55. C. Santos-Buelga, E. Francia-Aricha, S. Pascual-Teresa and J. C. Rivas-Gonzalo, *Zeitschr. Lebensm. Unters. Forsch.,* 1999, **209**, 411.
56. P. Sarni, H. Fulcrand, V. Souillol, J. M. Souquet and V. Cheynier, *J. Sci. Food Agric,* 1995, **69,** 385.
57. J. Bakker and C. F. Timberlake, *J. Agric. Food Chem.*, 1997, **45**, 35.
58. P. J. Cameira dos Santos, J. M. Brillouet, V. Cheynier and M. Moutounet, *J. Sci. Food Agric.,* 1996, **70**, 204.
59. V. Cheynier, T. Doco, H. Fulcrand, S. Guyot, E. Le Roux, J. M. Souquet, J. Rigaud and M. Moutounet, *Analusis*, 1997, **25**, M32.
60. S. Remy, H. Fulcrand, B. Labarbe, V. Cheynier and M. Moutounet, *J. Sci. Food Agric.,* 2000, **80**, 745.
61. N. Mateus, A. M. S. Silva, C. Santos-Buelga, J. C. Rivas-Gonzalo and V. de Freitas, *J. Agric. Food Chem.*, 2002, **50**, 2110.
62. C. Dallas, J. M. Ricardo-da-Silva and O. Laureano, *J. Agric. Food Chem.,* 1996, **44**, 2402.
63. N. E. Es-Safi, H. Fulcrand, V. Cheynier, M. Moutounet, M. Hmamouchi and E.M. Essassi, in *Polyphenols Communication 96*, eds J. Vercauteren, C. Cheze, M. Dumon and J. Weber, Groupe Polyphenols, Bordeaux, 1996, Vol. 2, p. 279.
64. H. L. Wildenradt and V. L. Singleton, *Am. J. Enol. Vitic.,* 1974, **25**, 119.
65. V. Atanasova, H. Fulcrand, V. Cheynier and M. Moutounet, *Anal. Chim. Acta,* 2002, **458**, 15.
66. N. E. Es-Safi, C. Le Guerneve, H. Fulcrand, V. Cheynier and M. Moutounet, *J. Agric. Food Chem.,* 1999, **47**, 5211.
67. N. E. Es-Safi, V. Cheynier and M. Moutounet, *J. Agric. Food Chem.,* 2000, **48**, 5946.
68. L. Jurd, *Am. J. Enol. Vitic.,* 1969, **20**, 197.
69. P. B. Bishop and C. W. Nagel *J. Agric. Food Chem.,* 1984, **32,** 1022.
70. L. Balas and J. Vercauteren, *Magn. Reson. Chem.,* 1994, **32**, 386.
71. C. Saucier, I. Pianet, M. Laguerre, D. Roux and Y. Glories, *J Magn. Reson. Anal.*, 1996, 237.
72. R. Julkunen-Tiitto and S. Sorsa, *J. Chem. Ecol.*, 2001, **27**, 779.
73. S. J. Cork and A. Krockenerger, *J. Chem. Ecol.,* 1991, **17**, 123.
74. S. Guyot, N. Marnet, D. Laraba, P. Sanoner and J. F. Drilleau, *J. Agric. Food Chem.*, 1998, **46**, 1698.
75. D. Cattell and H. Nursten, *Phytochemistry,* 1977, **16,** 1269.
76. L. Foo and J. Karchesy, *Phytochemistry*, 1989, **28**, 1743.

77. M. T. Escribano-Bailon, Y. Guttierez-Fernandez, J. C. Rivas-Gonzalo and C. Santos-Buelga, *J. Agric. Food Chem.*, 1992, **40**, 1794.
78. I. McMurrough, D. Madigan and M. R. Smyth, *J. Agric. Food Chem.*, 1996, **44**, 1731.
79. J. Delcour and D. Janssens de Varebeke, *J. Inst. Brew.*, 1985, **91**, 37.
80. B. Sun, M. Spranger and J. M. Ricardo da Silva, in *Polyphenols Communication 96*, eds J. Vercauteren, C. Cheze, M. Dumon and J. Weber, Groupe Polyphenols, Bordeaux, 1996, Vol. 1, p. 169.
81. A. E. Hagerman and L. G. Butler, *J. Agric. Food Chem.*, 1980, **28**, 947.
82. J. K. Nitao, B. A. Birr, M. G. Nair, D. A. Herms and W. J. Mattson, *J. Agric. Food Chem.*, 2001, **49**, 2207.
83. L. Foo and L. Porter, *J. Sci. Food Agric.*, 1981, **32**, 711.
84. R. Hemingway and G. J. McGraw, *Wood Chem. Technol.*, 1983, **3**, 421.
85. L. Porter, in *Methods in Plant Biochemistry: Plant Phenolics*, ed. J. B. Harborne, Academic Press, London, 1989, Vol. 1, p. 389.
86. S. Peng and C. Jay-Allemand, *J. Chem. Ecol.*, 1991, **17**, 123.
87. C. Prieur, J. Rigaud, V. Cheynier and M. Moutounet, *Phytochemistry*, 1994, **36**, 781.
88. M. P. Bradshaw, P. D. Prenzler and G. R. Scollary, *J. Agric. Food Chem.*, 2001, **49**, 934.
89. S. Ossipova, V. Ossipov, E. Haukioja, J. Loponen and K. Pihlaja, *Phytochem. Anal.*, 2001, **12**, 128.
90. L. Y. Foo, Y. Lu, A. B. Howell and N. Vorsa, *Phytochemistry*, 2000, **54**, 173.
91. J. M. Souquet, V. Cheynier, F. Brossaud and M. Moutounet, *Phytochemistry*, 1996, **43**, 509.
92. A. Degenhardt, U. H. Engelhardt, A. S. Wendt and P. Winterhalter, *J. Agric. Food Chem.*, 2000, **48**, 5200.
93. B. Sun, C. Leandro, J. M. Ricardo da Silva and I. Spranger, *J. Agric. Food Chem.*, 1998, **46,** 1390.
94. S. Vidal, D. Cartalade, J. M. Souquet, H. Fulcrand and V. Cheynier *J. Agric. Food Chem,* 2002, **50**, 2261.
95. A. G. G. Lea and C. F. Timberlake, *J. Sci. Food Agric.*, 1974, **25**, 1537.
96. G. Derdelinckx and J. Jerumanis, *J. Chromatogr.*, 1984, **285**, 231.
97. J. M. Ricardo da Silva, J. Rigaud, V. Cheynier, A. Cheminat and M. Moutounet, *Phytochemistry,* 1991, **30**, 1259.
98. Y. S. Bae, L. Y. Foo and J. J. Karchesy, *Holzforschung,* 1994, **48**, 4.
99. A. Yanagida, T. Kanda, T. Shoji, M. Ohnishi-Kameyama and T. Nagata, *J. Chromatogr. A*, 1999, **855,** 181.
100. T. Shoji, A. Yanagida and T. Kanda, *J. Agric. Food Chem.*, 1999, **47**, 2885.
101. C. Powell, M. N. Clifford, S. C. Opie, M. A. Ford, A. Robertson and C. L. Gibson, *J. Sci. Food Agric.*, 1992, **63**, 77.
102. M. Clifford, and C. Powell, in *Polyphenols Comunications 96*, eds J. Vercauteren, C. Cheze, M. Dumon and J. Weber, Groupe Polyphenols: Bordeaux, 1996, Vol. 1, p. 217.
103. A. G. H. Lea, *J. Sci. Food Agric.*, 1978, **29**, 471.

104. J. Rigaud, M. T. Escribano-Bailon, C. Prieur, J. M. Souquet and V. Cheynier, *J. Chromatogr. A*, 1993, **654**, 255.
105. J. F. Hammerstone, S. L. Lazarus, A. E. Mitchell, R. Rucker and H. H. Smith, *J. Agric. Food Chem.*, 1999, **47**, 490.
106. V. Cheynier, J. M. Souquet, E. Le Roux, S. Guyot and J. Rigaud, *Methods Enzymol., 1999,* **299**, 178.
107. E. Le Roux, T. Doco, P. Sarni, Y. Lozano and V. Cheynier, *Phytochemistry,* 1988, **48,** 1251.
108. S. Guyot, C. Le Guernevé, N. Marnet and J. F. Drilleau, in *Plant Polyphenols 2, Chemistry, Biology, Pharmacology, Ecology*, eds G. G. Gross, R. W. Hemingway and T. Yoshida, Kluwer Academic/Plenum Publishers, New York, 1999, p. 211.
109. Z. Peng, Y. Hayasaka, P. C. Iland, M. Sefton, P. Hoj and E. J. Waters, *J. Agric. Food Chem.,* 2001, **49**, 26.
110. A. G. H. Lea, P. Bridle, C. F. Timberlake and V. L. Singleton, *Am. J. Enol. Vitic.,* 1979, **30**, 289.
111. A. Degenhardt, U. H. Engelhardt, C. Lakenbrink and P. Winterhalter, *J. Agric. Food Chem.*, 2000, **48**, 3425.
112. A. Degenhardt, S. Hofmann, H. Knapp and P. Winterhalter, *J. Agric. Food Chem.*, 2000, **48**, 5812.
113. B. Labarbe, V. Cheynier, F. Brossaud, J. M. Souquet and M. Moutounet, *J. Agric. Food Chem.,* 1999, **47**, 2719.
114. C. Saucier, M. Mirabel, F. Daviaud, A. Longieras and Y. Glories, *J. Agric. Food Chem.*, 2001, **49**, 5732.
115. L. J. Porter, L. N. Hrstich and B. G. Chan, *Phytochemistry,* 1986, **25**, 223.
116. V. Cheynier, B. Labarbe and M. Moutounet, *Methods Enzymol.*, 2001, **335**, 82.
117. J. L. Goldstein and T. Swain, *Phytochemistry*, 1963, **2**, 371.
118. I. McMurrough and McDowell *Anal. Biochem.,* 1978, **91**, 92.
119. D. Treutter, C. Santos-Buelga, M. Gutmann and H. Kolodziej, *J. Chromatogr. A*, 1994, **667**, 290.
120. N. Vivas, Y. Glories, L. Lagune, C. Saucier and M. Augustin, *J. Int. Sci. Vigne Vin.*, 1994, **28**, 319.
121. R. B. Broadhurst and W. T. Jones, *J. Sci. Food Agric.,* 1978, **29**, 788.
122. L. G. Butler, M. L. Price and J. E. Brotherton *J. Agric. Food Chem.*, 1982, **30**, 1087.
123. M. Price, S. Van Scoyoc and L. G. Butler, *J. Agric. Food Chem.*, 1978, **26**, 1214.
124. C. W. Nagel and Y. Glories, *Am. J. Enol. Vitic.,* 1991, **42**, 364.
125. S. K. Sarkar and R. E. Norwarth, *J. Agric. Food Chem.*, 1976, **24**, 317.
126. P. Ribéreau-Gayon, and E. Stonestreet, *Chim. Anal.*, 1966, **48**, 188.
127. R. S. Thompson, D. Jacques, E. Haslam and D. J. N. Tanner, *J. Chem. Soc., Perkin Trans. 1*, 1972, 1387.
128. T. Matsuo, K. Tamaru and S. Itoo, *Agric. Biol. Chem.,* 1984, **48**, 1199.
129. L. Y. Foo and L. Porter, *J. Chem. Soc., Perkin Trans. 1*, 1978, 1186.

130. M. R. Koupai-Abyazani, A. D. Muir, B. A. Bohm, G. H. N. Towers and M. Y. Gruber, *Phytochemistry,* 1993, **34**, 113.
131. T. Tanaka, R. Takahashi, I. Kouno and G. I. Nonaka, *J. Chem. Soc., Perkin Trans. 1*, 1994, 3013.
132. N. E. Es-Safi, H. Fulcrand, V. Cheynier and M. Moutounet *J. Agric. Food Chem.*, 1999, **47**, 2088.
133. Z. Czochanska, L. Y. Foo and L. Porter, *Phytochemistry*, 1979, **18,** 1819.
134. S. Matthews, I. Mila, A. Scalbert, B. Pollet, C. Lapierre, C. Herve du Penhoat, C. Rolando and D. M. X. Donnelly, *J. Agric. Food Chem.*, 1997, **45**, 1195.
135. Z. Shen, E. Haslam, C. P. Falshaw and M. J. Begley, *Phytochemistry,* 1986, **25**, 2629.
136. J. Rigaud, X. Perez-Ilzarbe, J. M. Ricardo da Silva and V. Cheynier, *J. Chromatogr.*, 1991, **540**, 401.
137. J. Kennedy and G. P. Jones, *J. Agric. Food Chem.*, 2001, **49**, 1740.
138. Y. Cai, F. J. Evans, M. F. Roberts, J. D. Phillipson, M. H. Zenk and Y. Y. Glebas, *Phytochemistry*, 1991, **30**, 2033.
139. G. Nonaka, F. Sakai and I. Nishioka, *Phytochemistry*, 1984, **23**, 1753.
140. S. Morimoto, G. I. Nonaka and I. Nishioka, *Chem. Pharm. Bull.,* 1986, **34**, 633.
141. G. W. McGraw, P. J. Steynberg and R. W. Hemingway, *Tetrahedron Lett.*, 1993, **34**, 987.
142. D. Ferreira, E. Brandt, H. Van Rensburf, R. Bekker and R. Nel, in *Plant Polyphenols 2. Chemistry, Biology, Pharmacology, Ecology*, eds G. G. Gross, R. W. Hemingway and T. Yoshida, Kluwer Academic/Plenum Publishers New York, 1999, p. 163.
143. R. K. Gupta and E. Haslam, *J. Chem. Soc. Prekin Trans. 1*, 1978, 892.
144. Z. Czochanska, L. Y. Foo, R. H. Newman, L. J. Porter, W. A. Thomas and W. T. Jones, *J. Chem. Soc., Chem. Commun.*, 1979, 375.
145. F. Balza, Z. Abramowski, G. H. N. Towers and P. Wiriyachitra, *Phytochemistry*, 1989, **28**, 1827.
146. J. A. Kennedy, Y. Hayasaka, S. Vidal, E. J. Waters and G. P. Jones, *J. Agric. Food Chem.*, 2001, **49**, 5348.
147. S. Guyot, T. Doco, J. M. Souquet, M. Moutounet and J. F. Drilleau, *Phytochemistry.* 1997, **44**, 351.
148. M. L. Porter, C. G. Krueger, D. A. Wiebe, D. G. Cunningham and J. D. Reed, *J. Sci. Food Agric.,* 2001, **81**, 1306.
149. M. Ohnishi-Kameyama, A. Yanagida, T. Kanda and T. Nagata, *Rapid Commun. Mass Spectrom.*, 1997, **11**, 31.
150. Y. Yang and M. Chien, *J. Agric. Food Chem.*, 2000, **48**, 3990.
151. B. Gabetta, N. Fuzzati, A. Griffini, E. Lolla, R. Pace, T. Ruffillli and F. Peterlongo, *Fitoterapia*, 2000, **71**, 162.
152. H. Fulcrand, S. Guyot, E. Le Roux, S. Remy, J. M. Souquet, T. Doco and V. Cheynier, in *Plant Polyphenols 2. Chemistry, Biology, Pharmacology, Ecology*, eds G. G. Gross, R. W. Hemingway, T. Yoshida and S. J. Branham, Kluwer Academic/Plenum Publishers, New York, 1999, p. 223.

153. H. Fulcrand, S. Remy, J. M. Souquet, V. Cheynier and M. Moutounet, *J. Agric. Food Chem.*, 1999, **47**, 1023.
154. C. Guerra, C. Saucier, G. Bourgeois, C. Vitry, O. Busto and Y. Glories, *Proc. In Vino Anal. Sci.*, 1997, 124.
155. C. G. Krueger, N. C. Dopke, P. M. Treichel, J. Folts and J. D. Reed, *J. Agric. Food Chem.,* 2000, **48**, 1663.
156. Y. Takahata, M. Ohnishi-Kameyama, S. Furuta, M. Takahashi and I. Suda, *J. Agric. Food Chem.* 2001, **49**, 5843.
157. R. Self, J. Eagles, G. C. Galetti and I. Mueller-Harvey, *Biomed. Environ. Mass Spectrom.*, 1986, **13**, 449.
158. D. Barofsky, in *Chemistry and Significance of Condensed Tannins*, eds R. Hemingway and J. Karchesy, Plenum Press, New York, 1988, p. 175.
159. V. Williams, L. J. Porter and R. W. Hemingway, *Phytochemistry.* 1983, **22**, 569.

CHAPTER 14

The Analysis and Characterization of Chlorogenic Acids and Other Cinnamates

MICHAEL N. CLIFFORD

1 Introduction

Cinnamic acids are a series of *trans* phenyl-3-propenoic acids, differing in their ring substitution. These compounds are widely distributed as conjugates in plant material including many foods and beverages. The cinnamates are rarely found as free acids in unprocessed plant material, but they may be analysed after deconjugation and the contents be reported as free acids rather than as the original conjugates. The commonest is caffeic acid (3,4-dihydroxycinnamic acid): ferulic (3-methoxy-4-hydroxy), sinapic (3,5-dimethoxy-4-hydroxy) and *p*-coumaric (4-hydroxy) also are widespread and a few others occur much less frequently, for example 2-hydroxycinnamic and 3,4-dimethoxycinnamic which are unusual also in that they occur naturally as free acids in commercial green coffee beans.[1]

The best known conjugate is 5-caffeoylquinic acid (IUPAC numbering[2]), commonly referred to as chlorogenic acid (CGA), a term which should now be used only to refer to the family of related quinic acid conjugates. There have been several substantial reviews covering the chemistry, complex trivial nomenclature and quantification of CGA,[3,4] and their significance for the consumer[5] as well as the dietary burden of cinnamates in general.[6–8] This review focuses on analytical and physical aspects.

2 The Cinnamate Conjugates

The cinnamate conjugates may be divided in to two major categories.

The Chlorogenic Acids

Classically, chlorogenic acids (CGA) are a family of esters formed between certain *trans* cinnamic acids and (−)-quinic acid (1L-1(OH),3,4/5-tetrahydroxycyclohexanecarboxylic acid).[2] This isomer of quinic acid may also be referred to as (1*R*:3*R*:4*S*:5*R*)-3/1,4,5-tetrahydroxycyclohexane-1-carboxylic acid. Until recently, all characterized CGA contained quinic acid in the chair conformation with axial hydroxyls on carbons 1 and 3, and equatorial hydroxyls on carbons 4 and 5. Using the preferred IUPAC numbering[2] as illustrated in Figure 1, the commonest individual chlorogenic acid (CGA) is 5-*O*-caffeoylquinic acid (5-CQA). It is usually the only one commercially available and is still often called chlorogenic acid or 3-CQA (pre-IUPAC numbering) but this use of both terms should be discouraged.

These classic CGA may be subdivided by the identity, number and position of the acyl residues. The following subgroups are known:

- the relatively widespread mono-esters of caffeic acid [caffeoylquinic acids (CQA)], *p*-coumaric acid *p*-coumaroylquinic acids (*p*CoQA), and ferulic acid feruloylquinic acids (FQA);
- di-esters (diCQA), tri-esters (triCQA), and the single tetra-ester of caffeic acid (tetraCQA), the latter two groups being characteristic of Asteraceae;[9–13]
- di-esters of ferulic acid (diFQA) in *Brachiaria* spp. (Poaceae);[14]
- mixed di-esters of caffeic and ferulic acid [caffeoylferuloylquinic acids (CFQA)] which are characteristic of robusta coffee (*Coffea canephora*, Rubiaceae),[15–18] caffeic and sinapic acid [caffeoylsinapoylquinic acids (CSiQA)] as found in *Gardenia jasminoides* (Rubiaceae)[19] or ferulic and *p*-coumaric acid *p*-coumaroylferuloylquinic acids (*p*CoFQA) in *Brachiaria* spp. (Poaceae);[14]
- mixed esters involving various permutations of between one and three residues of caffeic acid with one or two residues of a dibasic aliphatic acid (*e.g.* glutaric, oxalic, succinic) which are characteristic of Asteraceae,[20–23] or various permutations of caffeic, sinapic and 3-hydroxy-3-methylglutaric acid in *Gardenia jasminoides* (Rubiaceae).[19] A mixed ester containing one residue of isobutyric acid and two residues of caffeic acid has been reported in *Buphthalmum salicifolium* (Asteraceae).[24]

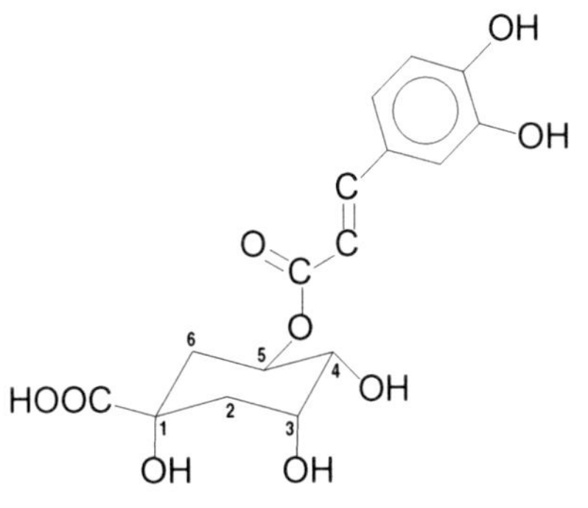

Figure 1 *The structure of 5-caffeoylquinic acid showing the preferred IUPAC numbering for the quinic acid residue*

It is convenient to extend this list of classic CGA to include the galloyl conjugates of quinic acid and the cinnamoyl conjugates of quinic acid derivatives such as shikimic acid,[25–28] methyl or butyl quinate,[29,30] 4-deoxyquinic acid[31] and the recently characterized *muco*-quinic acid[32] and *iso*-quinic acid.[33] These isomers differ from the norm described above by having respectively an equatorial hydroxyl on carbon 3 or an axial hydroxyl on carbon 4 (see Figure 2). The *p*CoFQA and diFQA of *Brachiaria* spp.[14] are so far unique in containing a 'normal' quinic acid residue but in the uncommon chair form with the carboxyl axial. Consequently, the substituents at carbons 1 and 3 are equatorial and the free hydroxyls at carbons 4 and 5 are axial.[14] This chair form is encountered also in the chlorogenic lactones found in roasted coffee (see later).

Other Cinnamate Conjugates

Cinnamic acids may be conjugated to many molecules other than quinic acid and its close relatives, although sometimes as very minor components of limited distribution. Further information on these less common conjugates is available in the previous reviews.[7,8] This second category includes:

- esters of other hydroxy acids, particularly α-hydroxyhydrocaffeic (rosmarinic), malic and tartaric, but including galactaric, glucaric, gluconic, hydroxycitric, methoxyaldaric, phenylpyruvic and tartronic;
- amides of amino compounds including aromatic amino acids, choline, anthranilic acids and diamines;
- esters of polysaccharides, simple sugars, sugar alcohols including glycerol and *myo*-inositol and glycosides including those of anthocyanins, flavonols and diterpenes;
- esters of lipids including alkanols, alkandiols, ω-hydroxy-fatty acids and sterols; and
- glycosides.

Cinnamate Transformation Products

With the notable exceptions of coffee roasting and wine making, cinnamate behaviour during food preparation and processing has been relatively little

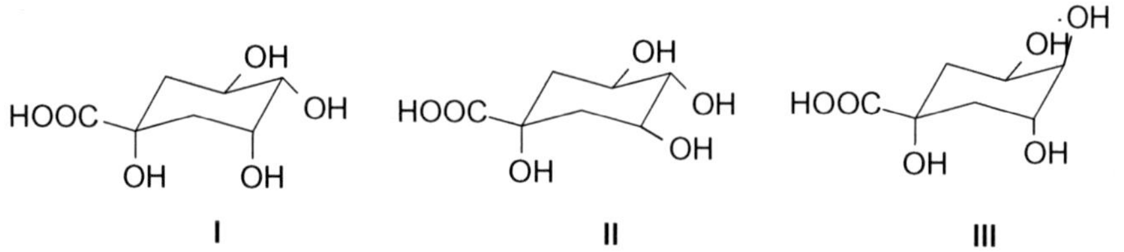

Figure 2 *The structures of the various quinic acids associated with the naturally occurring chlorogenic acids. I = quinic acid; II = muco-quinic acid; III = iso-quinic acid*

studied. Nevertheless, these studies have led to the characterization of several novel cinnamate transformation products that also deserve brief mention.

During the manufacture of black tea, the gallic acid residue of theogallin (5-GQA) has been reported to interact with the B-ring of EGCG and is partially transformed to a 2,2′-linked pale-brown adduct known as theogallinin.[34] However, Harbowy and Balentine,[35] citing this report from Hashimoto *et al.*,[34] describe the theogallinin as a benztropolone derived from the interaction of (−)-epicatechin and 5-GQA.

Coffee roasting causes a progressive destruction and transformation of CGA with some 8–10% being lost for every 1% loss of dry matter. During the early stages, CGA isomerization occurs accompanied by partial hydrolysis yielding quinic acid and the various cinnamic acids.[4,5] The released cinnamates may be decarboxylated, degraded to a range of simple phenols and further transformed. CGA in which position 5 is not substituted may lactonize forming caffeoyl and feruloyl-1,5-γ-quinides referred to collectively as chlorogenic lactones (CGL).[36] Free quinic acid epimerizes extensively forming the eight possible isomers of quinic acid, the eight 1,5-γ-quinides and the four 1,4-δ-quinides. Several of these exceed 0.1% in dark roast coffees,[37–40] but it is not known whether roasted coffee contains any epimeric CGA, or CGL other than the 1,5-γ-quinides referred to above. These lactones (4-CQL, and particularly 3-CQL) are rapidly hydrolysed to 4-CQA and 3-CQA. This hydrolysis occurs during domestic coffee brewing and could be an unwanted feature of some analytical extraction procedures.[41]

2-*S*-Glutathionyl-caftaric acid, its *o*-quinone and the *o*-quinone of caftaric acid have been found in grape must and white wines.[42] Cinnamic acid decarboxylation to the corresponding vinyl-phenol followed by cycloaddition to anthocyanins (*e.g.* malvidin) results in novel pigments characterisitc of matured red wines and ports.[36,43,44]

3 Analysis and Characterization

Structure and Hydrophobicity

Cinnamic acids are comparatively poorly soluble in the common solvents but the associated conjugates are generally soluble in water, aqueous alcohols and aqueous acetonitrile. The quinic acid carboxyl in its usual equatorial conformation has a p*K* value in the range 3.40–3.60 and at low pH values CGA will partition into ethyl acetate. For the classic quinic acid conjugates, greater solubility in water is associated with the presence of a greater number of more easily solvatable free equatorial hydroxyl groups. The sterically hindered free axial hydroxyl groups solvate less readily and impart a hydrophobic character to the molecule, as do the acylating residues.

Accordingly, the 1-acyl and 3-acyl isomers with two free equatorial hydroxyl groups and one free axial hydroxyl group are much more soluble in water (one order of magnitude) than the 4-acyl and 5-acyl isomers (one free equatorial and two free axial hydroxyl groups). All mono-acyl derivatives are more water-soluble than any of the diacyl quinic acids. Of these, 1,3-diCQA (two free equatorial hydroxyl groups) is much more soluble than 4,5-diCQA (with two free axial hydroxyl groups). 3,4-DiCQA is more hydrophilic than 3,5-diCQA. These structural features also govern the retention time of such compounds on reversed-phase HPLC packings,[17,45] and this aspect is discussed more fully in the section below dealing with HPLC.

Extraction from Plant Tissues

Efficient extraction from solid samples requires short diffusion pathways and large items must be ground or homogenized. For efficient extraction from coffee beans a particle size below 0.5 mm is desirable and this is most easily achieved in a hammer mill if the beans are first frozen either overnight at say −20 °C or in liquid nitrogen before grinding.

For commodities with low water contents (*e.g.* coffee beans) efficient extraction is achievable by refluxing for an appropriate number of times in 70% v/v aqueous methanol. The number of extraction cycles required is best determined by inspection – *i.e.* extract repeatedly and either analyse each extract by HPLC or measure on each extract some convenient parameter (*e.g.* A_{320}) so as to determine when extraction is effectively complete. As a guide, some 500 mg of ground green coffee can be extracted in four or five 20 min cycles each using some 30 ml of 70% v/v aqueous methanol. Note that green coffee beans contain some 50–100 g kg^{-1} CGA.[7] The CGA content determined on such extracts should have a coefficient of variation (CV) no greater than ~10%.[18]

For commodities comparatively high in water (*e.g.* apples) it is desirable to take into account the effect of the water in the sample on the final concentration of methanol in the solvent used for the first extraction cycle. Accordingly for apples it is better to use undiluted methanol for the first cycle, followed by 70% v/v aqueous methanol for subsequent cycles. As a guide, some 6 g of homogenized apple can be extracted in four or five 20 min cycles with the first using 30 ml undiluted methanol and cycles two to five using some 30 ml of 70% v/v aqueous methanol. Note, that apples contain some 60–400 mg kg^{-1} CGA.[7] The CGA content determined on such extracts should have a CV no greater than ~10%.[46]

The precautionary use of undiluted methanol is particularly important when extracting artichokes (*Cynara scolymus)*. The characteristic 1,5-diCQA (cynarin) is rapidly converted into 1,3-diCQA during extraction in hot aqueous methanol (Clifford, unpublished). However, the use of undiluted methanol for subsequent extraction cycles, or its application to 'dry' samples such as green coffee beans, is not to be recommended as it impairs recovery of CGA.[47]

Some investigators prefer to use acetone as the primary solvent. After evaporation to dryness the residue can be partitioned with ethyl acetate–water (1:1 by volume). The CGA enter the organic phase.[48]

Extracts of this type are rich in non-phenolic materials, particularly carbohydrate. The co-extraction of these materials can be minimized by increasing the methanol content of the extraction solvent or by using 70% v/v 2-propanol. These changes may require the other parameters of the extraction protocol to be reoptimized. Whatever extraction solvent is employed, it is advantageous to use Carrez reagents[49] to clarify the extract followed by centrifugation/filtration prior to HPLC analysis. Carrez reagent A is prepared by dissolving 21.9 g zinc acetate dihydrate in water containing 3 g glacial acetic acid and diluting to 100 ml with water. Reagent B is prepared by dissolving 10.6 g potassium ferrocyanide trihydrate in 100 ml water. Add sequentially 1 ml of each to the bulked extracts with thorough mixing prior to diluting to volume and centrifuging/filtering as convenient.

Extraction from Animal Tissues

With an increasing emphasis on studies of biological effects there is an increasing need for validated methods to extract such compounds and their metabolites from urine, plasma and other tissues. In such situations, the analyte concentration is much lower than in the plant tissues referred to previously, and the protein content is comparatively high. There have been studies of the binding of 5-CQA to gelatin[50] and human serum albumin (HSA).[51] The binding to HSA is endothermic, entropically driven and causes a conformational change as judged by the quenching of tryptophan fluorescence. Unless it is intended specifically to measure the 'free' concentrations, it is essential that any method adopted will release these analytes from protein complexes. A selection of extraction procedures is described in the following paragraphs.

Caffeic acid, ferulic acid and 5-CQA have been determined in rat plasma.[52] Plasma (50 μl) was added to 50 μl sodium acetate buffer (0.1 M, pH 5) and 900 μl methanol–acetic acid (100:5 v/v). The mixture was vortexed for 30 s, sonicated for 30 s, again vortexed for 30 s and centrifuged for 5 min at 4 °C and 5000 × *g*. The supernatant after dilution with water (1:1 v/v) was injected directly on to the HPLC. The electrochemical detection limits were in the region of 1–2 pmoles on column.

Ferulic acid and various diferulic acids have been analysed in plasma and intestinal contents of rats.[53] Blood samples were spiked with 3-hydroxycinnamic acid (1 μg) and cinnamic acid (1 μg) as internal standards. Plasma was obtained by centrifugation at 2000 × *g* for 10 min at 4 °C, followed by removal of protein by addition of three volumes of methanol per ml of plasma and centrifugation at 800 × *g* for 10 min at 4 °C. The protein pellet was resuspended twice in two volumes of methanol and centrifuged at 800 × *g* for 10 min at 4 °C. The pooled methanol phases were rotary-evaporated under vacuum at 40 °C and the dry residue dissolved in 3 ml phosphate-buffered saline (pH 7). This was treated with 5×10^5 units per litre of Sigma *E. coli* type IX-A β-glucuronidase (EC 3.2.1.31) and 2.5×10^4 units per litre of Sigma *Helix pomatia* sulfatase (EC 3.1.6.1) for 1 hour at 37 °C.[54] The reaction mixture was acidified to pH < 2 with HCl and extracted (×3) with an equal volume of ethyl acetate and the bulked organic phase evaporated to dryness. The residue was dissolved in a final volume of 0.4 ml of methanol–water (50:50 v/v) and centrifuged at 13 000 × *g* for 10 min before direct injection (100 μl). Despite these precautions, recoveries from the original blood sample were only some 40% of the spike. The limit of detection was some 90–130 pmole of pure standards on column for HPLC–DAD (280 nm), corresponding to some 60–90 nM in plasma. The sensitivity can be increased approximately ×10 by using LC–MS with selected ion moitoring.

The recovery of caffeic acid from rabbit urine has been reported.[55] Internal standard in acetonitrile (metoclopramide 100 $\mu g\,ml^{-1}$) was added to urine (2:1 v/v) and acidified with glacial acetic acid (0.002:1 v/v), vortexed for 30 s and centrifuged at 1945 × *g* for 10 min. The supernatant was diluted with 1 ml 0.4% phosphoric acid (pH 2) and injected directly. The limit of quantification at 320 nm was 360 pmole on column (signal to noise ~3 and CV < 10%).

It should be noted also that cinnamates are known to undergo a variety of metabolic transformations such as side chain hydrogenation and β-oxidation, accompanied by methylation or conjugation with glucuronate, sulfate[8] and glycine.[56] For analytical purposes, enzymic removal of sulfates and glucuronides is relatively straightforward (see below) but release from amino acid conjugates is not so easy. Traditionally, hot strong acid is used, but with the risk of destroying the cinnamic acid as observed during the characterization of caffeoyl-tyrosine in green coffee beans.[57] The tyrosine residue was released satisfactorily, but caffeic acid in low yield was accompanied by six degradation products. Some micro-organisms, including *E. coli*, are known to produce a benzoylglycine amidohydrolase (hippuricase) enzyme.[58] Carboxypeptidase A is able to hydrolyse some amino acid conjugates.[59]

Sample Clean-up by Solvent Partition and Solid-phase Adsorption

Various authors have reported clean-up/fractionation procedures based upon the use of various solid phases, especially polyamide. For example, chlorogenic lactones have been preferentially concentrated in chlorogenic acid-rich extracts of roasted coffee beans.[41] The polyamide (Macherey and Nagel MN–SC–6 in a glass column 22 × 3 cm) was conditioned with methanol and water (250 ml each) and 50 ml of coffee extract applied. The column was washed with 750 ml water and the lactones eluted with 500 ml methanol. Polyamide has also been used to separate hydroxycinnamates from anthocyanins, quinic acid conjugates from glucosides and glucose esters[60] and glucaric and galactaric acid conjugates of citrus fruits from glucose esters.[61] For the citrus extracts, elution with water removed contaminating sugars and carboxylic acids. Hydroxycinnamoyl glucoses were recovered in methanol, whereas conjugates with glucaric and galactaric acids were released from the polyamide with formic acid.[61] Chlorogenic acids are adsorbed from aqueous solutions at pH 2 onto Amberlite XAD-2 and, after thorough washing with acidified water (pH 2) followed by distilled water, may be recovered using methanol.[62] Cinnamic acids released from conjugates by saponification and acidification to pH 2 have been concentrated on C_{18} cartridges previously conditioned with 0.01% HCl. Contaminants were removed by washing with water and the cinnamic acids recovered in methanol containing 1% HCl.[63]

Flavanols and proanthocyanidins, and probably other neutral flavonoids, can be removed from acidic cinnamate conjugates by partitioning into ethyl acetate at pH 7.[64]

Artefacts

During such extraction and work-up procedures various artefacts may be produced. There is always a risk of oxidation and some workers favour the incorporation of ascorbate or metabisulfite as an antioxidant.[65] The 1-acyl CGA are comparatively heat labile.[14] *trans* Cinnamates may be converted into *cis* especially by exposure to UV light and these geometric isomers (as well as the positional isomers) may resolve on cellulose TLC plates and some HPLC stationary phases,[66] and relatively weak signals characteristic of *cis* isomer contamination have been seen in ^{1}H NMR spectra of '*trans*' isomers.[67] Cinnamates may skip around the hydroxyls on the quinic acid (acyl migration) or even be hydrolysed completely at the more extreme pH values.[65] At low pH values alkyl chlorogenates and alkyl cinnamates may be formed from free acids in alcoholic solutions, whereas at high pH values the alkyl cinnamate may be generated directly from the conjugate by transesterification.

Hydrolysis/Saponification

CGA and similar may be readily saponified. For example, treating 10 ml of citrus juice with 10 ml of 2 M NaOH in the dark for 4 hours at room temperature gave complete recovery of the hydroxycinnamic acids.[63] The quinic acid residue can be released from CGA by overnight saponification at room temperature in a final concentration of 0.5 M NaOH.[68] Enzymic hydrolysis has been achieved in 10–30 min at 37 °C using hydroxycinnamoylquinate esterase isolated from commercial *Aspergillus niger* pectinase by ammonium sulfate fractionation (80% saturation),[28] or similar.[69] Another suitable enzyme may be isolated from rat intestines.[53] One of the commercial enzymes hydrolysed the *trans* isomers of caftaric, coutaric and fertaric acids, but not the *cis* isomers.[69] Acid hydrolysis, especially as used to release flavonoid aglycones from glycosides, may cause loss of cinnamic acids and other artefacts that can apparently interfere with 5-CQA quantification.[65] It is thus difficult to 'simplify' analysis by simultaneous hydrolysis of cinnamate conjugates and flavonoid conjugates to a comparatively limited number of aglycones. In the author's experience it is better to analyse the more complex mixture of conjugates.

Availability of Authentic Standard Compounds

Lack of authentic commercial standards remains a major problem. Generally only 5-caffeoylquinic acid (5-CQA) is readily available, but this may well be offered in catalogues under non-IUPAC nomenclature as 3-caffeoylquinic acid/chlorogenic acid (for example Aldrich C4,420-6; Apin Chemicals Ltd 01032c; Fluka 25700;

Sigma Chemical Co. C3878). Apin also offers isochlorogenic acid, described in their catalogue as 5-CQA and which *presumably* is IUPAC 3-CQA. In the past, Roth (Karlsruhe) offered a preparation called *iso*chlorogenic acid, a mixture of 3,4-dicaffeoylquinic acid with the 3,5- and 4,5-isomers. Roth has sometimes offered cynarin. This compound is often described as 1,5-dicaffeoylquinic acid, but is more likely to consist predominantly of the 1,3-dicaffeoylquinic acid isomer that forms by spontaneous isomerization during isolation from artichokes using aqueous solvents. This isomerization can largely be circumvented if artichoke is extracted in 97% methanol using a solids to liquid ratio to ensure that sample water does not lower the final methanol concentration below about 90% (Clifford, unpublished).

Fortunately, there is very easy access to plant material that is rich in certain chlorogenic acids and provides distinctive and unequivocal fingerprints, that if used for spiking along with HPLC–DAD (or similar) are quite adequate for routine characterization of many unknowns. Many of these extracts are so rich in certain compounds that it is very easy to collect tens or hundreds of micrograms using even manual analytical HPLC. If required these can then be further characterized by MS and NMR and by treatment with tetramethylammonium hydroxide (TMAH) as described below. In the author's experience the materials shown in Table 1 are rich, easily available and convenient sources of 'standards'.

Using commercial 5-CQA and/or a few micrograms of reasonably pure in-house isolated 'standards', it is very easy to 'synthesize' a range of structurally associated isomers and simultaneously confirm the putative identity of the isolate by facile treatment with TMAH.[45] This procedure is detailed in the next section.[17]

Procedure for TMAH Characterization and Transformation of Chlorogenic Acids

This procedure can be carried out on a very small scale, for example using less than 10 μg of an unknown substance.[17] It is recommended that anyone wishing to

Table 1 *Convenient sources of CGA that are not available commercially*

• 3-CQA	Maté (*Ilex paraguariensis*), or leaves of other common *Ilex* spp.
• 4-*p*CoQA	Traditional English cider.[a]
• 5-FQA	Green (unroasted) arabica (*Coffea arabica*) coffee beans.[b]
• 1,3-diCQA	Artichoke (*Cynara scolymus*) extracted in water or dilute methanol.[103]
• 1,5-diCQA	Artichoke extracted in 97% methanol.
• 3,4-diCQA	Arabica or robusta coffee beans.[b]
• 3,5-diCQA	Arabica or robusta coffee beans,[b] or Maté, leaves of other common *Ilex* spp., or artichoke.
• 4,5-diCQA	Arabica or robusta coffee beans.[b]

[a] Note that American cider is apple juice and culinary varieties of apple are less rich in this component than cider varieties.

[b] Note that robusta (*Coffea canephora*) coffee beans are richer than arabica, but give a more complex chromatogram that might be more difficult to interpret by the inexperienced.

exploit this procedure should first practice with commercial 5-CQA so as to define precisely the optimum time and temperature of reaction and reagent ratios under their laboratory conditions. Accordingly, the following instructions should be taken as a starting point that might require some modest optimization.

- Add with immediate mixing commercial TMAH (X μl, 20% ethanolic solution) to 10X μl of an aqueous or methanolic solution containing not less than 2.5 μg 5-CQA.
- If a yellow colour does not appear immediately (attributable to co-extracted acids when dealing with a fraction prepared in-house) use additional aliquot(s) of TMAH.
- Allow the reaction to proceed at room temperature (*ca.* 20 °C) for 5 min (monoacyl CGA) or 3 min (diacyl CGA).
- Terminate the reaction by the addition, with immediate mixing, of 2X μl 3.5 M acetic acid. If additional aliquots of TMAH have been used, increase the acetic acid accordingly. Neutralization has been achieved when the yellow colour attributable to the phenolate anions has been abolished.

With 5-CQA as the starting material, this procedure will generate approximately equal concentrations of 3-CQA, 4-CQA, residual 5-CQA and methyl caffeate – see Figure 3. 4-CQA forms before 3-CQA and, if necessary, by using shorter treatments with TMAH, it is possible to distinguish between these two products by observing which appears first. This is an important check, since the 4-acyl and 5-acyl isomers are comparatively difficult to separate by HPLC. On early non-end-capped C_{18} phases the 4-acyl isomer always preceded the 5-acyl isomer: many more modern phases either fail to resolve these two isomers or the 5-acyl isomer elutes between the 3-acyl and 4-acyl isomers as discussed below.

Exactly the same products are obtained with 3-CQA or 4-CQA (and presumably 1-CQA) as the starting material. The analogous outcome is achieved with any *p*CoQA or FQA (and presumably with sinapoylquinic acids, although the author has not had an opportunity to examine these compounds). Since the pertinent methyl cinnamate is a significant product it is very easy to determine which cinnamic acid(s) is involved – the methyl esters are either commercially available or easily synthesized by refluxing the free acid in methanol containing *p*-toluenesulfonic acid.

DiCQAs give more complex products. 1,3- or 1,5-DiCQA are inter-convertible and yield also 3,4-, 3,5- and 4,5-diCQA as well as the monoacyl CQA and methyl caffeate. In contrast, the 3,4-, 3,5- and 4,5-diCQA isomers seem not to produce significant amounts of 1,3- or 1,5-diCQA. The author has never seen a peak that is unequivocally 1,4-diCQA but, as discussed below, this may be due to failure to resolve from 3,4-diCQA. This TMAH treatment was used to characterize the CFQA of green robusta coffee beans[17] as mixed diesters by releasing essentially equal amounts of CQA and FQA, plus methyl caffeate and methyl ferulate.

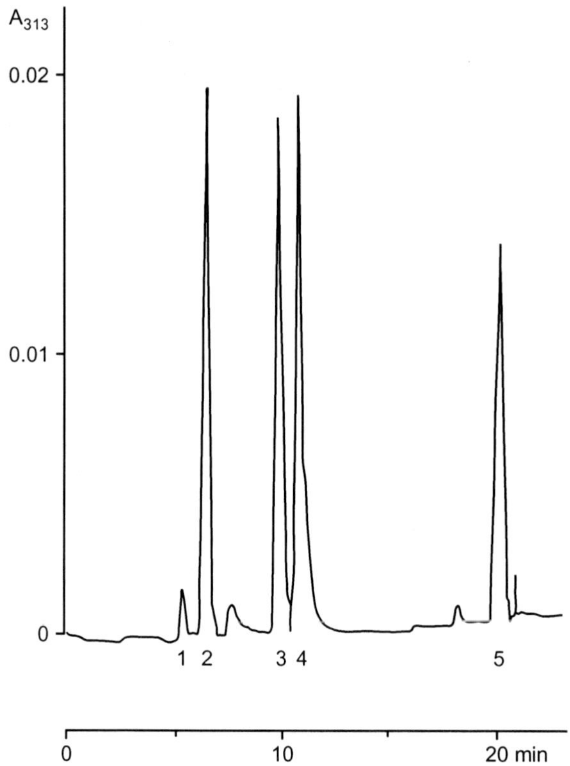

Figure 3 *Chromatogram of 5-caffeoylquinic acid treated with tetramethylammonium hydroxide for 5 minutes at 20 °C. Peak 1 = 1-CQA (tentative); peak 2 = 3-CQA; peak 3 = 4-CQA; peak 4 = 5-CQA; peak 5 = methyl caffeate. Conditions as described in the text*
(Reprinted from *Food Chem.*, 1989, **33**, 115–123, © 1989 with permission from Elsevier Science)

Ab initio Synthesis

The procedures available for the synthesis of chlorogenic acids have been reviewed.[4] Some recent advances have much simplified the procedures for the four monocaffeoylquinic acids,[70,71] and for some feruloyl-*myo*-inositols.[72,73]

Chromatography

Thin-layer Chromatography

On cellulose plates with *n*-butanol–acetic acid–water (4:1:1 by volume) as solvent *trans* and *cis* isomers separate as pairs of spots. Whether *cis* isomers were present in the intact plant tissue is a moot point since it is generally accepted that the plant synthesizes only the *trans* isomer.[74] Exposure to UV light facilitates the isomerization and thus one might reasonably expect some of the *cis* isomer to be a *normal* component of, for example, a cup of coffee, but *possibly* an artefact in many analytical extracts. Activated silica gel plates with toluene–ethyl formate–97% formic acid (5:5:1 by volume) will resolve the CQA, FQA and diCQA in coffee extracts. The diCQA remain close to the origin and are preceded by 5-CQA, 5-FQA, 4-CQA and 3-CQA, this latter compound having an R_f value of ~0.45.[75] Chloroform–methanol–water (1:1:0.1 by volume) is an alternative and complementary solvent system for 2D TLC.[14]

With care, the colour displayed under fluorescence may be used to distinguish between *p*-coumaric, caffeic and ferulic conjugates. However, it is desirable that a comparison is made with the colours displayed by standards of the commercially available free acids on a plate exposed to the same solvent system. The sequential use of the molybdate reagent and the metaperiodate reagent allows the caffeic and ferulic conjugates to be distinguished unequivocally on TLC plates. Only caffeic conjugates turn yellow with molybdate: ferulic conjugates also turn yellow with metaperiodate.[76,77] *p*-Coumaric conjugates do not show any colour change with either of these reagents. This approach was used during the characterization of caffeoyl-tyrosine.[57] Sinapic conjugates would not respond to molybdate. With metaperiodate there would be a positive response, possibly redder than seen with ferulic conjugates.[76] A phenol–sulfuric acid reagent used to detect sugar conjugates also gives a positive response with tartaric acid conjugates.[69] Its response with the quinic acid conjugates was not reported.

HPLC

Apart from the paper by Nagels *et al.*,[78] there have been few reports of normal-phase HPLC for the separation of chlorogenic acids. The vast majority of papers have reported the use of reversed-phase packings with elution in aqueous alcohol or acetonitrile, usually at low pH values or with phosphate buffer to suppress the ionization of the acidic analytes. Possibly the most suitable stationary phases are the non-end-capped C_{18} column packings such as the original Spherisorb ODS1. Extensive studies[45] established that the retention time of quinic acid conjugates on this packing was a predictable function of the position(s), number and identity of the acylating residue(s). The relative retention times of the common chlorogenic acids on this column packing are summarized in Table 2. This predictable behaviour of the CGA was used in combination with the TMAH inter-esterification and isomerization procedure (described more fully above) during the characterization of trace amounts of the coffee bean CFQA.[17]

It may be that the equivalent and similarly useful relationships can be established for the behaviour of CGA on other stationary phases, but it should not be assumed that these will necessarily be identical. For example, on some end-capped C_{18} materials (*e.g.* Spherisorb ODS2), it is difficult to resolve the 4-acyl and 5-acyl isomers.[18] If they do resolve, as for example on Waters Symmetry C_{18}[48] or Beckman ODS-Ultrasphere,[36] then at least for the CQA and FQA the 5-acyl isomer elutes before the 4-acyl isomer. The sequence of elution of the diCQA (3,4-di, 3,5-di, 4,5-di) was, however, as reported for the Spherisorb ODS1.

On Spherisorb ODS2 (250 × 4 mm with 5 mm guard column), *o*-quinones of *trans* caftaric acid and its 2-*S*-glutathionyl adduct elute later than the unoxidized phenols (~7% and ~56% acetonitrile for caftaric acid and ~13% and ~60% for the glutathione adducts).[42]

Möller and Herrmann[66] reported the separation of *cis* and *trans* isomers of several CGA using Lichrosorb RP-18 and a methanolic gradient containing 2% acetic acid. These workers commented that the sequence of the geometric isomers varied for individual CGA. The *cis* and *trans* isomers of the tartaric acid conjugates are resolved on Nucleosil 120 C_{18}.[79] In the author's experience such resolution is often not achieved on reversed-phase packings although it is not unusual to see a 'shoulder' on an otherwise sharp peak, at about 50% maximum peak height, that probably indicates partial separation of the geometric isomers.

Known difficult separations on reversed-phase packings include 3-FQA from 4-CQA, 4-FQA from 5-*p*CoQA, 1,5-diCQA from 3,5-diCQA, individual CFQA (*e.g.* 3-C, 4-FQA from 3-F, 4-CQA) and probably 1,4-diCQA from 3,4-diCQA. If possible where such compounds are under consideration additional criteria should be used when making these assignments. Errors have occurred,[15,80] and have been discussed elsewhere.[17]

1,5-DiCQA and 3,5-diCQA are difficult to resolve and may easily be confused or overlooked, especially as in some plant species (*e.g.* some Asteraceae, including artichoke) they are known to occur together. In contrast, 1,4-diCQA is poorly known. Indeed there are no modern reports of its occurrence. The author would predict that its behaviour would be very similar to that of 3,4-diCQA and possibly therefore 1,4-diCQA has been overlooked on some occasions.

It is now known that some plants synthesize quinic acid isomers (*e.g. iso*-quinic and *muco*-quinic),[32,33] and there may be other quinic acid isomers and conjugates awaiting discovery. Accordingly, it must also be recognized that conjugates of these relatively unknown quinic acid isomers may well co-elute with some of the better known conjugates and misidentification may follow on the assumption that it is the well-known compound that has been encountered. A similar complication arises with tartaric acid conjugates. These are probably best known as components of grapes and wines where caffeoyltartaric (caftaric), *p*-coumaroyltartaric (coutaric) and feruloyltartaric (fertaric) acid are found.[7,8,81,82] Dicaffeoyltartaric (chicoric) acid is found in lettuce, endive, chicory and dandelion.[7,8] In all of the foregoing the cinnamic acid is conjugated to L-(+)-tartaric acid (2*R*:3*R*-tartaric acid). In contrast, in spinach, *p*-coumaric acid is conjugated to *meso*-tartaric acid.[83] According to Ferreres *et al.*,[84] when using MeOH–formic acid (19:1) caffeoyltartaric acid elutes from a Merck RP18 packing (12 × 0.4 cm) in 9% methanol, followed by 5-CQA and dicaffeoyltartaric acid in 14% and 23% methanol respectively. The effect, if any, on the retention time/hydrophobicity of tartaric acid stereochemistry has not been reported. Clearly a chemotaxonomic awareness could be helpful in anticipating such occurrences, but caution must still be exercised. HPLC conditions for the separation of various ferulic acid dimers released by hydrolysis from cell wall polysaccharide have been reported.[85] Due to their hydrophobicity more than 70% acetonitrile is required for elution from an Inertsil ODS-2-Ik5 packing (20 × 0.46 cm). An alternative procedure has employed a Phenomenex Luna C_{18} 5 μm packing and a complex gradient involving acetonitrile–methanol mixtures acidified with TFA (1 mM).[53]

Analysts must be aware of such structure–behaviour relationships and should always be alert to the possibility of encountering unusual conjugates. Indeed, it is interesting to note that an early indication of the existence of a *muco*-quinic acid conjugate was the realization that a compound with mass spectral and UV characteristics consistent with a CQA eluted from a reversed-phase packing appreciably earlier than authentic 3-CQA (Haribal, personal communication).

The effect of changes in the quinic acid conformation on chromatographic behaviour is clearly illustrated also by the following examples. In a system where 3-CQA eluted before 4-CQA, 4-CQL eluted before 3-CQL (Clifford, unpublished). While both compounds have two free hydroxyls on the quinide moiety, chair inversion as a prerequisite for lactonization results in 3-CQL having one axial and one equatorial hydroxyl, whereas 4-CQL has two equatorial hydroxyls, making it more hydrophilic. The novel chair form associated with the 1,3-diFQA and 1,3-*p*CoFQA also has a profound effect on their elution from reversed-phase packings. The carboxyl equatorial 1,3-diCQA required approximately 19% aqueous acetonitrile[17] to elute from a non-end-capped C_{18} packing and, on the basis of the data in Table 2, the author would have predicted carboxyl axial 1,3-diFQA to elute in ~23% aqueous acetonitrile, all other factors being the same.

Table 2 *The relative retention times for various chlorogenic acids*

	Subfamily					
	CQA and diCQA	*pCoQA*		*FQA*		
Position of acylation	A_1	A_2	*B*	A_3	*C*	*Mean ± s.d. of* A_1*, B and C*
1-Acyl	0.64[a]	0.76[a]	0.63[a]	0.89[a]	0.68[a]	0.65 ± 0.03
3-Acyl	0.68	0.85	0.68	0.94	0.72	0.69 ± 0.02
4-Acyl	0.94	1.10	0.91	1.19	0.92	0.92 ± 0.02
5-Acyl	1.00	1.22	1.00	1.30	1.00	1.00
Cinnamic acid	1.08	1.36	1.12	1.51	1.13	1.11 ± 0.01
Methyl cinnamate	1.70	2.13	1.75	2.20	1.70	1.72 ± 0.03
1,3-Diacyl	1.22					
1,4-Diacyl	–					
1,5-Diacyl	1.67					
3,4-Diacyl	1.59					
3,5-Diacyl	1.67					
4,5-Diacyl	1.78					

[a] Identity not confirmed

A_1, A_2 and A_3 are relative to 5-CQA.

B is relative to 5-*p*CoQA.

C is relative to 5-FQA.

Chromatographic conditions: 15 cm × 4.6 mm analytical column packed with 3 μm Spherisorb ODS1 plus a 5 cm × 4.6 mm guard column packed with 5 μm Spherisorb ODS1. Gradient 6% v/v aqueous acetonitrile containing 0.5% v/v formic acid to 40% v/v aqueous acetonitrile containing 0.5% v/v formic acid linearly in 35 min.

Because this compound exists unexpectedly as the carboxyl axial conformer, it eluted in ~33% aqueous acetonitrile,[14] albeit probably from a column packing (unspecified) different from that used for 1,3-diCQA. Factors affecting the quinic acid conformation are discussed further in the NMR section of this review.

It is possible to make a preliminary distinction between *p*-coumaroyl, caffeoyl and feruloyl derivatives by comparing the relative absorption at 254 nm and 313 nm. The A_{313}:A_{254} ratios obtained on an old filter instrument were 0.07–0.09, 0.50–0.55 and 0.40–0.44 respectively.[4] This method can be used successfully with modern HPLC–dual wavelength or HPLC–DAD procedures, although the ability to distinguish between caffeoyl and feruloyl derivatives is impaired if the peaks are impure and/or there is appreciable base-line noise. The commercially available cinnamic acids should be used for calibration.

The use of LC–MS is discussed later.

Spectroscopic and Other Physical Properties

The solubility and relative hydrophobicity of the CGA have been discussed briefly above. A compilation of melting point data, crystal forms, partition coefficients and polarimetric data for CGA published prior to 1985 has been prepared.[4]

UV spectroscopy

For routine purposes, UV spectra can be informative. Caffeic acid and its derivatives have maxima in the regions 328–332 nm and 240–245 nm, with a shoulder in the range 290–300 nm. *p*-Coumaric acid derivatives have their primary maximum in the range 310–315 nm, but *trans* ferulic acid and sinapic acid are very similar to *trans* caffeic acid. Caffeic acid derivatives can be distinguished by their interaction with molybdate (primary $\lambda_{max} \sim 370$ nm), boric acid (~350 nm), or aluminium chloride (~360 nm). For a pure compound, the molar absorption coefficient clearly indicates how many acylating residues are present. For methanol as solvent, the log E value for monoacyl compounds falls in the range 4.2–4.3 and increases by some 6% for each additional acyl residue.

NMR Spectroscopy

Modern NMR instruments should have no difficulty in characterizing the simpler cinnamic acid conjugates, since NMR magnets of 7.05 T (⩾300 MHz) will even resolve the protons at C-2 and C-6 of the quinic acid residue,[67] especially if [$^2H_4$]methanol is used as solvent. The simplest approach to characterizing an unknown is to record the ^1H and ^{13}C spectra for the common cinnamic acids and the common associated moieties to which these might be attached (*e.g.* quinic acid, tartaric acid, tyrosine, *etc.*). Most of these substances are commercially available and this approach has been used successfully in the characterization of, for example, caffeoyl-tyrosine.[57] The conjugate spectrum will, in essence, be the sum of the constituent spectra, except for the carbons involved in the conjugation and any associated protons. Monoacylation produces a substituent chemical shift for the proton on the same carbon of some 1.1–1.6 ppm[67,86] with the effect being weaker for substitution at C-3 compared with C-5. DiCQA show larger substituent chemical shifts and it has been suggested that the magnitude is approximately the sum of the effects for the monoacyl derivatives,[87] but this has not been borne out precisely by the results from all investigations.[45] Acylation normally causes paramagnetic shifts in the olefinic protons (H-7′ and H-8′) of the cinnamic acid residue.

In many cases the proton spectrum alone will be sufficient for characterization. There are numerous published spectra for the commoner conjugates that can be consulted if necessary, *e.g.* see refs. 10, 17, 45, 88, 89. Spectra are also available for tartaric acid conjugates,[84] amino acid conjugates,[57,90,91] sugar conjugates,[92] acylated flavonoids[93–96] and some of the less common chlorogenic acids.[10,11,14,21,22,24,31,48]

One-dimensional ^{1}H and ^{13}C spectra are available also for chlorogenic lactones[41] and for the various stereoisomeric quinic acids and the γ- and δ-lactones.[38,89] There are fewer two-dimensional spectra available, but those for 1,3-diCQA,[97] 1,3-diFQA (COSY),[14] 3-caffeoyl-*muco*-quinic acid (ECOSY, DQCOSY, geHMQC and geHMBC),[32] dicaffeoyl, succinoyl-quinic acid (NOESY and HMBC)[22] and various mono- and diacyl-CGAs (TOCSY, HMBC)[67] are particularly useful.

Coupling constants (J) are particularly useful in assessing the conformation and assigning stereochemistry to quinic acid derivatives. For unstrained conformations where the substituent (*e.g.* caffeoyl residue) is axial, the proton–proton vicinal coupling constant is typically some 2–3 Hz for both axial–equatorial and equatorial–equatorial conditions. Similarly, when the substituent is equatorial, then the proton–proton coupling constants are some 10 Hz for axial–axial and some 5 Hz for axial–equatorial. The slightly low J values obtained by Haribal *et al.*[32] led them to conclude that in 3-caffeoyl-*muco*-quinic acid the quinic acid residue has a skewed or boat conformation.

The geminal coupling constant for the two protons at C-2 ($J_{2ax,2eq}$) can similarly be informative regarding the conformation. For the carboxyl axial 1,3-diFQA, the value is a relatively unstrained 12–13 Hz. For the carboxyl equatorial 1,3-diCQA, the corresponding value is a somewhat strained 16 Hz. According to Wenzl *et al.*[14] there is evidence from anisotropic ring-current effects of an interaction between the two axial caffeoyl moieties in 1,3-diCQA[67] which is presumably sufficient to stabilize this conformation. In the case of 1,3-diFQA, the presence in the cinnamoyl substituents of the two additional methyl residues makes such an interaction impossible and inversion occurs to yield the lower-energy carboxyl axial conformation, placing the two bulky feruloyl residues in equatorial positions.[14] Pauli *et al.*[67] have drawn attention to the usefulness of the $^{4}J_{2eq/6eq}$ *W*-coupling. High values are indicative that the quinic acid C-2, C-1 and C-6 are in the same plane – 3,5-diCQA had a lower value for the $^{4}J_{2eq/6eq}$ *W*-coupling constant than the other compounds studied (3-CQA, 5-CQA, 3-FQA, 1,3-diCQA and 4,5-diCQA).

Some published data are summarized in Tables 3 and 4 to illustrate the

Table 3 *^{1}H NMR Chemical shifts for protons at C-3, C-4 and C-5 in the quinic acid residue of various CGA relative to free quinic acid. Data from Clifford et al. (1989)*[45] *and Wenzl et al. (2000)*[14]

	Chemical shift (ppm)					
	C-3, quartet		*C-4, double doublet*		*C-5 triple doublet*	
Compounds	*δ*	*Δδ*	*δ*	*Δδ*	*δ*	*Δδ*
Quinic acid	4.22	–	3.70	–	4.02	–
3-CQA	**5.30**	**+1.08**	3.67	−0.03	4.11	+0.09
4-CQA	4.60	+0.38	**4.83**	**+1.13**	4.27	+0.25
5-CQA	4.37	+0.17	3.87	+0.07	**5.35**	+1.33
5-*p*CoQA	4.25	+0.03	3.71	+0.04	**5.25**	**+1.23**
5-FQA	4.14	−0.08	3.76	+0.06	**5.19**	**+1.17**
1,3-diCQA	**5.74**	**+1.52**	4.01	+0.31	4.15	+0.13
3,4-diCQA	**5.75**	**+1.53**	**5.25**	**+1.55**	4.25	+0.23
3,5-diCQA	**5.50**[a]	**+1.28**	4.06	+0.36	**5.45**[a]	**+1.43**
4,5-diCQA	4.40	+0.18	**5.15**	**+1.45**	**5.55**	**+1.53**
1,3-diFQA[b]	5.18	**(+0.96)**[c]	3.64	(−0.03)[c]	3.69	(−0.33)[c]

Spectra obtained at room temperature on a Brucker AC300 operating at 300.14 MHz. Samples were dissolved in D_2O containing tetramethylsilane standard.

[a] Overlapping signals, shifts approximate.

[b] Data from Wenzl *et al.* (2000).[14]

[c] Changes in chemical shift (Δδ) with acylation are only approximate as they were obtained in different experiments.

Table 4 *^{1}H NMR Chemical shifts, number of protons and assignment, multiplicities and coupling constants (J Hz) for protons in the quinic acid residue of carboxyl axial 1,3-diCQA*[a] *and carboxyl equatorial 1,3-diFQA*[b]

	Chemical shift (ppm), number of protons and assignment, multiplicities and coupling constants (J Hz)	
Position	*Carboxyl equatorial 1,3-diCQA*[104]	*Carboxyl axial 1,3-diFQA*[14]
C-2	2.99 ppm; 1 H_{ax}; *dd*; $J = 16, 4$ Hz	2.28 ppm; 1 H_{ax}; *t*; $J = 13, 13$ Hz
C-2	3.50 ppm; 1 H_{eq}; *dd*; $J = 16, 4$ Hz	2.00 ppm; 1 H_{eq}; *dd*; $J = 12, 3$ Hz
C-3	6.03 ppm; 1 H_{eq}; *q*; $J = 4, 4, 3$ Hz	5.18 ppm; 1 H_{ax}; *dt*; $J = 10, 4, 4$ Hz
C-4	4.19 ppm; 1 H_{ax}; *dd*; $J = 8, 3$ Hz	3.64 ppm; 1 H_{eq}; *m*
C-5	4.84 ppm; 1 H_{ax}; *dt*; $J = 8, 10, 4$ Hz	3.69 ppm; 1 H_{eq}; *m*
C-6	2.57 ppm; 1 H_{ax}; *dd*; $J = 14, 10$ Hz	H_{ax} covered by DMSO
C-6	3.20 ppm; 1 H_{eq}; *dd*; $J = 14, 4$ Hz	1.85 ppm; 1 H_{eq}; *d*; $J = 12$ Hz

[a] Spectrum recorded at 90 MHz in pyridine-d_5.

[b] Spectrum recorded at 300 MHz in DMSO-d_6.

Table 5 *MS data for selected CGA obtained with various modes of MS*

		Mode								
Compound	*Molecular mass*	*APCI +ve ion*[48]	*APCI −ve ion*[99]	*Ion trap +ve ion*[100]	*Ion trap −ve ion*[100]	*ES-MS −ve ion*[102,105]	*TS-MS +ve ion*[36,41]	*FAB-MS −ve ion*[106]	*FAB-MS +ve ion NBA*[14,28]	*FAB-MS +ve ion glycerol*[14]
3-CQA	354	355		355 (100%)	353 (100%) 191 (20%) 179 (6.5%)	353, 179				
4-CQA	354	355		355 (100%)	353 (100%) 191 (27%) 179 (5.5%)					
5-CQA	354		353, 191	355 (100%)	353 (100%) 191 (12%)	353, 179	355, 193		355	
5-FQA							369			
4-CQL							337			
3-CQL							337			
Unspecified diCQA	516		515, 516, 517, 353, 354, 355, 336, 191, 179, 162	517 (100%)	515 (100%) 353 (12%) 191 (7%) 179 (13%)	515, 353, 191, 179				
Unspecified diCQA	516			517 (100%)	515 (100%) 353 (19%) 191 (9%) 179 (14%)					

continued overleaf

Table 5 *(continued)*

		Mode								
Compound	*Molecular mass*	*APCI +ve ion*[48]	*APCI −ve ion*[99]	*Ion trap +ve ion*[100]	*Ion trap −ve ion*[100]	*ES-MS −ve ion*[102,105]	*TS-MS +ve ion*[36,41]	*FAB-MS −ve ion*[106]	*FAB-MS +ve ion NBA*[14,28]	*FAB-MS +ve ion glycerol*[14]
3,4-di, 3,5-di and 4,5-diCQA							517			
1,3-diFQA	544								177 (100%) 545 (5%) 567 (37%) 583 (4%) 589 (19%) 605 (4%)	177 (100%) 545 (14%) 567 (23%) 583 (5%)
Caffeoyl-tryptophan							367			
Unspecified 5-CSA									337	
p-Coumaroyl-malic acid	280							279, 163, 133		
Caffeoyl-malic acid	296							295, 179, 133		
Feruloyl-malic acid	310							309, 193, 133		
Sinapoyl-malic acid	340							339, 223, 133		

structure diagnostic value of the downfield shifts induced by acylation and the value of the proton–proton coupling constants in identifying the quinic acid conformation.

Mass Spectroscopy

Published spectra are available for many chlorogenic acids and related compounds.[14,48,98–102] Producing and interpreting such data would appear to be relatively straightforward, but sometimes unexpected signals are observed. For example, a diCQA (molecular mass 516) in negative ion APCI–MS gave signals at 515, 516 and 517 amu, with the signal at 516 being the largest.[99] These additional signals were not observed with either an ion trap [100] or ES-MS[102] and, in general, molecular ions (either with or without prior silylation) are normally observed in both positive and negative ion mode, with fragmentation and dehydration being observed at least for certain compounds in certain operating modes. Conjugates usually show fragments corresponding to the cinnamoyl residue and in some cases also for quinic acid. Cinnamic acids show a distinct decarboxylation. Water loss is also evident in some spectra and care may therefore be needed to distinguish between a quinic acid conjugate losing 18 amu as water, a preformed quinide (chlorogenic lactone) and the equivalent shikimic acid conjugate. Some illustrative data are presented in Table 5. There remains a need for MS^n data on the fragmentation of positional isomers since this might assist in distinguishing substances that are otherwise difficult to resolve, *e.g.* 1,5-diCQA and 3,5-diCQA, or 3-C, 4-FQA and 3-F, 4-CQA, *etc.*

References

1. P. B. Andrade, R. Leitao, R. M. Seabra, M. B. Oliveira and A. Ferreira, *Food Chem.*, 1998, **61**, 511.
2. IUPAC, *Biochem. J.*, 1976, **153**, 23.
3. M. N. Clifford, *Food Chem.*, 1979, **4**, 63.
4. M. N. Clifford, 'Chlorogenic Acids' in *Coffee 1. Chemistry*, R. J. Clarke and R. Macrae (eds), Elsevier Applied Science, London, 1985, pp 153–202.
5. M. N. Clifford, 'The Nature of Chlorogenic Acids – Are They Advantageous Compounds in Coffee?', in *Dix-septième Colloque*, ASIC, Paris, 1998, p. 79.
6. J. Radtke, J. Linseisen and G. Wolfram, *Z. Ernahrungswiss.*, 1998, **37**, 190.
7. M. N. Clifford, *J. Sci. Food Agric.*, 1999, **79**, 362.
8. M. N. Clifford, *J. Sci. Food Agric.*, 2000, **80**, 1033.
9. B. N. Timmermann, J. J. Hoffmann, S. D. Jolad, K. H. Schram, R. E. Klenk and R. B. Bates, *J. Nat. Prod.*, 1983, **46**, 365.
10. I. Merfort, *Phytochemistry*, 1992, **31**, 2111.
11. E. Scholz, M. Heinrich and D. Hunkler, *Planta Med.*, 1994, **60**, 360.
12. G. Peluso, V. de Feo, F. de Simone, E. Bresciano and M. L. Vuotto, *J. Nat. Prod.*, 1995, **58**, 639.

13. I. Agata, S. Goto, T. Hatano, S. Ishibe and T. Okuda, *Phytochemistry*, 1997, **33**, 508.
14. P. Wenzl, A. L. Chaves, J. E. Mayer, I. M. Rao and M. G. Nair, *Phytochemistry*, 2000, **55**, 389.
15. H. Iwahashi, H. Morishita, N. Osaka and R. Kido, *Phytochemistry*, 1985, **24**, 630.
16. H. Morishita, H. Iwahashi and R. Kido, *Phytochemistry*, 1986, **25**, 2679.
17. M. N. Clifford, B. Kellard and G. G. Birch, *Food Chem.*, 1989, **34**, 81.
18. K. J. Balyaya and M. N. Clifford, *J. Food Sci. Technol. Mysore*, 1995, **32**, 104.
19. M. Nishizawa, R. Izuhara, K. Kaneko and Y. Fujimoto, *Chem. Pharm. Bull.*, 1988, **36**, 87.
20. T. Tsuchida, M. Suzuki and M. Kurogi, *Nippon Shokuhin Kogyo Gakkaishi, J. Japan. Soc. Food Sci. Technol.*, 1968, **13**, 199.
21. Y. Maruta, J. Kawabata and R. Niki, *J. Agric. Food Chem.*, 1995, **43**, 2592.
22. Y. Chuda, H. Ono, M. Ohnishi-Kameyama, T. Nagata and T. Tsushida, *J. Agric. Food Chem.*, 1996, **44**, 2037.
23. Y. Chuda, M. Suzuki, T. Nagata and T. Tsushida, *J. Agric. Food Chem.*, 1998, **46**, 1437.
24. J. Heilmann, E. Müller and I. Merfort, *Phytochemistry*, 1999, **51**, 713.
25. M. Fukuoka, *Chem. Pharm. Bull.*, 1962, **10**, 3219.
26. V. P. Maier, D. M. Metzler and A. F. Huber, *Biochem. Biophys. Res. Commun.*, 1964, **14**, 124.
27. J. B. Harborne, C. A. Williams and J. Greenham, *Phytochemistry*, 1974, **13**, 1557.
28. P. M. Goupy, P. J. A. Varoquaux, J. J. Nicolas and J.-J. Macheix, *J. Agric. Food Chem.*, 1990, **38**, 2116.
29. D. Barron, M. Kaouadji and A.-M. Mariotte, *Z. Naturforsch.*, 1984, **39**, 167.
30. A. Rumbero-Sanchez and P. Vazquez, *Phytochemistry*, 1991, **30**, 311.
31. P. C. Stevenson, J. C. Anderson, M. Blaney and M. S. J. Simmonds, *J. Chem. Ecol.*, 1993, **19**, 2917.
32. M. Haribal, P. Feeny and C. C. Lester, *Phytochemistry*, 1998, **49**, 103.
33. Y. Wang, M. Hamburger, J. Gueho and K. Hostettmann, *Helv. Chim. Acta*, 1992, **75**, 269.
34. F. Hashimoto, G.-I. Nonaka and I. Nishioka, *Chem. Pharm. Bull.*, 1992, **40**, 1383.
35. M. E. Harbowy and D. A. Balentine, *CRC Crit. Rev. Plant Sci.*, 1997, **16**, 415.
36. K. Schrader, A. Kiehne, U. H. Engelhardt and H. G. Maier, *J. Sci. Food Agric.*, 1996, **71**, 392.
37. B. M. Scholz and H. G. Maier, *Z. Lebens. Unters. Forsch.*, 1990, **190**, 132.
38. B. M. Scholz-Böttcher, E. Ludger and H. G. Maier, *Liebigs Ann. Chem.*, 1991, 1029.
39. B. M. Scholz-Böttcher, PhD Thesis, 1991.
40. B. M. Scholz-Böttcher and H. G. Maier, 'Isomers of Quinic Acid and

Quinides in Roasted Coffee: Indicators for the Degree of Roast?', in *Quatorzième Colloque Scientifique International sur le Café*, Association Scientifique Internationale du Café, Paris, 1992, p. 220.
41. C. Bennat, U. H. Engelhardt, A. Kiehne and F. M. Wirries, *Z. Lebens. Unters. Forsch.*, 1994, **199**, 17.
42. V. Cheynier, J. Rigaud and M. Moutounet, *J. Chromatogr.*, 1989, **472**, 428.
43. H. Fulcrand, P.-J. Cameira-dos-Santos, P. Sarni-Manchado, V. Cheynier and M. Moutounet, 'New Anthocyanin-Derived Wine Pigments', in *Polyphenols 96. XVlle Journeés Internationales Group Polyphénols*, J. Vercauteren, C Chèze, M. C. Dumon and J. F. Weber, (eds), INRA Editions, Paris, 1996, Vol. 2, p. 259.
44. H. Fulcrand, P.-J. Cameira-dos-Santos, P. Sarni-Manchado, V. Cheynier and J. Favre-Bonvin, *J. Chem. Soc., Perkin Trans. 1*, 1996, 735.
45. M. N. Clifford, B. Kellard and G. G. Birch, *Food Chem.*, 1989, **33**, 115.
46. L. A. Mitchell, PhD Thesis, University of Surrey, 2000.
47. M. N. Clifford, 'The Influence of Extraction Method and Analytical Method on the Chlorogenic Acid Content of Green Coffee Beans', in *Onzième Colloque International sur la Chimie des Cafés Verts, Torréfiés et Leurs Dérivés*, Association Scientifique Internationale du Café, Paris, 1985, p. 252.
48. N. Nakatani, Si Kayano, H. Kikuzaki, K. Sumino, K. Katagiri and T. Mitani, *J. Agric. Food Chem.*, 2000, **48**, 5512.
49. H. Egan, R. S. Kirk and R. Sawyer, *Pearson's Chemical Analysis of Foods*, Churchill Livingstone, London, 1981.
50. M. Naish, M. N. Clifford and G. G. Birch, *Ann. N. Y. Acad. Sci.*, 1998, **855**, 823.
51. B. K. Muralidhara and V. Prakash, *Int. J. Peptide Protein Res.*, 1995, **46**, 1.
52. K. Azuma, K. Ippoushi, M. Nakayama, H. Ito, H. Higashio and J. Terao, *J. Agric. Food Chem.*, 2000, **48**, 5496.
53. M. F. Andreasen, P. A. Kroon, G. Williamson and M. T. Garcia-Conesa, *Free Radical Biol. Med.*, 2001, **31**, 304.
54. T. Ohta, T. Nakano, Y. Egashira and H. Sanada, *Biosci. Biotechnol. Biochem.*, 1997, **61**, 1942.
55. Y. S. Uang and K. Y. Hsu, *Biopharm. Drug Dispos.*, 1997, **18**, 727.
56. R. R. Scheline, *Mammalian Metabolism of Plant Xenobiotics*, Academic Press, London, 1978.
57. M. N. Clifford, B. Kellard and E. Ah-Sing, *Phytochemistry*, 1989, **28**, 1989.
58. S. F. Park, *Lett. Appl. Microbiol.*, 1999, **28**, 285.
59. M. Lee and D. H. Kim, *Bioorg. Med. Chem.*, 2000, **8**, 815.
60. M. Winter and K. Herrmann, *J. Chromatogr.*, 1984, **315**, 243.
61. B. Risch and K. Herrmann, *Z. Lebens. Unters. Forsch.*, 1988, **187**, 530.
62. B. M. Silva, P. B. Andrade, P. Valentão, G. C. Mendes, R. M. Seabra and M. A. Ferreira, *Food Chem.*, 2000, **71**, 281.
63. P. Rapisarda, G. Carollo, B. Fallico, F. Tomaselli and E. Maccarone, *J. Agric. Food Chem.*, 1998, **46**, 464.
64. A. Schieber, P. Keller and R. Carle, *J. Chromatogr. A*, 2001, **910**, 265.

65. A. M. Nuutila, K. Kammiovirta and K.-M. Oksman-Caldentey, *Food Chem.*, 2002, **76**, 519.
66. B. Möller and K. Herrmann, *J. Chromatogr.*, 1982, **241**, 371.
67. G. F. Pauli, F. Poetsch and A. Nahrstedt, *Phytochem. Anal.*, 1998, **9**, 177.
68. M. N. Clifford and P. S. Staniforth, 'Critical Comparison of Six Spectrophotometric Methods for Measuring Chlorogenic Acids in Green Coffee Beans', in *Huitième Colloque Scientifique International sur le Café*, Association Scientifique Internationale du Café, 1979, p. 109.
69. J. D. Baranowski and C. W. Nagel, *Am. J. Enol. Vitic.*, 1981, **32**, 5.
70. M. Sefkow, *Eur. J. Org. Chem.*, 2001, 1137.
71. M. Sefkow, A. Kelling and U. Schilde, *Eur. J. Org. Chem.*, 2001, 2735.
72. A. Hosoda, E. Nomura, A. Murakami, K. Koshimizu, H. Ohigashi, K. Mizuno and H. Taniguchi, *Bioorg. Med. Chem. Lett.*, 2000, **10**, 1439.
73. A. Hosoda, E. Nomura, K. Mizuno and H. Taniguchi, *J. Org. Chem.*, 2001, **66**, 7199.
74. A. J. Parr and G. P. Bolwell, *J. Sci. Food Agric.*, 2000, **80**, 985.
75. M. N. Clifford, PhD Thesis, University of Strathclyde, 1971.
76. M. N. Clifford, *J. Chromatogr.*, 1973, **86**, 222.
77. M. N. Clifford and J. Wight, *J. Sci. Food Agric.*, 1976, **27**, 73.
78. L. Nagels, W. van Dongen, J. de Brucker and H. de Pooter, *J. Chromatogr.*, 1980, **187**, 181.
79. C. Bétes-Saura, C. Andrés-Lacueva and R. M. Lamuela-Raventós, *J. Agric. Food Chem.*, 1996, **44**, 3040.
80. H. Morishita, *Wakayama Diagaku*, 1987, **36**, 69.
81. J. M. Souquet, B. Labarbe, C. Le Guerneve, V. Cheynier and M. Moutounet, *J. Agric. Food Chem.*, 2000, **48**, 1076.
82. M. Strasser and K. Gutzler, *Int. J. Food Sci. Technol.*, 2000, **35**, 81.
83. K. Tadera and H. Mitsuda, *Agric. Biol. Chem.*, 1971, **35**, 1431.
84. F. Ferreres, M. I. Gil, M. Castañer and F. A. Tomás-Barberán, *J. Agric. Food Chem.*, 1997, **45**, 4249.
85. A. J. Parr, K. Waldron, A. Ng and M. Parker, *J. Sci. Food Agric.*, 1996, **71**, 501.
86. E. L. Copeland, M. N. Clifford and C. M. Williams, *Monitoring Consumer's Intake of Tea*, SCI, London, 1994.
87. H. Morishita, H. Iwahashi, N. Osaka and R. Kido, *J. Chromatogr.*, 1984, **315**, 253.
88. J. Corse, R. E. Lundin, E. Sondheimer and A. C. Waiss, *Phytochemistry*, 1966, **5**, 767.
89. J. Corse and R. E. Lundin, *J. Org. Chem.*, 1970, **35**, 1904.
90. S. Jin and M. Yoshida, *Biosci. Biotechnol. Biochem.*, 2000, **64**, 1614.
91. M. Murata, H. Okada and S. Homma, *Biosci. Biotechnol. Biochem.*, 1996, **59**, 1887.
92. T. S. Kujala, J. M. Loponen, K. D. Klika and K. Pihlaja, *J. Agric. Food Chem.*, 2000, **48**, 5338.
93. A. Degenhardt, S. Hofmann, H. Knapp and P. Winterhalter, *J. Agric. Food Chem.*, 2000, **48**, 5812.

94. O. P. Johansen, O. M. Andersen, W. Nerdal and D. W. Aksnes, *Phytochemistry*, 1991, **30**, 4137.
95. F. Ferreres, E. Esteban, R. Carpena Ruiz, M. A. Jimenez and F. A. Tomás-Barberán, *Phytochemistry*, 1995, **39**, 1443.
96. S. S. Kang, Y. S. Chang and J. S. Kim, *Chem. Pharm. Bull. (Tokyo)*, 2000, **48**, 1242.
97. I. Horman, R. Badoud and W. Ammann, *J. Agric. Food Chem.*, 1984, **32**, 538.
98. R. G. Bailey, H. E. Nursten and I. McDowell, *J. Sci. Food Agric.*, 1994, **66**, 203.
99. M. A. Aramendia, I. M. Garcia, F. Lafont, J. Lizaso, J. M. Marinas and F. J. Urbano, *Rapid Commun. Mass Spectrom.*, 2000, **14**, 1019.
100. M. Carini, G. Aldini, S. Furlanetto, R. Stefani and R. M. Facino, *J. Pharm. Biomed. Anal.*, 2001, **24**, 517.
101. P. Cremin, S. Kasim-Karakas and A. L. Waterhouse, *J. Agric. Food Chem.*, 2001, **49**, 1747.
102. F. A. Tomás-Barberán, M. I. Gil, P. Cremin, A. L. Waterhouse, B. Hess-Pierce and A. A. Kader, *J. Agric. Food Chem.*, 2001, **49**, 4748.
103. J. Slanina, E. Táborská, H. Bochorává, I. Slaninová, O. Humpa, W. E. Robinson and K. H. Schram, *Tetrahedron. Lett.*, 2001, **42**, 3383.
104. M. N. Clifford, *Phytochemistry*, 1986, **25**, 1767.
105. M. M. Pedrosa, M. Muzquiz, C. Garcia-Vallejo, C. Burbano, C. Cuadrado, G. Ayet and L. M. Robredo, *J. Sci. Food Agric.*, 2000, **80**, 459.
106. W. Brandl, K. Herrmann and L. Grootjahn, *Z. Naturforsch.*, 1984, **39c**, 515.

CHAPTER 15

Analysis of Anthocyanins

JULIÁN C. RIVAS-GONZALO

1 Introduction

These compounds are found in nature in flowers as well as in fruits, to which they provide a broad spectrum of colour ranging from orange to blue. They may also occur in roots, stems, leaves and bracts, accumulating in the vacuoles. Anthocyanins are vacuolar pigments found in various plant tissues. In flowers they are principally located in epidermal cells, and rarely in the mesophyll. However, in rye leaves, they are restricted to the mesophyll cells.[1]

The functions of the anthocyanins in plants are similar to the general functions of all flavonoids: antioxidant functions, photoprotection, defence mechanisms, and other ecological functions (symbiotic phenomena). Since they give colour to different parts of plants, they also play an interesting role in reproductive mechanisms: found in flowers, they serve to attract polinators, and in seeds and fruits to attract seed disseminators. It has also been observed that cyanide-3-glucoside inhibits larval growth in tobacco worms (*Heliothis viriscens*), and thus anthocyanins may be considered agents of biological control. It has also been suggested that anthocyanins may serve as taxonomic markers in plants, but up to now research has been carried out only on some levels of species within specific families.

Anthocyanins have been attracting greater interest as food colouring agents as a result of the present tendency toward the use of natural colours. To this end, researchers have been looking for sources in the plant kingdom that can provide sufficient quantities of stable anthocyanins. At the same time, the observed biological effects of these compounds have resulted in an interest in incorporating them in foods. All of the above have led to the development of methods of analysis in order to achieve maximum efficiency in identifying, isolating, and quantifying anthocyanins for biological testing from plant sources as well as from animal fluid and tissue sources.

The aim of this chapter is to present information about the methods for analysing anthocyanins. Although the classic techniques for analysing anthocyanins that were well reviewed by Strack and Wray[2] in 1989 are still being used today, in recent years the generalized application of sophisticated techniques has led to great advances in anthocyanin identification. One of the main problems associated with working with anthocyanins is their instability in the presence of various chemical and physical conditions. Since these factors must usually be taken into consideration during anthocyanin analysis, we will review their effects.

2 Chemistry

Structure

They are present in nature in the form of heterosides whose aglycone (or anthocyanidin) is a derivative of the flavylium ion or 2-phenylbenzopyrilium, diversely hydroxylated. This aglycone is found bound to one or several sugar molecules which, in turn, may or may not be acylated with different organic acids. The presence of these hydroxyl groups on the rings, as well as one or several sugar molecules, makes these compounds quite soluble in water, ethanol and methanol.

2-Phenylbenzopyrilium is an oxonium ion in which the oxygen is tetravalent; oxonium ions are generally not very stable, but in the case of the pyrilium ion the aromatic character lends it stability. The true structure of this molecule is that of a resonance hybrid in which the positive charge is not located on a specific atom but spread out over several atoms, there being several possible structures of the oxonium–carbonium type (Figure 1).

According to the literature data, twenty-one anthocyanidins have been fully described, of which only six are widespread and usually found in fruits and vegetables (Table 1). They are: pelargonidin, cyanidin, peonidin, delphinidin, petunidin and malvidin. These are based on the same 2-phenylbenzopyrilium (flavylium) skeleton hydroxylated at the 3, 5, and 7 positions, and differing in the number and position of hydroxyl and methoxyl groups in the B-ring. Derived from these anthocyanidins, a wide range of anthocyanins exist which differ in the nature and number of sugars attached to the aglycone and the position of the attachment, and in the nature and number of aliphatic or aromatic acids attached to the sugar residues. The most commonly found sugars are xylose, arabinose and rhamnose, among pentoses, and galactose and glucose, among hexoses. They are

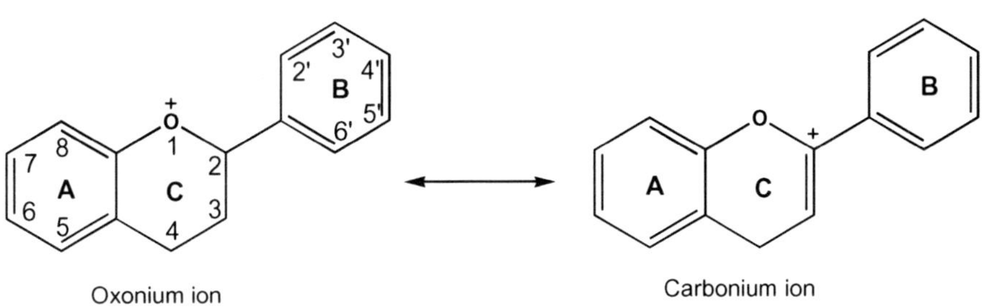

Figure 1 *Resonance structures of the flavylium cation*

Table 1 *Anthocyanidins present in nature*

Anthocyanidins	*Substitution pattern*
Common anthocyanidins	
Delphinidin	3,5,7,3′,4′,5′-OH
Cyanidin	3,5,7,3′,4′-OH
Pelargonidin	3,5,7,4′-OH
Petunidin	3,5,7,4′,5′-OH; 3′-OMe
Peonidin	3,5,7,4′-OH; 3′-OMe
Malvidin	3,5,7,4′-OH; 3′,5′-OMe
Rare anthocyanidins	
Hydroxylated structures	
Aurantinidin	3,5,6,7,4′-OH
6-Hydroxycyanidin	3,5,6,7,3′,4′-OH
6-Hydroxydelphinidin	3,5,6,7,3′,4′,5′-OH
Methoxylated structures	
5-Methylcyanidin	3,7,3′,4′-OH; 5-OMe
Capensinidin	3,7,4′-OH; 5,3′,5′-OMe
Europinidin	3,7,4′,5′-OH; 5,3′-OMe
Hirsutinidin	3,5,7,4′-OH;
Pulchellidin	3,7,3′,4′,5′-OH; 5-OMe
Rosinidin	3,5,4′-OH; 7,3′-OMe
3-deoxyanthocyanidins	
Apigeninidin	5,7,4′-OH
Luteolinidin	5,7,3′,4′-OH
Tricetinidin	5,7,3′,4′,5′-OH
Carajurin	6,7-OH; 5,4′-OMe
Carajurin-like compound I	6,7,4′-OH; 5′-OMe
Carajurin-like compound II	6,7,3′,4′-OH; 5′-OMe

linked to the aglycone by an α or β linkage of the free anomeric hydroxyl. Di- and tri-saccharide functional groups are also common, among which the most usual are rutinose (6-*O*-α-rhamnosyl-D-glucopyranose), sophorose (2-*O*-β-glucopyranosyl-D-glucopyranose), 2^G-xylosylrutinose and glucosylrutinose. Sugar moieties are always found attached to the hydroxyl group at position 3 of the anthocyanidin. When additional sugar groups exist, they occupy positions 5 and/or 7, and less frequently 3′ and 5′.

3-Deoxyanthocyanidins have been observed; carajurin plus another two carajurin-like anthocyanidins having only recently been identified,[3] which are the only ones naturally stable without substitution at position 3. Unusual anthocyanins found bearing sugar functional groups only at the 3′ position have also been described by Fossen and Andersen.[4] To our knowledge, no natural anthocyanins have been described with more than three hydroxyl groups substituted at the same time. Sugars may be substituted by aliphatic, hydroxybenzoic or hydroxycinnamic acids, the most usual being acetic, malonic, *p*-coumaric and caffeic acids.

Acylation with oxalic, succinic, malic, *p*-hydroxybenzoic, gallic, vanillic, syringic, protocatechuic, ferulic and sinapic acids is more restricted. The usual substitution position for the acyl residues is at 6″ of the sugar, although unusual 3″-acyl functional groups have also been reported in the genus *Allium*.[5] Anthocyanins have also been observed bearing a sugar molecule to which an acylated group is joined, which, in turn, bears another attached sugar molecule,[6] and there have even been described anthocyanins that bear sugar–acyl group–sugar–acyl group type chains.[7–9]

For a list of identified anthocyanins see Mazza and Miniati,[10] and regular updates published in successive editions of *The Flavonoids*, edited by Harborne.[11] Nevertheless, published lists are always incomplete, as new anthocyanins are continually being discovered in the plant kingdom. According to Andersen,[12] more than 560 anthocyanins have been described to date.

Stability

Anthocyanins are compounds that may fairly easily suffer reactions that alter their structures through the action of different agents due to the electronic deficiency of their flavylium nuclei. Anthocyanin stability increases with the number of methoxyls in the B ring and decreases as hydroxyls increase. Thus, among the most common anthocyaninidins, the most stable is malvidin, followed by peonidin, petunidin, cyanidin and delfinidin. In general, anthocyanins are more stable at an acid pH. Glycosylation and acylation of the sugars also increase stability and, therefore, the diglycosides are more stable than their corresponding monoglycosides.[13,14]

In aqueous dissolution, anthocynins are found in equilibrium in four basic structures, flavylium cation, quinonoidal anhydrobase, carbinol pseudobase and chalcone pseudobase. The proportion of these forms is determined principally by pH, with the flavylium ion predominating in very acid media. When the pH is raised, most of the flavylium ion, red in colour, changes into other anthocyanic forms which may be coloured or colourless, depending on whether the A and B rings are conjugated or not.[15]

The relative amount of each of the structural forms that coexists in equilibrium is a function of the pH of the medium and its functional groups. The distribution equilibrium as a function of pH at 25 °C among the different basic structural forms of malvidin-3-monoglucoside in aqueous solution was studied by means of kinetic, thermodynamic and spectroscopic techniques.[16] At low pH (<2) anthocyanins are mainly found in the form of flavylium cation (AH^+), red in colour. When the pH is increased, the flavylium cation quickly disappears making way for the deprotonation of the hydroxyl groups in positions 5, 7 and 4′, to produce various coloured forms called quinonoidal bases. These structures are bluish with maximum absorption at higher wavelengths than the cationic forms. At pH values between 3.5 and 4.5, flavylium cation and the quinonoidal base coexist, while at pH 4.5–6, only the neutral quinonoidal bases are present. As the pH value approaches neutrality, anionic and neutral quinonoidal bases coexist. If the pH goes above 8.5, the anhidrobases ionize completely. The proton transference reaction is established very quickly, but in aqueous media the bases formed are very unstable and the equilibria move quickly toward the formation of hydrated bases (covalent hydration reaction of the cation).

When pH values are between 3 and 6, hydration of the flavylium cation takes place, leading to the formation of colourless carbinol pseudobases. Nucleophilic attack by water generally occurs at the 2 carbon of the pyrilium nucleus. Even when other positions are suggested for the point of attack, the corresponding structures have been observed in natural anthocyanins, possibly due to deactivation originated by the 3-OH group. When position 4 of the pyrilium ring is occupied, stabilization of the cationic form occurs which renders hydration reactions difficult,[17] and this is one of the reasons for the greater stability of the pyranoanthocyanins recently described in wine.[18] While some authors suggest the necessity for the anthocyanin to be in cationic form in order to receive this kind of nucleophilic attack,[19,20] others maintain that the attack may occur even better on the anhydrobase forms.[17] Hydration reactions are much slower than protonation reactions and may take several seconds or even minutes to reach equilibrium.

In only slightly acid or neutral solutions, when the temperature of the solution is raised, the carbinol pseudobase is transformed, leading to an open chalcone. At room temperature and in a rather acidic medium, equilibrium takes several hours to reach. [16,21] Chalcone may be neutral and colourless or exist as ionized forms, yellow in colour; the neutral forms may in turn be *cis* or *trans*.[22] The *cis* forms may revert to flavylium cation when the pH is lowered, in a slow process, but it is believed that this transformation is very difficult to achieve in the case of the *trans*-chalcones.[23]

Monoglucoside anthocyanins have a greater tendency to give the chalcone form than the diglucosides under the same temperature conditions. Thus, at 25 °C, the amount of chalcone formed in the 3-monoglucoside anthocyanins is approximately double that of the corresponding 3,5-diglucosides.

The formation of chalcone-type structures is favoured by the absence of functional groups in the heterocycle, while the existence of a methyl at position 4 or a sugar at position 3 reduces their proportion in the equilibrium.[16]

The existence of chalcones has been observed in insufficiently acidic anthocyanin extracts which may cause quantitative errors in analysis and lead to yield losses in the process of isolating pure pigments. In this situation, it is necessary to acidify the extract and leave it for several hours to accomplish complete reversal of the chalcone forms to cationic form.

During analysis, it is important to control the temperature since the rate of degradation of anthocyanins increases as temperature rises, and is higher at higher pH values.[24–27]

Iacobucci and Sweeny[28] and Shenoy[25] found that light radiation accelerated the degradation of anthocyanins, probably due to the existence of autooxidative reactions. For this reason, it is a good idea to protect anthocyanin solutions from exposure to light when they are not going to be analysed right away. Van Buren *et al.*[29] studied the effect of light on wine anthocyanins, finding that acylated diglucosides were the most stable and that monoglucosides were the most unstable.

Bisulfite ions (HSO_3^-) attach at positions 2 or 4 of the flavylium cation, producing a colourless adduct, stable at the pH of wine. The reaction leads to discolouration of the anthocyanins, but lends stability at the same time since the complex is less sensitive to processes of hydration and polymerization, as the reactive positions are occupied.[30,31] This reaction is reversible to a variable extent, depending on the nature of the anthocyanin and the pH.[32] The adduct dissociates again to the original flavylium salt and SO_2 is released when the medium becomes acidified (pH = 1). This property, along with the effect of pH on colouration, is used for global anthocyanin analysis by spectrophotometry.

Compounds that have position 4 occupied cannot attach to SO_2. Thus, the complex pigments derived from anthocyanins formed in wine during ageing with position 4 occupied will be totally or partially resistant to action of bisulfite.[33,34]

The presence of ascorbic acid, either naturally present or added as an antioxidant, is negative for anthocyanin dissolutions, since in the presence of oxygen, both types of compounds mutually degrade.[35–38] The exact mechanism by which ascorbic acid induces degradation has not been clearly established, though it appears to be related to its autooxidation, since the reaction is inhibited in inert atmosphere[36] and in the presence of antioxidants, such as quercetin, which act to scavenge the free radicals generated in the process.[39] In some studies it has been stated that the products of anthocyanin degradation formed in the presence and absence of ascorbic acid are identical, indicating that the process must take a similar pathway in both cases.[40] It has also been shown that the addition of ascorbic acid at different stages of the process of making experimental red wines generally proves prejudicial for colour as it accelerates anthocyanin degradation, in spite of the presence of phenolic compounds that could act as preventative agents.[41]

Anthocyanins can form compounds with several di- or trivalent metals, such as tin,[42] copper,[43] aluminum or iron, leading to changes in colour, varying in function of the pH.[44–46] The trivalent cations may establish bonds with anthocyanins that possess *ortho*-dihydroxyl groups on ring B,[47] causing bathochromatic shift leading to a bluish colouration.[48] Thus, the addition of $AlCl_3$ has been used as an analytical test for differentiating cyanidin, delfinidin and petunidin from pelargonidin, malvidin and peonidin.[49]

3 Methods of Analysis

Extraction

The most common procedure for extracting anthocyanins is still by using methanol or ethanol solutions acidified with different proportions of HCl (0.1–1.0%). Using an organic medium has the advantage of inactivating the enzymes present in plant tissue, and of facilitating subsequent treatment of the extract obtained, as it is easily evaporated. Nevertheless, depending on the nature of the sample, it may be necessary to add water in order to achieve complete anthocyanin extraction.

For some applications, especially for use in food, it is preferable to use ethanol as it is less toxic, even though it is less efficient in extraction and more difficult to eliminate later. With ethanol, if the extraction process is carried out in the presence of heat, artefacts may be produced through esterification reactions with free carboxylic groups, for example in malonyl conjugate anthocyanins from maize.[50] It has been reported that during the extraction process and subsequent concentration, the HCl may produce hydrolysis of highly acylated anthocyanins, especially those with aliphatic acids. For this reason, in order to achieve good extractions with labile anthocyanins, gentler acid conditions such as those obtained with acetic, tartaric or citric acids are preferred.[2] Trifluoracetic acid (TFA) in different concentrations (0.5–3%) has been used to extract polyacylated anthocyanins, since, in spite of its high acidity, it is easily eliminated due to its high volatility during the subsequent concentration step. Methanol acidified with formic acid has also been used,[51] though it has been noted that this acid may lead to anthocyanin sterification *in situ* creating analytical artefacts.[52]

Acetone has also been used to extract anthocyanins from several plant sources.[53,54] In comparison with acidified methanol, this technique for extracting anthocyanins from red fruits allows an efficient and more reproducible extraction, avoids problems with pectins and permits a much lower temperature for sample concentration. However, pyranoanthocyanins have been detected when acetone was used in the extraction. The cycloaddition of acetone to anthocyanins is slow at room temperature, but increases significantly when the temperature is raised.[55]

After filtration and decantation, the extracts may be vacuum concentrated at a temperature no higher than 30 °C, in order to minimize degradation. After the alcohol is eliminated it is preferable to lyophylize the extract to achieve a less aggressive concentration. When the existence of lipidic material or liposoluble substances (waxes, chlorophylls, *etc.*) is suspected, before vacuum evaporation, water is added and, once all the alcohol has been eliminated, the aqueous solution is cleaned with an organic solvent such as hexane.

Very recently, analysis of anthocyanins has been carried out on biological samples with the objective of studying their bioavailability and kinetics in human beings after oral ingestion. This requires certain particular modifications in the stages of analysis because of the nature of the samples and because of the low concentrations in which the anthocyanins will be found. For blood analysis, the method described by Miyazawa *et al.* is used,[56] though there have been slight modifications in the protocols as described by different authors.[57–60] First, plasma is obtained through centrifugation at 1500–1800*g* for 10–15 min at 4 °C. The plasma is immediately acidified, generally with TFA, and subjected to a clean-up with ODS solid-phase extraction cartridge. After eliminating hydrosoluble compounds, polar lipids and neutral lipids with 7% formic acid, dichloromethane, and benzene respectively, the anthocyanins are eluted with methanol containing 5% TFA. The methanol extract is evaporated until dryness and the residue is redissolved in 5% TFA aqueous solution for the HPLC–DAD analysis.

In order to avoid interference of high molecular weight macromolecules (*e.g.* seroproteins) and to concentrate the anthocyanin extract before analysis, Murkovic *et al.* have proposed using restricted access phase during the clean-up step.[61] The packing material used has a hydrophilic outer surface (diol silica) and a hydrophobic inner surface (C_{18}). The proteins are not retained in the column while the anthocyanins interact with the hydrophobic surface. The anthocyanins are eluted with a water–formic acid–acetonitrile solution and then transferred to an RP18 analytical column for determination.

Isolation and Purification

It is difficult to separate anthocyanin mixtures since the basic skeletons of the different aglycones differ very little one from another, whether in molecular weight or in their functional groups, so that they have very similar physico-chemical properties. Because of this, chromatographic techniques are the choice, as much for separating the various anthocyanin fractions as for obtaining pure pigments that will allow for subsequent identification and quantification.

At first, the technique employed was *paper chromatography* (PC), which permitted not only isolation, but also identification, and quantitation of certain anthocyanins.[62–65] In accordance with the nature of the sample and the desired pigments, different mobile phases have been utilized, and these have permitted good resolutions for some pigment mixtures.[66] The drawbacks of this technique are that it doesn't permit large quantities of pure anthocyanins to be obtained, and that it needs very long development times, especially with very apolar eluents.

Thin layer chromatography (TLC) has advantages over PC in that it requires a lower quantity of pigments for analysis, requires much shorter times for elution and, in general, achieves better resolution. This technique offers results directly comparable to PC and, like PC, doesn't permit pure anthocyanins to be obtained in large quantities. Due to the variability in R_f values, which makes it difficult to compare them in lists,[67] it becomes necessary to use reference compounds in most cases. Strack and Wray compiled chromatographic conditions and applications in anthocyanin analysis.[2]

In spite of their drawbacks, the two techniques above, due to their low cost and the constant development of better supports and mobile phases, are still being used as routine techniques in many laboratories.

Column chromatography (CC) techniques began to be used because of the need to obtain pure compounds in sufficient quantities for subsequent identification and characterization, and to be used as reference substances in qualitative and quantitative analyses. An efficient column chromatography technique ought to be capable of separating the impure anthocyanins (sugars, organic acids, flavonoles, products of the decomposition of polymers, *etc.*), that are habitually present in extracts, as well as satisfactorily resolving the natural mixtures of these compounds. Whenever possible, these two functions should be carried out by the same resin.[68]

Ever since 1936 when Karrer and collaborators (cited in ref. 69) used alumina and calcium sulfate to separate anthocyanidins, many different chromatographic supports have been tried. Thus, Somers used Sephadex columns, but, due to the small differences in molecular weight among anthocyanins, to its adsorption to the gel through hydrogen bonds, and to the interference of the sugars and organic acids in the separation, it wasn't very successful.[70,71] In 1976 Golstein (cited in ref. 72), using chromatography in polyethylene glycol (PEG) dimethacrylate gel managed to separate, besides other phenolic compounds, different anthocyanidins and their diglucosides, with an acceptable degree of resolution.

Ion-exchange resins were frequently used to separate interferers such as sugars, acids, *etc.*, and thus obtain purer anthocyanin extracts, but they have not been shown useful for separating anthocyanins among themselves.[73,74] Currently, Amberlite GC-50[75] and Amberlite XAD-7[76,77] ion-exchange resins are often used as a first purification step. Further isolation is carried out by gel chromatography using Sephadex LH-20 or Toyopearl columns.[78]

Methods have recently been developed for separating wine pigments based on cation-exchange chromatography in the absence and presence of excess bisulfite.[79] Thus, authors using sulfoxyethyl cellulose, and taking advantage of the different bleaching ability for anthocyanins and their derivatives, have separated anthocyanins and anthocyanin derivatives which form bisulfite addition products from pyranoanthocyanins, 4-substituted anthocyanins, which do not form thosc adducts.

Polyamide shows good anthocyanin retention and allows elimination of sugars, organic acids, soluble polysaccharides, proteins and cations, using a simple washing with water. Some anthocyanin mixtures have been resolved well with this type of column, but in most cases the resolution is mediocre and it becomes necessary to use additionally paper or thin layer chromatography for a better separation and purification. Strack and Mansell managed to obtain some acylated derivatives of pure anthocyanins using polyamide.[80] These same authors observed that, as long as adequate mobile phase is used, the best resolution is obtained for anthocyanin mixtures from the same aglycone, as it appears that the nature of the sugars has more effect on retention than the ring B hydroxyl functional groups.

Fuleki and Francis attempted to develop a quantitative purification in polyamide columns,[81] but they observed excessive diffusion in the bands, as well as a loss of some pigments through degradation.

Starting in the 1970s, laboratories began using insoluble polyvinylpyrrolidone (PVP, Polyclar AT) which permits elimination of the impurities that accompany the anthocyanins in plant extracts, as well as separation of these compounds among themselves.[14,68,82] Although the mechanism has not been clearly elucidated, it seems that hydrogen bridges form between their proton acceptor groups and the free hydroxyl groups of the anthocyanins,[14,83] which permit a good separation between them, given that the ring B functional groups are different for each of these compounds. This characteristic has, in some cases, permitted the separation of different anthocyanins within specific groups (3-monosides, 3-biosides, 3,5-diglucosides, acylated derivatives). However, and just as occurs with polyamide, the best resolution is obtained for anthocyanin mixtures containing different derivatives of the same aglycone. Another researcher, Glories,[84] taking as a basis results obtained by Hrazdina,[14] developed a method for separating monomeric anthocyanins from those found attached to tannins using insoluble PVP.

However, when the objective is to resolve complex mixtures of anthocyanins, the degree of glucosylation and acylation of the aglycones interferes with the absorption of the pigments on the PVP and so the separation is not satisfactory.[72,85] Another drawback of this resin is that it strongly retains anthocyanin polymers and anthocyanins attached to tannins, which makes it difficult to re-use.

Since none of the above resins permits complete resolution of complex mixtures of anthocyanins, some authors have opted for using several together. Thus, Hrazdina and Franzese[86] managed to fractionate and isolate anthocyanins from Ives grapes using first polyamide, and then re-chromatographying the fractions obtained with PVP. In this way, they succeeded in isolating the diglucoside anthocyanins, as well as their coumarates and those corresponding to the three monoglucosides. However, though the results obtained using the combination of these two resins were acceptable, certain complex anthocyanin mixtures required a subsequent separation using paper chromatography.[87] Along the same lines, Moutounet and Chudzikiewicz,[88] using a Sephadex LH 20 column plus another Polyclar AT column in series, managed to separate monomeric anthocyanins and three groups of polymerized or combined wine anthocyanins.

Another solution that has been suggested, and which offers rather satisfactory results, is to use resin mixtures. Thus, Bourzeix *et al.* designed a method to separate monomer, red polymer, and brown polymer anthocyanins from each other in wine samples.[89] They used, as stationary phase, a resin mixture composed of: Polyclar AT (20%) + Silicagel G for fine layer chromatography (10%) + Silicagel 60 for column chromatography (70%). This mixture has the advantage of permitting elution of the polymerized compounds of the wine, as opposed to what occurs using PVP and permits, as well, a greater eluting speed, with considerable savings in time and thus less degradation of the anthocyanins. This column has been used successfully for isolating pure anthocyanins as standards for quantitative analysis in complex samples or as reagents for model systems.[90–93]

In CC, the best resolution is achieved with reverse-phase columns of the type Lichroprep-RPS,[94] which offer a quick, efficient separation of each and every one of the anthocyanins present in grapes. The main drawback to these resins is their high cost.

The technique of choice today is *high performance liquid chromatography* (HPLC), since it permits simultaneous separation, identification and quantification of anthocyanin compounds, without requiring excessive purity in the extracts. Nevertheless, solid-phase extraction (SPE) on C_{18} cartridges or Sephadex is commonly used for the initial purification of the crude anthocyanin extracts before carrying out the HPLC. The anthocyanins are bound strongly to these adsorbents through their unsubstituted hydroxyl groups and are separated from unrelated compounds by using a series of solvents of increasing polarity.

The first references to anthocyanin analyses using HPLC techniques were in 1975 (Manley and Subiak, cited in ref. 85), but until the work of Wulf and Nagel no one managed to establish acceptable working conditions.[95] Given the diverse polarity of the different anthocyanins, analysis with reversed-phase HPLC is generally carried out establishing a gradient, linear or otherwise, between a sufficiently acidified aqueous phase and an organic phase (methanol, acetonitrile, or acetone), the proportion of which continually increases throughout the analysis. Because each laboratory has tended to develop an analytical method for a specific separation requirement, it is impossible to describe a single standard procedure. Da Costa *et al.* discuss the work of numerous laboratories that have used this technique for analysing anthocyanins in food.[96]

The main problem with reversed-phase columns is that they may deteriorate at pH levels below 2, due to hydrolysis of the silica gel bonds and the octyl or octadecyl bonds. However, it is necessary to work at such low pH values, since it is only under these conditions that a high percentage of anthocyanins in the form of flavylium is found. Using higher pH values causes the different chromatographic peaks to broaden, or even split when equilibrium is reached between the flavylium salt, the quinonoidal base and the pseudobase, resulting in the existence of more than one chemical form at the same time. Yet, in spite of all this, different authors have worked at these pH levels without observing any unusual deterioration of the columns. A comparison of the separation performances of several newly developed pH-stable stationary reversed phases are described for the analysis of anthocyanins in red wines.[97]

Excellent results have been achieved using preparative HPLC to purify anthocyanins from raw plant material extracts.[75,78,98]

High-speed countercurrent chromatography (HSCCC) and *high-performance centrifugal partition chromatography* (HPCPC) have been used very recently for separating and purifying anthocyanins from plants and red wine. As an all-liquid chromatographic technique, CCC uses no solid support and allows complete recovery of the sample, since no irreversible adsorption on active surfaces can occur.[99] The biphasic mixture of *tert*-butyl methyl ether–*n*-butanol–acetonitrile–water (2:2:1:5) acidified with 0.1% TFA was found to be a suitable solvent system in HSCCC, and gradients of the ternary system EtOAc–*n*-BuOH–water acidified with 0.2% TFA in HPCPC technique have been applied for anthocyanin preparative separation.[99–102]

Capillary electrophoresis (CE) is a relatively new analytical tool with excellent mass sensitivity, high resolution, low sample consumption and minimal generation of solvent waste. Very few reports have appeared in the literature on the application of capillary zone electrophoresis methods (CZE) for separating anthocyanins.[103–105] In the case of anthocyanins, the applicability of this technique is limited, in part, by the fact that anthocyanins are not stable under alkaline conditions. This is why anthocyanins are separated using an acidic running buffer.[106] The use of micellar electrokinetic chromatography (MEKC) with SDS solution in a phosphate buffer at pH 7.0 produces excellent results for separating anthocyanins from elderberry juice.[107]

In spite of the advantages cited, CE does not offer the range for separating complex samples that can be achieved with LC. Because the CE method uses much smaller samples, it does not offer any advantages over LC with respect to sensitivity.[96]

Detection and Identification

UV-Vis Spectroscopy

Spectral properties are often very useful for characterizing anthocyanins, especially for identifying anthocyanidin type.[69,90,95,108–110] Anthocyanins present absorption maxima in the visible region (510–528 nm, in the case of those from grapes) as well as in UV region (270–280 nm). Absorbance in the visible is due to the structure of the central heterocycle and the conjugation of the two benzene cycles and is influenced by the pH level and by the solvent. Ring B functional groups may have some influence as well, so that the anthocyanins that possess two functional groups in that cycle (cyanidin and peonidin) have absorption maxima 11 nm lower than those with three functional groups (delfinidin, petunidin and malvidin). These differences remain the same for aglycones and for mono- and di-glucosides. Thus, the presence of methoxyl groups on ring B provokes a certain movement of colour toward the red, lending, at the same time, greater stability to the molecule. Also, delfinidin, petunidin and cyanidin, unlike malvidin and peonidin, undergo bathochromic shift in the presence of $AlCl_3$, a fact that has been used occasionally as a criterion for differentiating between them.

Glycosylation produces a displacement of the absorption maximum in the visible region of some 10 nm toward lower wavelengths, in relation to the corresponding aglycone and, at the same time, decreases the molar absorption coefficient. Diglycoside anthocyanins present absorption maxima toward lower wavelengths than the corresponding monoglycosides. In contrast, the presence of a shoulder in the region of visible light at about 440 nm in the 3-monoglycosides, which is not seen in the 3,5-diglycosides, allows differentiation between the two types of anthocyanins. However, pelargonidin 3-monoglucoside does present a marked shoulder in visible light at around 440 nm. Furthermore, the use of the $A_{440nm}/A_{\lambda max}$ ratio has been also suggested for the characterization of anthocyanins providing that the absorbance values are included in the 0.6–1 range, since this ratio increases considerably when A_{max} of the solution is less than 0.6.[111]

The acylation of sugars provokes slight displacements in the absorption maximum in the visible spectrum toward higher wavelengths. Absorbance in the UV region is due to phenolic groupings and is not seen to be affected by pH. Glycosylation does not provoke changes in absorbance, though the presence of acyl groups may cause the appearance of an additional maximum (or shoulder) in the UV spectrum. Thus, esterification with *p*-coumaric acid increases absorption at approximately 308–313 nm and with caffeic acid at 326–329 nm. With acetic acid no modifications in this range of the spectrum are produced.

The introduction of diode array detection systems (DAD) and their utilization jointly with HPLC has opened new perspectives for anthocyanin separation, identification and quantification in that they made it possible to acquire on-line UV-Vis spectra during chromatographic analysis.[90,110] Today, DAD is the usual technique used for analysing these compounds (see Chapter 5).

Spectrofluorometry

3,5-Diglucosides of malvidin and peonidin emit fluorescence when illuminated with UV light, a property that may be used for their identification, especially when attempting to detect their presence in TLC.[112] Moreover, the 3,5-diglucoside of malvidin (malvin), through oxidation and subsequent adjustment of pH to alkaline values, leads to the formation of a fluorescent derivative, a reaction that may be used to detect its presence in wines,[113] and which has been used in some cases for quantifying it, even in the complex anthocyanin mixtures such as are present in wine.[114,115] Though the method is good for qualitative objectives, it doesn't offer reliable quantitative results due to interference of the matrix in the fluorimetric reaction. Separation of samples by CC, using Polyclar AT as stationary phase and acidic methanol and ethanol as eluents, permits obtaining quantitative results for malvin comparable to those obtained using HPLC.[116]

Mass Spectrometry

Mass spectrometry is a technique that permits anthocyanin identification by determinination of the mass of the molecular ions in the sample and identification of the fragments from the rupture of these compounds through the application of higher ionization energies. In the 1980s, the introduction of fast atom bombardment mass spectrometry (FAB-MS) meant a great advance for structural identification of chemical compounds and especially for anthocyanin analysis.[2]

The application of HPLC–DAD–MS using electrospray ionization (ESI) has facilitated the identification of the anthocyanins in numerous plant samples such as, for example, black carrots,[117] grapes,[118] strawberries,[119] corn cobs[50] and red wines.[120,121]

The study of anthocyanin fragmentation highlights the fact that they have repeating fragmentation patterns. Recognition of these patterns may prove useful for identifying the original compound. The nomenclature has been described for flavonoid fragmentation as well as for that of sugars when bonded to an aglycone.[122]

Anthocyanidin in its flavylium cation form is relatively stable and does not have an easily broken linkage site, and it is unusual to find, in soft ionization conditions, fragments of this structure. The presence of the flavylium cation makes large ionization energies in positive mode unnecessary when working with electrospray probes and, consequently, the spectra of the anthocyanins and their anthocyanidins present a m/z signal which corresponds to a molecular ion with very little fragmentation. To provoke fragmentation, electrospray ionization in combination with CID (collision-induced dissociation) and/or tandem mass spectrometry (MS–MS) are used. Oliveira *et al.* have studied the fragmentation pattern of four of the most commonly found anthocyanidins in nature, cyanidin, delphinidin, peonidin and malvidin,[122] observing that they fragmented very similarly and that the differences they presented were mainly due to the ring B functional groups. Hayasaka and Asenstorfer,[123] using a combination of tandem nanoelectrospray plus spectrometry techniques, have determined the elimination masses from aglycone cations of wine anthocyanins and have described the fragmentation profiles of other wine anthocyanin derivatives.

These authors have also described the fragmentation behaviour of peonidin-3-glucoside and malvidin-3-glucoside, cyanidin-3,5-diglucoside, malvidin-3,5-diglucoside and cyanidin-3-rutinoside. The conclusions these authors arrived at are similar to those presented by Giusti *et al.* when they experimented with different anthocyanins from various sources.[124]

In all cases, the rupture of the glycosidic bonds occurs between the flavylium ion ring and the sugars directly attached to it. This breaking is accompanied by the transference of a hydrogen atom attached to oxygen from the carbohydrate to the ions formed. The fragments formed through loss of the sugar units contain only remaining aglycone, which is the more abundant ion in the mass–mass spectrum.

The number and type of fragments formed will vary depending on whether the substance is monoglucoside, diglucoside or bioside. With anthocyanins substituted on position 3 with a single sugar, the ion mainly detected will correspond to aglycone and residually the peak at *m*/*z* of the molecular ion. In the case of a diglucoside with sugars on positions 3 and 5, several ions will be detected in the spectrum, the ion corresponding to aglycone alone, the ions corresponding to anthocyanin substituted on 3 and on 5 and the anthocyanin molecular ion.

If the two sugars are the same, only one ion appears, apart from the molecular ion and the aglycone ion, so that the elimination sequence cannot be characterized. In the case of a 3-bioside, only one fragment will be formed, since the glucosidic bond between the two sugar units does not usually break. There is an exception: when the bioside on 3 is a rutin derivative, there will be a rupture in the glycosidic linkage, leading to an additional fragment, corresponding to the loss of the rhamnose. This 1–6 linkage between rhamnose and glucose permits free rotation and greater accessibility of the gas used in fragmentation.[124] In this type of compound, the formation of an ion has also been observed due to loss of the rhamnose terminal and of oxygen from the glucosidic bond.

In the case of acylated anthocyanins, there will not be any breaking of the ester bonds and the sugar and acid will be lost together. Giusti *et al.*, using the fragmentation pattern of acylated anthocyanins, made a rough determination of the location of the acylating groups in radish and red cabbage anthocyanins.[124]

Gas chromatography combined with mass spectrometry has been used in some cases, though this proves to be a difficult and costly application due to the structural lability of the anthocyanins with regard to electron impact ionization techniques, and due to the necessity of derivatizing them.[72,125,126]

NMR

These techniques have been used for identifying anthocyanins, and they undoubtedly permit correct identification. Nevertheless, we must take into account that the complexity of these compounds, due, above all, to the presence of sugars in the molecule, makes it necessary to use instruments capable of good resolution in order to achieve correct assignments of the proton and carbon signals in the NMR spectra.

On the other hand, it is necessary to use acidified media such as DCl or CF_3COOD, in order to displace the equilibria, so that only flavylium cation exists in the solution, since, otherwise, the spectra would become extraordinarily complicated because of the coexistence of several chemical forms.

The use of 1D and 2D ^1H and ^{13}C NMR spectroscopy with large signal suppression methods enables reliable assignments for ^1H and ^{13}C resonance signals of individual anthocyanins in CD_3OD,[127] which will serve as a tool for the identification of a particular fraction in LC analysis. (See Chapter 6).

Apart from the drawbacks already mentioned, these techniques require relatively large quantities of product in order to obtain spectra sufficiently discernible from background noise, which is a serious problem in using them with anthocyanins, which are found in small quantities in plant media. In addition, it is necessary to work with very pure compounds, which implies very precise initial isolation. Nevertheless, the NMR technique, in homo- and hetero-nuclear experiments, such as COSY, TOCSY, NOESY, HMQC and HMBC, is commonly used for identifying anthocyanins and their derivatives in natural plant sources and in industrial products as well.

Resonance Raman Spectroscopy

This technique uses a laser ray as a source of light. When this light falls upon a sample, it essentially scatters with no change in frequency, Rayleigh scattering; however, a small portion of the incident protons, upon diffusing, suffer a change in frequency, Raman scattering which is characteristic of the sample and does not depend on the excitation frequency.

Ordinary Raman spectroscopy has drawbacks in that it requires very elevated compound concentrations and that the spectrum obtained will correspond to those of all the molecules present in the sample. In resonance Raman spectroscopy, this difficulty is overcome through the use of laser light, the frequency of which corresponds to the absorption maximum of the compound to be characterized, which, for anthocyanins, may be that relative to the flavylium cation. Resonance Raman spectra specific to the structure corresponding to the absorption band used to produce the effect of resonance are thus obtained. Some researchers studying anthocyanins have used this technique.[128]

Techniques of Hydrolysis

In the studies first carried out on anthocyanins, their structure was characterized through selective hydrolysis and subsequent identification of the resulting compounds. In spite of the development of the above-cited modern methods for identification, these techniques, due to their simplicity, are still commonly used as complements to the newer techniques.

Acid hydrolysis, which is usually carried out in a concentrated hydrochloric acid medium at boiling temperature, provokes the rupture of the heteroside linkages, leading to the appearance of the corresponding sugars and aglycones.[129] This technique may be carried out in one go or gradually, in the latter case leading to the appearance of intermediate products of hydrolysis, which facilitate the correct identification of the anthocyanins.[98,130] The aglycones obtained are extracted with amyl alcohol[129,131] or with 1-pentanol,[98] while the sugars remain in aqueous solution. Aglycone identification may be carried out through UV-vis spectroscopy,[14] TLC or PC,[131] or HPLC,[130] the latter being the technique of choice, since it permits greater resolution. Sugars may be identified by classic paper, thin layer, or gas chromatography techniques.

Alkaline hydrolysis provokes the rupture of the bonds between the anthocyanin heterosides and the acylating acids. These reactions are produced by adding NaOH or KOH at room temperature, in a nitrogen atmosphere, and in darkness. After hydrolysis, the acylating acids are extracted with an organic solvent, usually ether, while the anthocyanins remain in the aqueous phase. Analysis of these compounds is also carried out through spectroscopic and chromatographic techniques.[95,131,132] If the acylating acid is acetic acid, identification requires previously formation of *p*-nitrobenzyl acetate, which absorbs light in the UV region and can be detected spectroscopically[133] or chromatographically.[95]

Gao and Mazza developed a rapid method for complete chemical characterization of simple and acylated anthocyanins, based on a single acid hydrolysis to effectively cleave the intact pigments onto stable aglycones, sugars and acyl groups. Anthocyanidin and phenolic acid characterization were carried out through RP–HPLC and aliphatic acylating acids and the sugars were characterized by GLC analysis of trimethylsilyl derivatives of sugars and methyl esters of aliphatic acids.[134]

When alkaline hydrolysis is carried out using $Ba(OH)_2$ the heterocycle breaks and results in the appearance of phloroglucinol and a phenolic derivative of benzoic acid, which may be identified directly by their spectral properties[129] or by PC,[135] TLC or HPLC.[95]

Oxidation with H_2O_2 and subsequent hydrolysis with NH_4OH, followed by silanization of the resulting acid-sugar and gas chromatography analysis, have been used to determine the site where the acylating acid is attached.[132]

References

1. F. Delgado-Vargas, A. R. Jiménez and O. Paredes-López, *CRC Food Sci. Nutr.*, 2000, **40**, 173.
2. D. Strack and V. Wray, in *Methods in Plant Biochemistry*, Academic Press, New York, 1989, Vol. I, p. 325.
3. B. Devia, G. Llabres, J. Wouters, L. Dupont, M. T. Escribano-Bailón, S. de Pascual-Teresa, L. Angenot and M. Tits, *Phytochem. Anal.*, 2002, **13**, 114.
4. T. Fossen and O. M. Andersen, *Phytochemistry*, 1999, **50**, 1185.
5. O. M. Andersen and T. Fossen, *Phytochemistry*, 1995, **40**, 1809.

6. N. Saito, F. Tatsuzawa, K. Yoda, M. Yokol, K. Kasahara, S. Iida, A. Shigihara and T. Honda, *Phytochemistry*, 1995, **40**, 1283.
7. T. Goto, H. Tamura, T. Kawai, T. Hoshino, N. Harada and T. Kondo, *Ann. N.Y. Acad. Sci.*, 1986, **471**, 155.
8. T. Goto and T. Kondo, *Angew. Chem. Int. Ed.*, 1991, **30**, 17.
9. F. Tatsuzawa, T. Ando, N. Saito, K. Yoda, H. Kokubun, H. Watanabe, G. Hashimoto, K. Asakura, R. Hara and H. Seki, *Phytochemistry*, 1999, **52**, 351.
10. G. Mazza and E. Miniati, *Anthocyanins in Fruits, Vegetables and Grains*, CRC Press, London, 1993.
11. D. Strack and V. Wray, in *The Flavonoids: Advances in Research since 1986*, ed. J. B. Harborne, Chapman and Hall, London, 1994, p. 1.
12. O. M. Andersen, in *Polyphenols 2000*, eds S. Martens, D. Treutter and G. Forkmann, Tecnische Universität München, 2002, p. 49.
13. L. S. Diaz, F. Gasque and B. Lafuente, *Rev. Agroquim. Tecnol. Aliment.*, 1976, **16**, 509.
14. G. Hrazdina, *J. Agric. Food Chem.*, 1970, **18**, 243.
15. R. Brouillard and O. Dangles in *The Flavonoids: Advances in Research since 1986*, ed. J. B. Harborne, Chapman and Hall, London, 1994, p. 565.
16. R. Brouillard, in *Anthocyanins as Food Colours*, ed. P. Markakis, Academic Press, New York, 1982, p. 1.
17. D. Amic, J. Baranac and V. Vukadinovic, *J. Agric. Food Chem.*, 1990, **38**, 937.
18. C. Romero and J. Bakker, *J. Agric. Food Chem.*, 2000, **48**, 2135.
19. R. Brouillard, B. Delaporte and J. E. Dubois, *J. Am. Chem. Soc.*, 1977, **99**, 8461.
20. R. Brouillard and J. E. Dubois, *J. Am. Chem. Soc.*, 1977, **99**, 1359.
21. L. J. Chen and G. Hrazdina, *Experientia*, 1982, **38**, 1030.
22. R. Brouillard, *Bull. Liaison Groupe Polyphenols*, 1986, **13**, 14.
23. A. Cheminat and R. Brouillard,. *Tetrahedron Lett.*, 1986, **27**, 4457.
24. L. Havlikova and K. Mikova,. *Z. Lebensm. Unters. Forsch.*, 1985, **181**, 427.
25. V. R. Shenoy, *Current Sci.*, 1993, **64**, 575.
26. J. P. Calvi and F. J. Francis, *J. Food Sci.*, 1978, **43**, 1448.
27. E. Maccarone, A. Maccarone and P. Rapisarda, *J. Food Sci.*, 1985, **50**, 901.
28. G. A. Iacobucci and J. G. Sweeny, *Tetrahedron*, 1983, **39**, 3005.
29. J. P. Van Buren, J. J. Bertino and W. B. Robinson, *Am. J. Enol. Vitic.*, 1968, **19**, 147.
30. Y. Glories, *Conn. Vigne Vin*, 1984, **18**, 195.
31. T. C. Somers, *Phytochemistry*, 1971, **10**, 2175.
32. J. F. Schopfer and J. Aerny, *Bull. O. I. V.*, 1985, **58**, 515.
33. P. Sarni-Manchado, H. Fulcrand, J. M. Souquet, V. Cheynier and M. Moutounet, *J. Food Sci.*, 1996, **61**, 938.
34. C. Romero and J. Bakker, *J. Agric. Food Chem.*, 1999, **47**, 3130.
35. F. J. Francis, *Crit. Rev. Food Sci. Nutr.*, 1989, **28**, 273.
36. G. A. King, J. C. Sweeny, T. Radford and G. A. Iacobucci, *Bull. Liaison Groupe Polyphenols*, 1980, **9**, 121.

37. M. S. Poei-Langston and R. E. Wrostald, *J. Food Sci.*, 1981, **46**, 1218.
38. Z. Shi, M. Lin and F. J. Francis, *J. Food Sci.*, 1992, **57**, 758.
39. A. J. Shrikhande and F. J. Francis, *J. Food Sci.*, 1974, **39**, 904.
40. M. D. Ortega-Meder, Efectos del Ácido Ascórbico sobre la Estabilidad del 3-Monoglucósido de Malvidina en Disoluciones Modelo, Tesis Doctoral, Universidad de Salamanca, 1995.
41. M. T. Guerra-Sánchez Simón, Estudio del Ácido Ascórbico en Vinificación. Efecto sobre el Color de los Vinos Tintos, Tesis Doctoral, Universidad de Salamanca, 1997.
42. F. W. Salt and J. G. Thomas, *J. Appl. Chem.*, 1957, **7**, 231.
43. D. Somaatmadja, J. J. Powers and M. Handy, *J. Food Sci.*, 1964, **29**, 655.
44. S. Asen, K. H. Norris and R. N. Stewart, *Phytochemistry*, 1969, **8**, 653.
45. L. Jurd and S. Asen, *Phytochemistry*, 1966, **5**, 1263.
46. H. Pyysalo, *Suomen Kemistil*, 1973, **46**, 103.
47. M. Elhabiri, P. Figueiredo, K. Toki, N. Saito and R. Brouillard, *J. Chem. Soc., Perkin Trans. 2*, 1997, 355.
48. T. A. Geissman, E. C. Jorgenson and J. B. Harborne, *Chem. Ind. (London)*, 1953, **52**, 1389.
49. K. R. Markham in *The Flavonoids*, J. B. Harborne, T. J. Mabry and H. Mabry, eds., Chapman and Hall, London, 1975.
50. S. de Pascual-Teresa, C. Santos-Buelga and J. C. Rivas-Gonzalo, *J. Sci. Food Agr.*, 2002, **82**, 1003.
51. A. M. Leone, E. La Notte and G. Gambacorta, *Vignevini*, 1984, **11**, 17.
52. J. Bakker and C. F. Timberlake, *J. Sci. Food Agric.*, 1985, **36**, 1315.
53. M. M. Giusti, L. E. Rodriguez-Saona, J. R. Baggett, G. L. Reed, R.W. Durst and R. E. Wrolstad, *J. Food Sci.*, 1998, **63**, 219.
54. C. Garcia-Viguera, P. Zafrilla and F. A. Tomás-Barberán, *Phytochem. Anal.*, 1998, **9**, 274.
55. Y. Lu and Y. Foo, *Tetrahedron Lett.*, 2001, **42**, 1371.
56. T. Miyazawa, K. Nakagawa, M. Kudo, K. Muraishi and K. Someya, *J. Agric. Food Chem.*, 1999, **47**, 1083.
57. T. Tsuda, F. Horio and T. Osawa, *FEBS Lett.*, 1999, **449**, 179.
58. G. Cao, H. U. Muccitelli, C. Sánchez-Moreno and R. L. Prior, *Am. J. Clin. Nutr.,* 2001, **73**, 920.
59. H. Matsumoto, H. Inaba, M. Hishi, S. Tominaga, M. Hirayama and T. Tsuda, *J. Agric. Food Chem.*, 2001, **49**, 1546.
60. I. Suda, T. Ohi, M. Masuda, Y. Nishiba, S. Furuta, K. Matsugano, K. Sugita and N. Terahara, *J. Agric. Food Chem.*, 2002, **50**, 1672.
61. M. Murkovic, U. Adam and W. Pfannhauser, *Fresenius' J. Anal. Chem.*, 2000, **366**, 379.
62. P. Ribereau-Gayon, *Ann. Physiol. Veg.*, 1964, **6**, 119.
63. R. A. Fong, R. E. Kepner and A. D. Webb, *Am. J. Enol. Vitic.*, 1971, **22**, 150.
64. J. P. Van Buren, J. J. Bertino, T. J. Einset, G. W. Remailey and W. B. Robinson, *Am. J. Enol. Vitic.*, 1970, **21**, 117.
65. A. Vaccari and P. G. Pifferi, *Chromatographia*, 1978, **11**, 193.

66. F. J. Francis, in *Anthocyanins as Food Colours*, ed. P. Markakis, Academic Press, New York, 1982, p. 181.
67. J. B. Harborne, *Comparative Biochemistry of the Flavonoids*, Academic Press, New York, 1967.
68. C. G. Van Teeling, P. E. Cansfield and R. A. Gallop, *J. Chromatogr. Sci.*, 1971, **9**, 505.
69. P. Ribereau-Gayon, *Les Composés Phénoliques des Vegetaux*, Ed. Dunod, Paris, 1968.
70. T. C. Somers, *Nature,* 1966, **209**, 368.
71. T. C. Somers, *Vitis*, 1968, **7**, 303.
72. G. Hrazdina, in *The Flavonoids. Advances in Research*, eds J. B. Harborne and T. J. Mabry, Chapman and Hall Ltd, London, 1982, p. 135.
73. P. G. Pifferi and A. Vaccari, *Lebensm. Wiss. U. Technol.*, 1980, **14**, 85.
74. R. E. Simard and T. Fuleki, *Bull. Liaison Groupe Polyphenols*, 1980, **9**, 89.
75. M. Fiorini, *J. Chromatogr. A*, 1995, **692**, 213.
76. K. Torskangerpoll, T. Fossen and O. M. Andersen, *Phytochemistry*, 1999, **52**, 1687.
77. L. Cabrita, N. A. Froystein and O. M. Andersen, *Food Chem.*, 2000, **69**, 33.
78. N. Mateus, S. de Pascual-Teresa, J. C. Rivas-Gonzalo, C. Santos-Buelga and V. de Freitas, *Food Chem.*, 2002, **76**, 335.
79. R. E. Asenstorfer, Y. Hayasaka and G. P. Jones, *J. Agric. Food Chem.*, 2001, **49**, 5957.
80. D. Strack and R. L. Mansell, *J. Chromatogr*, 1975, **109**, 325.
81. T. Fuleki and F. J. Francis, *J. Food Sci.*, 1968, **33**, 26.
82. R. Wrolstad and B. J. Struthers, *J. Chromatogr.*, 1971, **55**, 405.
83. O. Lennart and O. Samuelson, *J. Chromatogr.*, 1974, **93**, 189.
84. Y. Glories, *Conn. Vigne Vin*, 1984, **18**, 253.
85. G. Hrazdina, Abstracts of papers of the American Chemical Society, Apr. 20, 1979.
86. G. Hrazdina and A. J. Franzese, *Phytochemistry*, 1974, **13**, 225.
87. G. Hrazdina, H. Iredale and L. R. Mattick, *Phytochemistry*, 1977, **16**, 297.
88. M. Moutounet and J. Chudzikiewich, *Bull. Liaison Groupe Polyphenols*, 1980, **9**, 145.
89. M. Bourzeix, N. Heredia, M. I. Estrella, J. L. Puechi and K. Fartsov, *Bull. Liaison Groupe Polyphenols*, 1980, **9**, 131.
90. E. Hebrero, C. Santos-Buelga and J. C. Rivas-Gonzalo. *Am. J. Enol. Vitic.*, 1988, **39**, 227.
91. J. C. Rivas-Gonzalo, S. Bravo-Haro and C. Santos-Buelga, *J. Agric. Food Chem.*, 1995, **43**, 1444.
92. E. M. Francia-Aricha, M. T. Guerra, J. C. Rivas-Gonzalo and C. Santos-Buelga, *J. Agric. Food Chem.*, 1997, **45**, 2262.
93. F. J. Heredia, E. M. Francia-Aricha, J. C. Rivas-Gonzalo, I. M. Vicario and C. Santos-Buelga, *Food Chem.*, 1998, **63**, 491.
94. L. Piergiovanni and G. Volonterio, *Riv. Vitic. Enol.*, 1980, **33**, 289.
95. L. W. Wulf and Ch. W. Nagel, *Am. J. Enol. Vitic.*, 1978, **29**, 42.

96. C. T. da Costa, D. Horton and S. A. Margolis, *J. Chromatogr. A*, 2000, **881**, 403.
97. B. Berente, M. Reichenbächer and K. Danzer, *Fresenius' J. Anal. Chem.*, 2001, **371**, 68.
98. O. M. Andersen, *J. Food Sci.*, 1985, **50**, 1230.
99. A. P. Foucault and L. Chevolot, *J. Chromatogr. A*, 1998, **808**, 3.
100. A. Degenhardt, U. H. Engelhardt, P. Winterhalter and Y. Ito, *J. Agric. Food Chem.*, 2001, **49**, 1730.
101. A. Degenhardt, H. Knapp and P. Winterhalter, *J. Agric. Food Chem.*, 2000, **48**, 338.
102. A. Degenhardt, S. Hofmann, H. Knapp and P. Winterhalter, *J. Agric. Food Chem.*, 2000, **48**, 5812.
103. P. Bridle, C. García-Viguera and F. A. Tomás-Barberán., *J. Liq. Chromatogr. Rel. Technol.*, 1997, **19**, 537.
104. C. T. da Costa, B. C. Nelson, S. A. Margolis and D. Horton. *J. Chromatogr. A,* 1998, **799**, 321.
105. T. Ichiyanagi, C. Tateyama, K. Oikawa and T. Konishi, *Biol. Pharm. Bull.*, 2000, **23**, 492.
106. J. Sádecka and J. Polonsky, *J. Chromatogr. A*, 2000, **880**, 243.
107. T. Watanabe, A. Yamamoto, S. Nagai and S. Terabe, *Anal. Sci.*, 1998, **14,** 839.
108. J. B. Harborne, *Biochem. J.*, 1958, **70**, 22.
109. C. F. Timberlake and P. Bridle, in *The Flavonoids*, eds J. B. Harborne, T. J. Mabry and H. Mabry, Chapman and Hall. London, 1975, p. 214.
110. E. Hebrero, C. García-Rodríguez, C. Santos-Buelga and J. C. Rivas-Gonzalo, *Am. J. Enol. Vitic.*, 1989, **40**, 283.
111. L. S. Díaz and R. Olave, *Rev. Agroquim. Tecnol. Aliment.,* 1981, **21**, 419.
112. Y. Gutiérrez-Fernández, *Determinación de Malvina en Vinos*, Tesis Doctoral, Universidad de Salamanca, 1983.
113. P. Dorier and L. P. Verelle, *Ann. Fals. Exp. Chim.*, 1966, **59**, 1.
114. V. H. Bieber, *Deustch Lebensmitt-Rundsch.*, 1967, **63**, 44.
115. *Recueil des Méthods Internationales d'Analyse des Vins et des Moûts*, Office International de la Vigne et du Vin, Paris, 1990.
116. Y. Gutiérrez-Fernández, J. C. Rivas-Gonzalo and C. Santos-Buelga, *Vitis*, 1993, **32**, 119.
117. W. E. Glassgen, H. U. Seitz and J. W. Metzger, *Biol. Mass Spectrom.*, 1992, **21**, 271.
118. D. Favretto and R. Flamini, *Am. J. Enol. Vitic.*, 2000, **51**, 55.
119. F. Lopes da Silva, S. de Pascual-Teresa, J. Rivas-Gonzalo and C. Santos-Buelga, *Eur. Food Res. Technol.*, 2002, **214**, 248.
120. N. Mateus, S. Pascual-Teresa, J. C. Rivas-Gonzalo, C. Santos-Buelga and V. de Freitas, *Food Chemistry*, 2002, **76**, 335.
121. A. M. Vivar-Quintana, C. Santos-Buelga and J. C. Rivas-Gonzalo, *Anal. Chim. Acta*, 2002, **458**, 147.
122. M. C. Oliveira, P. Esperança and M. A. Almoster Ferreira, *Rapid Commun. Mass Spectrom.*, 2001, **15**, 1525.

123. Y. Hayasaka and R. E. Asenstorfer, *J. Agric. Food Chem.*, 2002, **50**, 756.
124. M. M. Giusti, L. E. Rodriguez-Saona, D. Griffin and R. E. Wrolstad, *J. Agric. Food Chem.*, 1999, **47**, 4657.
125. J. Bakker, *Vitis*, 1986, **25**, 203.
126. K. Hostettmann and M. Hostettmann, in *The Flavonoids. Advances in Research*, ed. J. B. Harborne and T. J. Mabry, Chapman and Hall Ltd, London, 1982, p. 1.
127. J. Kosir and J. Kidric, *Anal. Chim. Acta*, 2002, **458**, 77.
128. J. C. Merlin, A. Statoua, J. P. Cornard, M. Saidi-Idrissi and R. Brouillard, *Phytochemistry*, 1994, **35**, 227.
129. L. S. Díaz and R. Olave, *Rev. Agroquim. Tecnol. Aliment.*, 1981, **21**, 419.
130. D. Strack, N. Akavia and H. Reznik, *Z. Naturfosch. C*, 1980, **35**, 533.
131. R. F. Albach, R. E. Kepner and A. D. Webb, *J. Food Sci.*, 1965, **30**, 69.
132. R. A. Fong, A. D. Webb and R. E. Kepner, *Phytochemistry*, 1974, **13**, 1001.
133. H. D. Durst, M. Milano, E. G. Kikta, S. A. Connelly and E. Grushka, *Anal. Chem.*, 1975, **47**, 1797.
134. L. Gao and G. Mazza, *J. Agric. Food Chem.*, 1994, **42**, 118.
135. L. F. Chen and B. S. Luh, *J. Food Sci.*, 1967, **32**, 66.

CHAPTER 16

Analysis and Purification of Flavanones, Chalcones and Dihydrochalcones

FRANCISCO A. TOMÁS-BARBERÁN,
ANGEL GIL-IZQUIERDO, FEDERICO FERRERES AND
MARÍA I. GIL

1 Introduction

Flavanones, chalcones and dihydrochalcones are biochemically related compounds of restricted occurrence. For this reason they are described as minor flavonoids despite sometimes being present in foods at a dietary significant concentration.[1] This is the case of flavanones in citrus fruits, chalcones in tomato and dihydrochalcones in apple.

Flavanones have a saturated C-ring, chalcones are unsaturated and along with dihydrochalcones have an open structure and a carbon skeleton numbered in a way different from other flavonoids (Figure 1).

These compounds are often responsible for relevant quality characteristics of plant-derived food products (taste, colour and appearance, *etc.*).[2] Thus, some flavanones have an intense bitter taste, being responsible for the characteristic taste of sour orange and grapefruit. In contrast, some dihydrochalcones that are obtained by chemical transformation of the bitter flavanones show an intense sweet taste and they are used as low-calorie sweeteners. Chalcones, together with aurones, are responsible for the pigmentation of many yellow and orange flowers, such as carnations and many flowers of the Compositae family.

Some of these compounds, such as citrus fruits flavanones, are ingested in the diet in very significant concentrations (150–300 mg per serving).[1] As orange juice is one of the main fruit-derived products in the diet of European countries,[3] flavanones are one of the main phenolic compounds ingested in the diet.

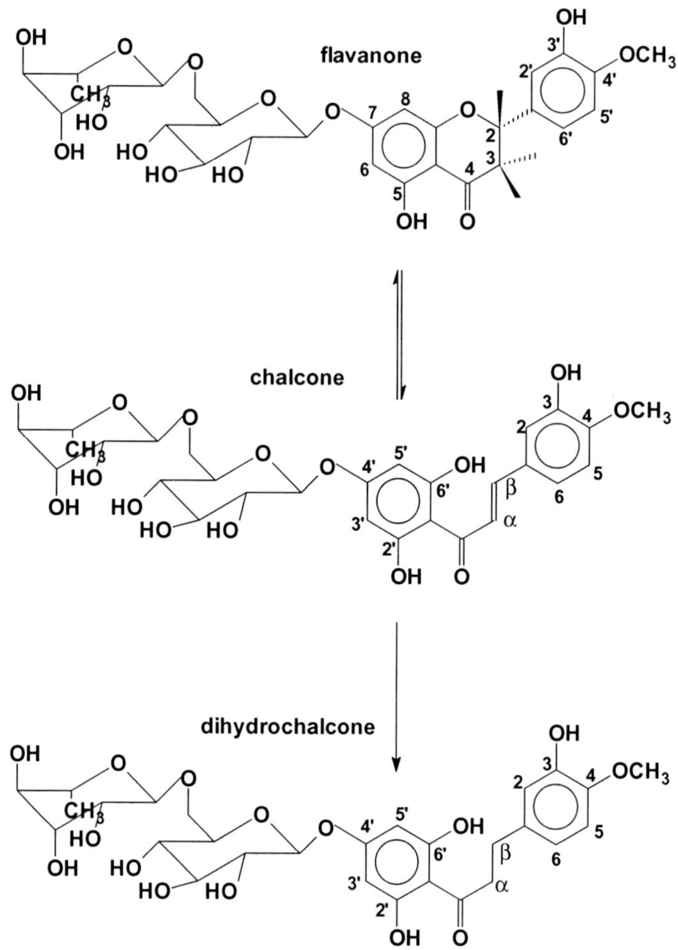

Figure 1 *Flavanone, chalcone and dihydrochalcone structures and numbering. Flavanone–Chalcone chemical interconversion*

2 Extraction

Some flavanones and chalcones are very difficult to dissolve in water and in the usual organic solvents used in flavonoid extraction from plant sources (methanol, ethanol or mixtures of these solvents with water). Flavanone solubility depends on the pH of the water solution. This is very small in acid, neutral or slightly alkaline conditions (Table 1) but it increases markedly at pH values above 10 (11–12). When trying to dissolve 500 mg of hesperidin in one litre of water, only 1.5% of hesperidin is dissolved at pH values between 1.5 and 10 (Table 1). Then at pH 12 70% is soluble, 25% remains insoluble and 5% is converted into the corresponding chalcone. At pH 14 all the hesperidin is dissolved and 45% is transformed to chalcone (Table 1). The chalcone is already present at all pH values as part of the insoluble fraction. The solubility of flavanones like hesperidin in the vacuole sap

Table 1 *Effect of pH on the water solubility of hesperidin and its transformation to chalcone*

pH	*% Soluble hesperidin*	*% Insoluble hesperidin*	*% Soluble hesperidin chalcone*	*% Insoluble hesperidin chalcone*
1.5	1.0 (0.0)	99.0 (0.0)	nd	tr
3.5	1.1 (0.0)	98.5 (0.1)	nd	0.5 (0.1)
6.0	1.0 (0.1)	97.8 (0.1)	nd	1.2 (0.0)
7.5	1.1 (0.0)	98.2 (0.0)	nd	0.7 (0.0)
8.5	1.5 (0.1)	97.5 (0.1)	tr	1.0 (0.0)
10.0	1.5 (0.1)	97.7 (0.0)	tr	0.9 (0.2)
12.0	70.2 (0.8)	24.1 (1.4)	4.6 (0.1)	1.1 (0.5)
14.0	54.2 (0.1)	nd	45.1 (0.1)	0.7 (0.0)
pH	*Soluble hesperidin (mg l^{-1})*	*Insoluble hesperidin (mg l^{-1})*	*Soluble hesperidin chalcone (mg l^{-1})*	*Insoluble hesperidin chalcone (mg l^{-1})*
1.5	4.9 (0.1)	496.6 (5.5)	nd	0.2 (0.0)
3.5	4.9 (0.1)	451.6 (6.1)	nd	2.2 (0.4)
6.0	4.4 (0.3)	441.3 (6.3)	nd	5.6 (0.0)
7.5	5.3 (0.0)	486.0 (0.7)	nd	3.6 (0.0)
8.5	7.5 (0.5)	488.5 (0.7)	tr	5.1 (0.0)
10.0	8.0 (0.1)	535.5 (41.8)	0.1 (0.0)	4.6 (1.0)
12.0	336.2 (11.9)	115.3 (4.1)	21.8 (0.8)	7.6 (0.8)
14.0	247.0 (1.8)	nd	205.5 (2.0)	3.4 (0.0)

nd = not detected; tr = trace. Values are means (n = 3). Standard deviations are in brackets.

of plant tissues is much higher than that observed after extraction of juice. Soon after extraction, flavanone precipitation starts, and this precipitation increases under some technological treatments and storage conditions.[4] Once the flavanone is precipitated, re-solubilisation is very difficult using either water or hydroalcoholic mixtures or even pure methanol or ethanol. In this case, the use of solvents like dimethyl sulfoxide (DMSO) or methanol–DMSO mixtures is essential for complete flavanone recovery.

This is essential for flavanone analysis and quantification, or in preparative chromatography of these compounds. In this case 100% of the flavanones and chalcones are soluble. However, when 100 μl of a stock solution of hesperidin in DMSO–methanol 1:1 (v:v) are added to 900 μl water at different pH values, the percentages of soluble and insoluble hesperidin differ from those obtained when trying to dissolve hesperidin directly in water. The results obtained are shown in Table 2. Under these conditions, at pH values between 1.5 and 10, the soluble hesperidin ranges between 20 and 30% while the rest is mainly in the insoluble fraction with very small amounts of the corresponding chalcone in the insoluble fraction. As in the case of direct solution in water, at pH 12, 90% of hesperidin is found soluble and the remaining 10% is found as soluble chalcone.

Table 2 *Effect of pH on the solubility of hesperidin and its transformation to chalcone. 100 μl of the stock solution (hesperidin in DMSO–MeOH (1:1, v:v)) were added to 900 ml of water equilibrated at the different pH values*

pH	*% Soluble hesperidin*	*% Insoluble hesperidin*	*% Soluble hesperidin chalcone*	*% Insoluble hesperidin chalcone*
1.5	32.1 (2.0)	67.6 (2.0)	0.02 (0.0)	0.3 (0.0)
3.5	20.0 (1.7)	79.6 (1.7)	0.1 (0.0)	0.3 (0.0)
6.0	21.4 (0.5)	77.8 (5.2)	tr	0.8 (0.2)
7.5	19.6 (3.6)	79.9 (3.6)	tr	0.5 (0.0)
8.5	25.9 (2.2)	73.6 (2.2)	0.1 (0.0)	0.5 (0.0)
10.0	31.4 (4.0)	68.0 (4.0)	0.2 (0.0)	0.4 (0.1)
12.0	90.0 (0.0)	0.8 (0.1)	9.2 (0.0)	nd
14.0	44.2 (0.3)	nd	55.8 (0.3)	nd
pH	*Soluble hesperidin (mg l^{-1})*	*Insoluble hesperidin (mg l^{-1})*	*Soluble hesperidin chalcone (mg l^{-1})*	*Insoluble hesperidin chalcone (mg l^{-1})*
1.5	163.2 (9.2)	343.7 (12.8)	0.1 (0.1)	1.5 (0.1)
3.5	101.4 (8.6)	403.4 (10.5)	0.4 (0.0)	1.7 (0.1)
6.0	107.4 (4.4)	390.3 (3.6)	tr	3.0 (0.3)
7.5	102.4 (9.0)	418.4 (23.4)	0.1 (0.0)	2.5 (0.1)
8.5	132.1 (12.3)	375.5 (8.0)	0.4 (0.0)	2.7 (0.1)
10.0	161.5 (19.4)	350.5 (23.2)	1.0 (0.2)	2.2 (0.2)
12.0	446.0 (6.4)	3.81 (0.5)	45.2 (0.7)	nd
14.0	180.2 (1.4)	nd	227.7 (1.4)	nd

nd = not detected; tr = trace. Values are means (n = 3). Standard deviations are in brackets.

On increasing the pH to 14, the transformation to chalcone is more favoured reaching around 55% in the soluble fraction (Table 2).

Not all the favanones have similar solubility properties. Hesperidin (hesperetin 7-rutinoside) is less soluble than narirutin (naringenin 7-rutinoside; 5,7,4′-trihydroxyflavanone 7-rhamnosyl 1–6 glucoside). When comparing naringenin and eriodyctiol (5,7,3′,4′-tetrahydroxyflavanone) glycosides the former are less soluble than the latter.

This characteristic solubility behaviour of some flavanones, like citrus flavanones, has been responsible for the very different quantitative data reported for the flavanone content of orange juice.[4] Flavanone solubility also matters for its biological significance as the solubility in biological fluids can affect their bioavailability and, therefore, their biological activity *in vivo*.[5]

Another important factor to consider when analysing flavanones and chalcones is that during their precipitation they follow exponential kinetics until reaching equilibrium. In the case of hesperidin dissolved in a stock solution of DMSO–methanol 1:1 (v:v) and diluted 9:1 (v:v) with water at pH 1.5 at room

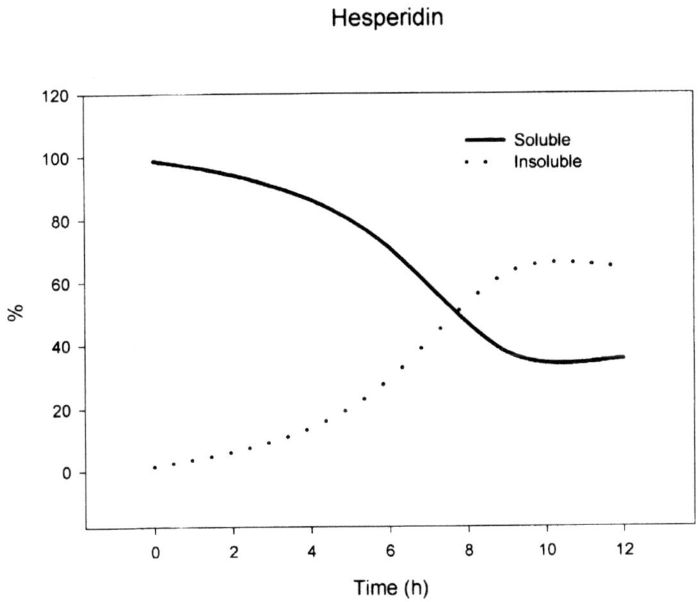

Figure 2 *Precipitation kinetics of hesperidin in water solution (pH 1.5). 100 μl of a stock solution of hesperidin in DMSO–MeOH 1:1 (v:v) were added to 900 μl of water pH 1.5 to reach a hesperidin final concentration of 500 mg l^{-1}*

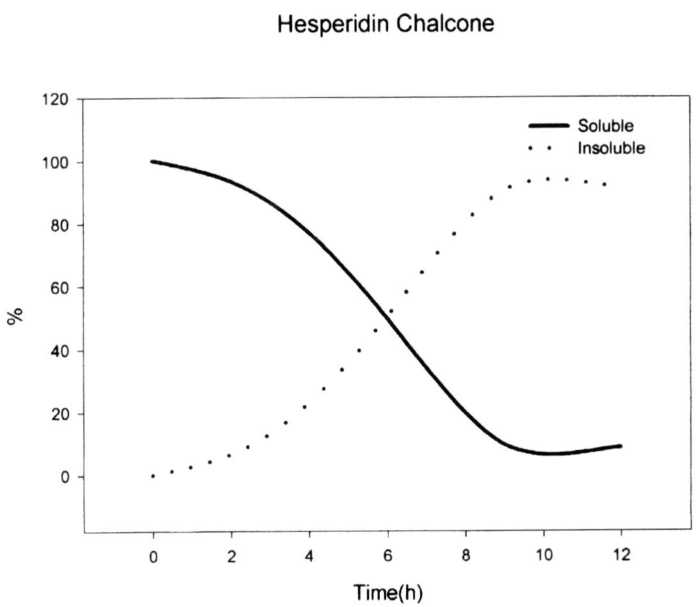

Figure 3 *Precipitation kinetics of hesperidin chalcone in water solution (pH 1.5). 100 μl of a stock solution of hesperidin in DMSO–MeOH 1:1 (v:v) were added to 900 μl of water pH 1.5 to reach a hesperidin final concentration of 500 mg l^{-1}*

temperature, the soluble hesperidin precipitates to reach equilibrium after 8 hours with 40% soluble and 60% in the insoluble fraction (Figure 2).

In the case of chalcones, and under the same conditions, something similar is observed, but in this case after 9 hours nearly all the chalcone is present in the insoluble fraction (Figure 3).

The solubility of dihydrochalcones is higher than those of the corresponding flavanones and chalcones. Thus, the solubility of hesperidin dihydrochalcone is around 500 $mg\,l^{-1}$ at room temperature, while that of the flavanone hesperidin is 100 times lower (around 5 $mg\,l^{-1}$) (Table 1). The low solubility of hesperidin contrasts with the values reported for soluble hesperidin in freshly prepared orange juice (250 $mg\,l^{-1}$) or that of pasteurised orange juice (40 $mg\,l^{-1}$) indicating that other juice constituents must have an effect increasing flavanone solubility.[4]

3 Chemical Transformation

The analysis of flavanones and chalcones is always difficult due to the chemical properties of these products that affect their solubility, as mentioned above, and the chemical interconversion between flavanones and chalcones. Chalcones and flavanones coexist at most pH values. At acidic pHs the flavanones are more favoured, while at alkaline pHs the chalcones are more favoured. At pH around 7, the chalcones are nearly absent (Tables 1 and 2).

In 1962 Seshadri reported the interconversion of chalcones and flavanones.[6] These two groups of compounds are isomeric and undergo interconversion readily (Figure 1). Acid or alkali acts as catalyst and the change can take place in either direction. Usually in acid medium the formation of the flavanone is more favoured and in alkaline medium the chalcone. A very important part in this interconversion is played by the presence of a hydroxyl in the 6′ position of the chalcone or the 5-position of the flavanone (Figure 1). In this case the flavanone becomes predominantly stable and the corresponding chalcone cannot be isolated free. This is the case for citrus flavanones (hesperidin, hesperetin and naringenin). If this hydroxyl group is absent, usually an equilibrium mixture is produced (*e.g.* liquiritigenin, butein, *etc.*). Flavanones are usually less soluble in alcohol and tend to separate first in fractional crystallisation.

As shown in Tables 1 and 2, the flavanones can be converted into chalcones by changing the pH of the solution. In model solutions of hesperidin, the transformation from flavanones to chalcones at slightly alkaline pH (8.5–9.0) is very small, and in these cases chalcones are mainly insoluble and are detected in the insoluble fraction. This transformation has some relevance as under the physiological conditions during the digestion of citrus fruit flavanones the transformation into chalcones has been reported to occur, although in very small amount.[5]

Dihydrochalcones do not undergo this transformation, as the lack of a double bond decreases the reactivity of the molecule and prevents the rearrangement to produce the flavanone. However, dihydrochalcones are readily produced from the corresponding chalcones by catalytic hydrogenation.[7,8]

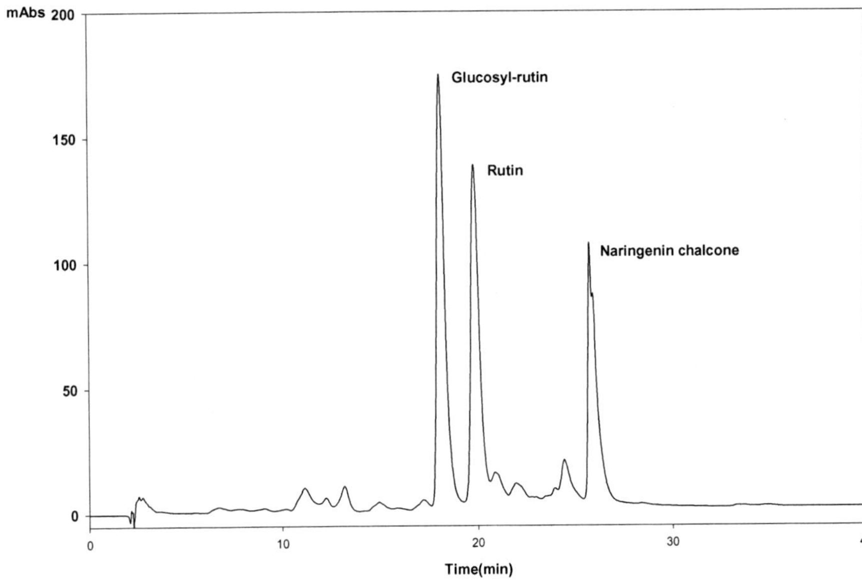

Figure 4 *HPLC chormatogram (340 nm) of a methanol extract from tomato peel*

4 Chromatographic Analysis

HPLC, as for most flavonoids, is the method of choice for the analysis of flavanones, chalcones and dihydrochalcones. The optimal wavelength for quantification depends on the compound as flavanones have a maximum at around 290 nm, chalcones at 340–360 nm and dihydrochalcones at *ca.* 280 nm. The chromatographic conditions used are similar to those of other flavonoids (reversed-phase columns with methanol–water–formic acid (or acetic acid) gradients or acetonitrile–water–formic acid (or acetic acid) gradients. The important issue in analysing these compounds is to take into account the solubility and chemical interconversion characteristics of these compounds that were mentioned above. DMSO or DMSO–methanol mixtures for sample preparation (extraction) and for injection (samples should be dissolved in these solvents to guarantee that all flavanones are completely dissolved) should be used for the complete recovery of these compounds. In addition it is important to take into account that chalcones under acid conditions are converted into flavanones. This transformation has been responsible for the overestimation of naringenin in tomato products after acid hydrolysis of the extracts, although in the tomato tissues naringenin-chalcone is the main compound present and naringenin is only a minor constituent (Figure 4).

5 Isolation and Purification

For preparative chromatography is also essential to make sure that flavanones are completely dissolved before chromatographic separation. A poor solubility of the flavanones in the sample solvent, or in the mobile phase used for the chromatographic separation, will produce a precipitation of these compounds in the column's head. This will lead to poor resolution, a decrease in solvent flow, and can even block completely the column and spoil both sample and stationary phase. A suitable alternative in this case is the use of solvent mixtures containing DMSO. This approach can be nicely used in the case of separation by molecular filtration using Sephadex LH-20 columns, in which both sample solvent and mobile phase can contain DMSO together with methanol. This type of solvent, however, cannot be used as mobile phase in other chromatographic columns such as reversed-phase and partition chromatography. In this case the only alternative is to dissolve the sample in DMSO–methanol and separate small amounts of sample under the usual chromatographic conditions.

Non-ionic polymeric resins such as different Amberlite XADs can be used to retain flavanones, chalcones and dihydrochalcones from complex extracts and then elute the phenolic compound fraction with methanol or ethanol. This has been previously reported for different flavonoid extracts,[9] and for the removal of bitter flavanones from citrus juices.[10,11] This method is useful for the adsorption of compounds dissolved in a water solution by filtration through an Amberlite XAD column, and then elution of the adsorbed compounds with an organic solvent as methanol or methanol–DMSO mixtures.

6 UV Spectroscopy

Flavanones

All flavanones give similar UV spectra in methanol as a result of their having little or no conjugation between the rings of the molecule. They are readily distinguished from flavones and flavonols by their UV spectra (see Chapter 5). Flavanones have their major absorption peak in the range 270–295 nm. Increasing oxygenation in the B ring has no noticeable effect on their UV spectra. The use of alkaline and metal reagents to detect the presence of specific free hydroxyls in the flavonoid nucleus has been extensively documented by Mabry *et al.*[12] in a classic work that is still very much in use. The UV spectra of flavanones in the presence of NaOMe exhibit bathochromic shifts of the main absorption band (BII) of about 35 nm for 5,7-dihydroxyflavanones. The shift is accompanied by an increase in the intensity of BII. Under alkaline conditions, however, some flavanones (in particular those lacking a free hydroxyl group at the 5-position) isomerise to chalcones, which have an entirely different UV spectrum. The detection of a free hydroxyl group at the 7-position in flavanones can be readily achieved by the effect of NaOAc on the UV spectrum. The presence or absence of a free 7-hydroxyl group may be readily determined from their UV spectra by comparing the positions of the major absorption peak (BII) in the methanol spectrum with that in the NaOAc spectrum. The shift is about 34–37 nm for the 5,7-dihydroxyflavanones and 51–58 nm for 7-hydroxyflavanones.[12] The detection of a free hydroxyl at position 5 of the flavanones can be achieved by the effect of the addition of $AlCl_3$/HCl in the methanol spectrum. In 5-hydroxyflavones, the

$AlCl_3$/HCl reagent also causes a consistent bathochromic shift (20–26 nm) of BII. The detection of B-ring *ortho*-dihydroxyl groups in flavonols and flavones is readily evidenced by the effect of the addition of NaOAc/H_3BO_3 or $AlCl_3$/HCl reagents on the methanol spectrum. In the case of flavanones, however, the *ortho*-dihydroxyl groups are not detectable because the B-rings in these flavonoids lack effective conjugation with the major chromophore.

Chalcones

The UV spectrum in methanol of chalcones is characterised by an intense BI and a diminished BII absorption. The maximum of BI usually occurs in the range 340–390 nm. BII is usually a minor peak in the 220–270 nm region. An increase in oxygenation usually results in bathochromic shifts in BI. Free 4-hydroxyl groups can be readily detected by the addition of NaOMe, as this causes a 60–100 nm bathochromic shift of BI with an increase in peak intensity. Chalcones without a 4-hydroxyl group but with either a free 2- or 4′-hydroxyl group also give, in the presence of NaOMe, a 60–100 nm bathochromic shift in BI but without an increase in peak intensity. The presence of B-ring *ortho*-dihydroxyl groups is readily detected by the 28–36 nm bathochromic shift in BI after the addition of NaOAc/H_3BO_3. This group is also detected by a 40–70 nm bathochromic shift of BI (relative to the BI position in the $AlCl_3$/HCl spectrum) on the addition of $AlCl_3$.

Dihydrochalcones

The information provided by the UV spectra of dihydrochalcones in methanol and after the addition of the alkaline and metal reagents is similar to that obtained in the case of flavanones. The lack of conjugation between rings means that the introduction of free hydroxyls on the B-ring has no detectable effect on the UV spectrum.

Their UV spectra in the presence of NaOMe exhibit bathochromic shifts of the main absorption band (BII) of about 40 nm in the case of phloridzin (4,2′,4′,6′-tetrahydroxy dihydrochalcone 2′-glucoside). The shift is accompanied by an increase in the intensity of BII. The detection of a free hydroxyl group at the 7-position in dihydrochalcones can be readily achieved by the effect of NaOAc on the UV spectrum. The presence or absence of a free 7-hydroxyl group may be readily determined from the UV spectra by comparing the positions of the major absorption peak (BII) in the methanol spectrum with that in the NaOAc spectrum. The shift is about 40 nm in the case of phloridzin. The detection of a free hydroxyl at position 2′ or 6′ of the dihydrochalcones can be achieved by the effect of the addition of $AlCl_3$/HCl in the methanol spectrum. In 2′-hydroxy- or 6′-hydroxy-dihydrochalcones, the $AlCl_3$/HCl reagent also causes a consistent bathochromic shift (20–28 nm) of BII.

7 ^{1}H NMR spectroscopy

^{1}H NMR analysis of these compounds is better performed using [$^2H_6$]DMSO as a solvent, as this avoids the solubility problems mentioned above and provides good signal resolution. An excellent review on the use of this technique in flavonoid structure determination has been published by Markham and Geiger.[13]

Flavanones

In flavanones, the ^{1}HNMR spectra show characteristic signals that are unequivocally distinguished by the typical coupling patterns of the C-ring protons. In this case the H-2 signal appears as a double doublet at around 5.4–5.5 ppm, the H-3 axial as a double doublet at 3.15–3.4 ppm and the H-3 equatorial as a double doublet at 2.6–2.7 ppm. The signals of the 3-axial protons are often obscured by sugar signals. Most naturally occurring flavanones show a phloroglucinol-type A-ring, and their 6 and 8 protons are not distinguishable, appearing as a doublet signal at 5.8–6.15 ppm. The resonances of the B-ring protons H-2′ and H-6′ occur at about 0.6 ppm higher field than do those of the corresponding flavones due to the saturation of the C-ring. The H-3′ and H-5′ signals are, however, only slightly shifted.

Chalcones

All the chalcone nucleus protons appear in the aromatic region of the spectrum. In this case the α and β protons are in resonance with the A- and B-aromatic rings, and they are shifted downfield to 7.8–7.9 ppm where they appear as doublets. The A- and B-ring proton resonances are close to those of the corresponding flavones.

Dihydrochalcones

These compounds are distinguished by the characteristic responses of the protons in the C-ring. Both α and β hydrogens appear as triplets, α-hydrogens at around 3.2–3.3 ppm and β-hydrogens at 2.7–2.8 ppm. The α proton signals are sometimes obscured by sugar signals in dihydrochalcone glycosides. The signals for the protons in the A- and B-rings are similar to those found in flavanones.

8 HPLC–ESI (Electrospray Ionisation) Mass Spectrometry

In the last few years HPLC chromatography coupled to MS detectors has become available for most laboratories and this is the method of choice to obtain the mass spectrum of flavonoids in crude extracts after HPLC separations. This technique is especially efficient in the analysis of complex flavonoid combinations with sugars, phenolic acids and aliphatic acids. This can be nicely applied to the identification of flavanones, chalcones and dihydrochalcones. In those instruments with an ion trap system, the selected ions can be selectively fragmented, providing information on the structure of the original compound.

In Figure 5 an HPLC chromatogram of the chemical conversion of hesperidin

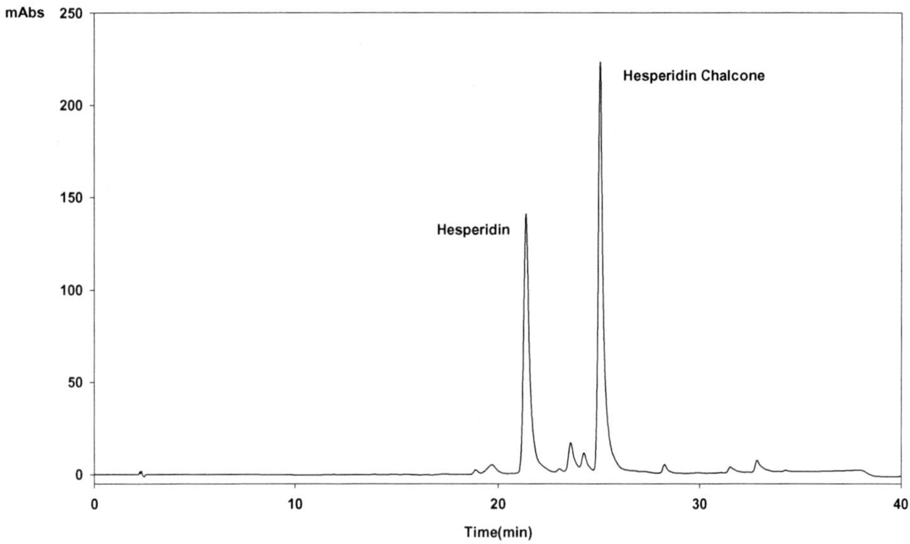

Figure 5 *HPLC chromatogram (290 nm) of hesperidin and its chalcone obtained by solution in water at pH 14*

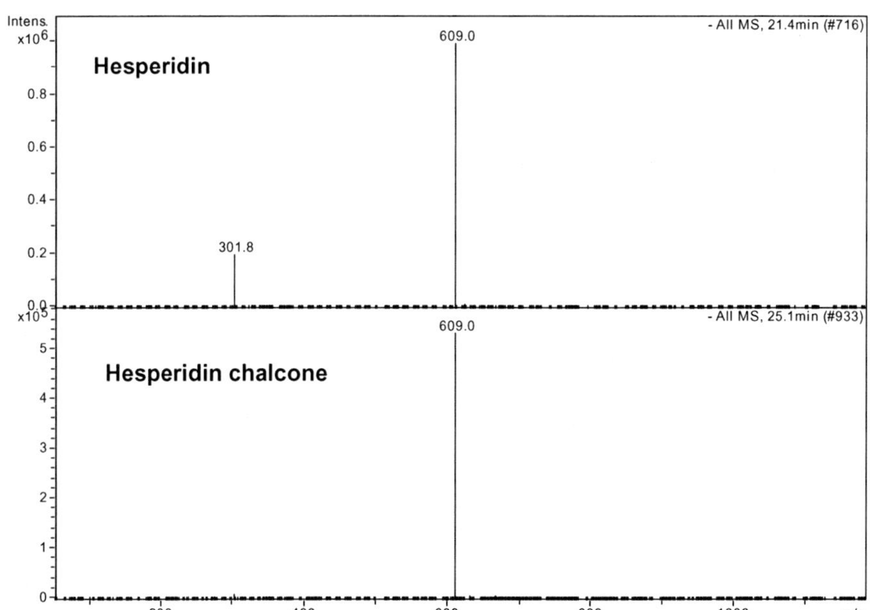

Figure 6 *HPLC–ESI–MS spectra (negative form) of hesperidin and its chalcone*

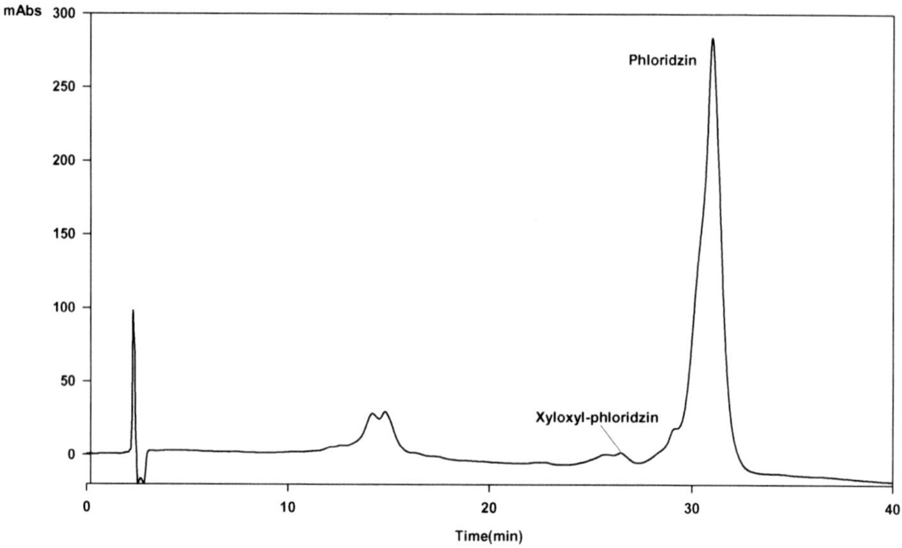

Figure 7 *HPLC chromatogram (280 nm) of apple seed methanol extracts*

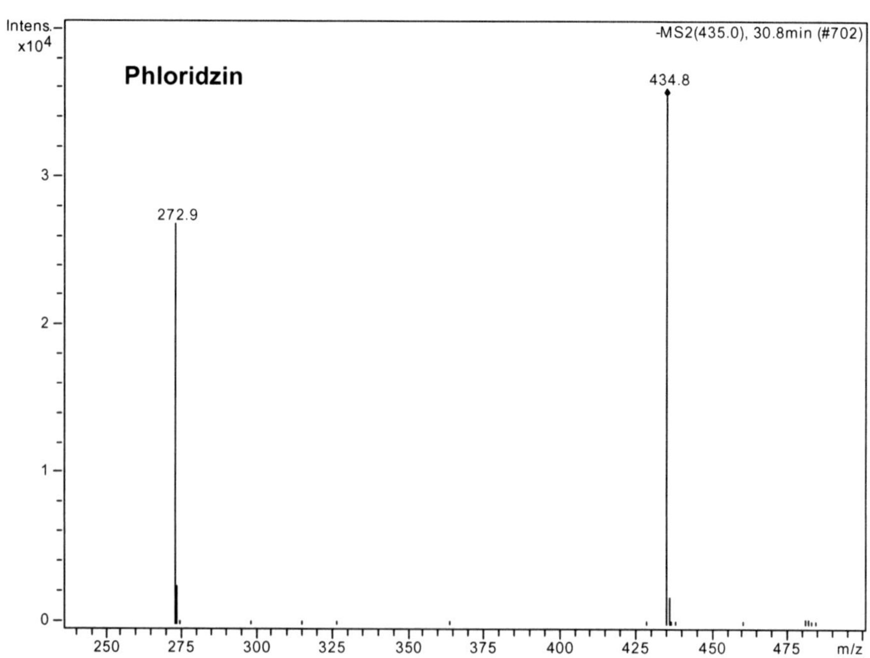

Figure 8 *HPLC–ESI–MS of phloridzin from apple seed extracts*

to its chalcone at pH 14 is shown. The HPLC–ESI–MS analysis in the negative form gives in the case of hesperidin an (M – H) ion at m/z 609 corresponding to the rhamnosyl-glucoside of hesperetin, and another fragment at m/z 301 corresponding to the aglycone hesperetin. In the case of hesperidin chalcone, only the peak of the (M – H) ion is observed, and the fragment of the aglycone is only observed after fragmentation of the isolated ion (Figure 6). This is a difference between flavanones and chalcones that should be studied in other flavanones in order to establish whether it is a common feature that can be used as a rule for differentiation between flavanones and the corresponding chalcones.

In Figure 7 an HPLC chromatogram of the methanol extract obtained from apple seeds is shown. In this case the dihydrochalcone phloridzin (phloretin 2′-glucoside; 4, 2′,4′,6′-tetrahydroxydihydrochalcone 2′-gucoside) is the main compound, although the corresponding 2′-xylosyl-glucoside is also present as a minor compound. Its HPLC–ESI–MS spectrum in the negative form shows an (M – H) ion at m/z 435 correponding to phloridzin, and the aglycone fragment at m/z 273 (M – 162) (Figure 8).

This method has been a great step forward in flavonoid analysis, in some ways similar to that produced when diode-array detection coupled to HPLC became available to most phytochemical laboratories during the eighties.

References

1. F. A. Tomás-Barberán and M. N. Clifford, *J. Sci. Food Agric.*, 2000, **80**, 1073.
2. F. A. Tomás-Barberán and J. C. Espín, *J. Sci. Food Agric.*, 2001, **81**, 853.
3. U. Justesen and P. Knuthsen, T. Leth, *Cancer Lett.*, 1997, **114**, 165.
4. A. Gil-Izquierdo, M. I. Gil, F. Ferreres and F. A. Tomás-Barberán, *J. Agric. Food Chem.*, 2001, **49**, 1035.
5. A. Gil-Izquierdo, P. Zafrilla and F. A. Tomás-Barberán, *Eur. Food Res. Technol.*, 2002, **214**, 155.
6. T. R. Seshadri, in *The Chemistry of Flavonoid Compounds*, ed. T. A. Geissman, Pergamon Press, Oxford, 1962, p. 156.
7. R. M. Horowitz and B. Gentili, *J. Agric. Food Chem.*, 1969, **17**, 696.
8. R. M. Horowitz and B. Gentili, *Food Sci. Technol.*, 1991, **48**, 97.
9. F. A. Tomás-Barberán, M. A. Blázquez, C. García-Viguera, F. Ferreres and F. Tomás-Lorente, *Phytochem. Anal.*, 1992, **3**, 178.
10. R. L. Johnson and B. V. Chandler, *Food Technol.*, 1988, 130.
11. P. E. Shaw and B. S. Buslig, *J. Agric. Food Chem.*, 1986, **34**, 837.
12. T. J. Mabry, K. R. Markham and M. B. Thomas, *The Systematic Identification of Flavonoids*, Springer-Verlag, Berlin, 1970.
13. K. R. Markham and H. Geiger, in *The flavonoids: Advances in Research since 1986*, ed. J.B. Harborne, Chapman and Hall, London, 1994, p. 441.

Subject Index